ADVANCING RULE OF LAW IN A GLOBAL CONTEXT

T0313237

PROCEEDINGS OF THE INTERNATIONAL CONFERENCE ON LAW AND GOVERNANCE IN A GLOBAL CONTEXT (ICLAVE 2017), NOVEMBER 1-2, 2017, DEPOK, INDONESIA

Advancing Rule of Law in a Global Context

Editors

Heru Susetyo, Patricia Rinwigati Waagstein & Akhmad Budi Cahyono
Universitas Indonesia, Indonesia

CRC Press
Taylor & Francis Group
Boca Raton London New York

CRC Press is an imprint of the
Taylor & Francis Group, an **informa** business
A BALKEMA BOOK

Published by:
CRC Press/Balkema
P.O. Box 447, 2300 AK Leiden, The Netherlands
e-mail: Pub.NL@taylorandfrancis.com
www.crcpress.com – www.taylorandfrancis.com

First issued in paperback 2021

ISBN 13: 978-1-03-223627-8 (pbk)
ISBN 13: 978-1-138-32782-5 (hbk)

DOI: 10.1201/9780429449031

Typeset by Integra Software Services Pvt. Ltd., Pondicherry, India

Publisher's Note
The publisher has gone to great lengths to ensure the quality of this reprint but points out that some imperfections in the original copies may be apparent.

Visit the Taylor & Francis Web site at
http://www.taylorandfrancis.com

and the CRC Press Web site at
http://www.crcpress.com

Table of contents

Foreword

The manuscripts published in this book originate from articles presented at the 1st International Conference on Law and Governance, (ICLAVE 2017), which took place at the Faculty of Law, Universitas Indonesia in 2017. The participants originate from Indonesia and some foreign countries. Invited speakers at this conference come from Malaysia, Thailand, People's Republic of China, Australia, Japan, Singapore, USA, and Indonesia.

This conference aims to provide an excellent international platform for academicians, researchers and budding students around the world to share their research findings and to find international linkage for future collaborations in the areas of Government policy-disparity between Law in the Book and Law in Society, Independent Judiciary, Corruption Eradication and Asset Recovery, Legal Protection for Women and Children, Rule of Law and Constitutional Democracy, Law and the Global Labor Market, Law and Management Problems in Multinational Corporations, Global Investment Governance, New Developments in Criminal Law, Penal Law and Criminal Justice, Law, Policy and Conflict Management on Environmental and Natural Resources, Law and Identities in Blurred Boundaries, Law and Intellectual Property Rights Protection, Protection to Indigenous People/ Local Communities, The Protection of Individual Identity and Privacy at Cyberspace, Competition Law and the Monopoly of State-owned Enterprises, Opportunities and Challenges of ASEAN Economic Community, Global Market Penetration by Micro, Small and Medium enterprises, Law, Radicalism and Terrorism, Climate Change Mitigation and Adaptation, The development of Family Law and Inheritance Law in shaping future global communities, and Islamic finance and economic opportunities.

We convey our utmost gratitude to all participants and professors who actively took part in this conference, to all editors and staff at the research and publication unit of the Faculty of Law, Universitas Indonesia, as well as to Universitas Indonesia, particularly DRPM and KPPRI.

Last but not least, our gratitude goes to the staff of CRC Press (Taylor & Francis Group), for the proceedings.

Depok, Indonesia
5 July 2019

Assoc. Prof. Heru Susetyo, PhD
Editor

Advancing Rule of Law in a Global Context – Susetyo, Rinwigati Waagstein & Budi Cahyono (eds)
© 2020 Taylor & Francis Group, London, ISBN 978-1-138-32782-5

Organizing Committee

Keynote Speaker
Prof. Jimly Ashiddiqie (Universitas Indonesia, Indonesia)

Invited Speakers
Prof. Harkristuti Harkrisnowo (Universitas Indonesia, Indonesia)
Prof. Sulistyowati lrianto (Universitas Indonesia, Indonesia)
Prof. Hikmahanto Juwana (Universitas Indonesia, Indonesia)
Prof. Abu Bakar Munir (University of Malaya, Malaysia)
Prof. Dr. David W. Tushaus (Missouri Western State University, United States of America)
Assoc. Prof. Dr. Dan W. Puchniak (National University of Singapore, Singapore)
Dr. H. Nadirsyah Hosen (Monash University, Australia)

Director
Prof. Dr. Topo Santoso, SH.,MH.
Dr. Ratih Lestarini, S.H., M.H.
Dr. Akhmad Budi Cahyono, S.H., M.H.

Chairman
Dr. Patricia Rinwigati Waagstein, S.H., M.I.L.

Vice-Chairman
Bono Budi Priambodo, S.H., M.Sc.

Secretariat
Qurrata Ayuni, S.H., MCDR
Abdul Karim Munthe, S.Sy., S.H., M.H.
Milson Kamil
Arrofi Maretha Rusian
Adi Prabowo, A.Md.Hk

Finance
Rizki Wulandari, SE
Alfien Qorrina

Logistics
Wardi, S.H., M.H
Ghunarsa Sujatnika, S.H., M.H.
Ali Abdillah, S.H., LL.M.

Transportation
Tirtawening Parikesit, S.H., M.Si
Ahmad Madison, S.H.

Accommodation
Iva Kasuma, S.H., M.Si.
Rizki Banyualam, SH. LL.M

Public Relations
Mohammad Novrizal, S.H., LL.M
Sri Supriyanti
Mutia Ariani
Irwan Hermawan
Yuli Wirdyaningsih

Program
Kris Wijoyo Soepandji, S.H., M.P.P.
Desrezka Gunti Larasati, S.H., LL.M.
Hari Prasetiyo, S.H., M.H.
Raafi Seif, SH.
Muhammad Hanif Pahlawan
Mutiara Elisabet
Ryu Kristoforus
Adara Skyla Sakinah
Muliani Wahab

Consumption
Sumarni, S.Sos

Articles Publication
Arie Afriansyah, SH., MIL, Ph.D

Advancing Rule of Law in a Global Context – Susetyo, Rinwigati Waagstein & Budi Cahyono (eds)
© 2020 Taylor & Francis Group, London, ISBN 978-1-138-32782-5

Scientific Committee

- Heru Susetyo, S.H., LL.M., M.Si., Ph.D. (Universitas Indonesia, Indonesia)
- Dr. Akhmad Budi Cahyono, S.H., M.H. (Universitas Indonesia, Indonesia)
- M. Ramdan Andri Gunawan Wibisana, S.H., LL.M., Ph.D. (Universitas Indonesia, Indonesia)

Editor Biographies

Heru Susetyo, Ph.D.

Heru Susetyo in an Associate Professor at the Faculty of Law, Universitas Indonesia, Jakarta – Indonesia, Executive Board of Victim Support Asia, Secretary General of the *Association of Lecturers of Victimology in Indonesia* (APVI) and Founder of the *Indonesian Victimology Society* in 2011, Member of Jakarta Province Research Council (2018 – 2022) as well as member of the former executive committee of the World Society of Victimology (2009 – 2015). He is currently teaching human rights, victimology, law and social welfare, children protection law, Law and Society and Legal Research Method.

Heru Susetyo graduated from Universitas Indonesia in 1996 (bachelor of law) and 2003 (master of social work) and also obtained a master degree in International Human Rights Law (LL.M) from Northwestern Law School, Chicago (2003) and a PhD in Human Rights and Peace Studies from Mahidol University, Thailand (in 2014). Currently, he is conducting research on *Victimization to the Family of Terrorism Suspects* as an external PhD researcher at INTERVICT (International Victimology Institute), Tilburg University, The Netherlands. Heru Susetyo has also written numerous articles in various newspapers, magazines and online media and also published papers in several journals and books related to human rights, victimology, terrorism, social work and international criminal law.

He has also been responsible for handling research and publication affairs at the Law Faculty of the University as Manager of Research and Publication (since 2018), Editor in Chief of *Jurnal Hukum dan Pembangunan* FHUI (since 2018) and Editor in Chief of Journal of Islam and Islamic Law Studies (since 2007). He can be contacted by email at hsusetyo@ui.ac.id and heru-s@lawyer.com. His own blog : http://herususetyo.com

Dr. Patricia Rinwigati Waagstein

Dr. Patricia Rinwigati Waagstein, S.H., MIL is a lecture at the Faculty of Law, University of Indonesia. She has also been a practicing attorney and consultant who has worked in Timor Lest, USA, Sweden, and ASEAN countries for over fifteen years. She has extensive experience and expertise in human rights, business and human rights issues, criminal justice, and international law. She holds a Bachelor degree in law from the University of Indonesia (1996), a Master's Degree in Public International Law from Lund University, Sweden (1999) and a Doctoral Degree in law from Uppsala University, Sweden (2009). In addition to lecturer, she is currently the director of the Djokosoetono Research Center at the Faculty of Law, University of Indonesia. Areas of research interest: human rights, humanitarian law, criminal law, and international law.

Dr. Akhmad Budi Cahyono

Akhmad Budi Cahyono is a senior lecturer of civil law at the Faculty of Law, Universitas Indonesia. He earned a bachelor's degree from Universitas Indonesia in 1998, a Master's degree from Universitas Indonesia in 2002, and a doctoral degree from Universitas Indonesia in 2015. He also conducted Research in completion of his dissertation at Leiden

University, the Netherlands. His qualifications are civil law in general, contract law, and tort law. He has been a lecturer since 1998 and published three private law books. The first one was published in 2005 with the title "Hukum Perdata: Suatu Pengantar" (Introduction of Private Law), the second one published in 2008 with the title "Mengenal Hukum Perdata" (Understanding of Private Law) and the last book published in 2014 with the title "Modul Hukum Perdata" (Private Law Module).

Acknowledgements

The Editors and organizers of the 1st ICLAVE 2017 express their utmost gratitude to the Dean of Universitas Indonesia, Faculty of Law (2013-2017), Prof Dr Topo Santoso. Also to the (late) Dean (2017-2019), Prof Dr Melda Kamil and to Universitas Indonesia for taking the initiative, provide constant support and financial assistance so that this manuscripts are suitable for publication by CRC Press, Taylor & Francis Group.

Advancing Rule of Law in a Global Context – Susetyo, Rinwigati Waagstein & Budi Cahyono (eds)
© 2020 Taylor & Francis Group, London, ISBN 978-1-138-32782-5

The limitation of investigator authority for illegal narcotics offences

Riza Alifianto Kurniawan & Sapta Aprilianto
Faculty of Law, Universitas Airlangga, Indonesia

ABSTRACT: Narcotics crime has become a priority for law enforcement in Indonesia. The granting of special powers in the investigation of narcotics crime for INP (Indonesian National Police) and BNN (National Anti-Narcotics Agency) provides more effective law enforcement opportunities, but the extent of full investigation authority provides an opportunity for the abuse or misuse of authority if such authority is not regulated or evaluated. The evaluation of the authority of the investigator in the Indonesian criminal justice system is done through pretrial mechanisms. Special powers require specific legal and regulatory basis so that pretrial institutions can be used to test or protect the rights of suspects undergoing the investigation process at the level of full investigation and prosecution. The research method used is to test the concept of the authority of the investigator of narcotic crime by using the law and the general principles of good governance.

Keywords: controlled buying, controlled delivery, discretion

1 INTRODUCTION

Narcotics crime is one of the offences set out in a special law. Specific arrangements in a separate legal product indicate a narcotic crime can be categorized as a serious criminal offence. (Atmasasmita, 2010) It can be seen that the narcotics criminal penalty threat on average is more than five years, and there is the existence of accumulation of the principal penalty.

The Law on Narcotics is very aggressive in enforcing the law against the perpetrators of narcotics abuse. The authority of the investigator in the narcotic criminal case is very wide and has the opportunity to cause irregularities. One of the abuses of investigators' authorities of narcotics crime is a controlled buy and controlled delivery. The granting of this investigative authority as a follow-up in the disclosure of narcotic criminal cases is very difficult to disclose. Controlled buy and controlled delivery in some cases have a positive impact on opening a network of narcotics abuse crimes and organized crime. The authority of this investigation has the potential to be misused by INP and BNN investigators in trafficking or in case engineering. Therefore, to prevent abuse of suspect rights, it is necessary to limit the authority of the investigators.

The authority of the investigator in a narcotic crime is vast when compared to the authority of an ordinary criminal investigator. One of the widespread authorities possessed by narcotics investigators is the investigation techniques of controlled buy and controlled delivery. The two authorities in this investigation technique are the means given by Law No. 35 of 2009 on narcotics as the authority to further trap the narcotics perpetrators. It is very difficult to catch the narcotics crime's perpetrators because the nature of this criminal act is very organized. All perpetrators have their own crime networks that are very difficult to detect by law enforcement officers.

Controlled buy and controlled delivery techniques are recognized as part of the authority of the investigator based on Law No. 35 of 2009 article 79. The explanation regarding these two authorities becomes unclear because such investigative authority is highly subjective depending on the investigator's judgment. Article 79 states that controlled buy and controlled delivery are made on the basis of a written order from the superior of the investigator. A written order from full investigator to impose controlled buy and controlled delivery authority provides an opportunity for infringement of the suspect's right.

Violations that may occur in controlled buy and controlled delivery are an engineering crime committed by an investigator. This technique can be interpreted as a form of abuse of power by investigators and cannot be measured objectively. The mechanisms perpetrated by the suspect in protecting his rights are through pretrial institutions set out in the Criminal Procedure Code. The pretrial process is actually a mechanism to test the full investigation authority owned by BNN in particular to examine the legitimacy of controlled buy and controlled delivery technique.

Means that can be used in controlling the investigation techniques of controlled buy and controlled delivery is by applying the principle of good governance in the administration of criminal full investigation cases (Sasangka and Rosita, 2003). A very common term known for this principle is the general principle of good governance. This principle provides a benchmark for investigators when exercising their authority not to infringe upon others. The unwritten provisions in the principle of good governance in the investigation provide uncertainty of interpretation or immeasurable discretion. This study seeks to provide a legal void in the setting of full investigation techniques such as controlled buy and controlled delivery to safeguard the professionalism of law enforcement of narcotic crimes which does not violate human rights.

Problems

1. What are the limits of controlled buy and controlled delivery authority under the Law on Narcotics?
2. How is the protection of the suspect's rights in the application of controlled buy and controlled delivery?

2 DISCUSSION

2.1 *The limitation of controlled buy and controlled delivery authority*

A preliminary criminal investigation is the beginning of a process in criminal procedural law to determine the material truth of the alleged occurrence of a criminal event (Harahap, 2005). The preliminary investigation in Law No. 8 of 1981 on Criminal Procedure Law means as an act of investigating officer or preliminary investigator in determining whether reports or complaints received from the public are a criminal act or not (Criminal Procedures Code-KUHAP 1981). In the preliminary investigation phase, law enforcement officers only assess whether a public report or complaint is a criminal act regulated by law or simply a civil or administrative Act that is not a crime. This preliminary investigation stage is the beginning of a stage in the procedural proceedings in order to be subject to criminal punishment or not against a suspect.

Criminal procedure law has several stages of a sustainable process in the disclosure of a criminal case. These stages are preliminary investigation, investigation, prosecution and examination process in court (Atmasasmita, 2010). These stages have their own meanings and procedures involving the investigation apparatuses such as INP, KPK (The Corruption Eradication Commission), and BNN. Prosecution apparatus such as prosecutors, and judges as apparatuses who have the authority to issue judicial decisions.

The stages of examination under Law No. 8 of 1981 on Criminal Procedures Code or can be referred to as KUHAP, provide the authorities that can be used by each law enforcement apparatus to disclose a criminal case. The authority generally authorized by the law against the investigator is the authority to arrest, authorize witnesses, authorize detention, the authority to conduct a letter examination, the authority to tapping or intercept, the authority to foreclosure the goods and instruments of crime, and to the seizure of certain goods.

The authority of the criminal investigator as described in the preceding paragraph is closely linked to the process and evidence collection techniques. The definition of investigation in the KUHAP (Criminal Procedure Code) is all actions to collect evidence to clarify the crime committed and find the suspect. Moving from the notion of such kind of full investigation, thus the full investigation has a very wide authority, and even this authority can violate the rights of others. However, as long as such authority is sourced from a written legal rule or law, it can be legally denied or applied by the investigator.

The vast authority of the investigator is closely related to the evidentiary system adopted by the Indonesian Criminal Procedure System. Indonesian's criminal procedure law system adheres to a limited system of evidence (Negatief Wettelijk Bewijstheorie) (Harahap, 2010). In proving the crime and wrongdoing of the defendant, the judge does not just stick to the instruments of real evidence specified in the law. The judge has a subjective assessment of the defendant through beliefs sourced from the instruments of real evidence given by law. So in determining the defendant's fault or issuing the verdict, the judge holds to the real evidence given by law and his conviction to determine the defendant guilty or not.

The authority of the investigator in a narcotic crime is vast when compared to the authority of an ordinary criminal investigator. One of the widespread authorities possessed by narcotics investigators is the investigation techniques of controlled buy and controlled delivery. The two authorities in this investigation technique are the means given by Law No. 35 of 2009 on narcotics as the authority to further trap the narcotics perpetrators. It is very difficult to catch the perpetrators of narcotics crime because the nature of this criminal act is very organized. All perpetrators have their own crime networks that are very difficult to detect by law enforcement officers.

Controlled buy and controlled delivery techniques are recognized as part of the authority of the investigator based on Law No. 35 of 2009 article 79. The explanation regarding these two authorities becomes unclear because such an investigator's authority is highly subjective depending on his/her judgment. Article 79 states that controlled buy and controlled delivery are made on the basis of a written order from the superior of the investigator. A written order from the investigator to apply the controlled buy and controlled delivery provides an opportunity for infringement of the suspect's rights.

Violations that may occur in controlled buy and controlled delivery are an engineering crime committed by an investigator. This technique can be interpreted as a form of abuse of power by investigators and cannot be measured objectively. The mechanisms perpetrated by the suspect in protecting his rights are through pretrial institutions set out in the Criminal Procedure Code. The pretrial process is actually a mechanism to test the investigation authority owned by BNN in particular to examine the legitimacy of controlled buy and controlled delivery.

A Tool that can be used in controlling for full investigation techniques of controlled buy and controlled delivery is by applying the principle of good governance in the administration of criminal investigations. A very common term known for this principle is the general principle of good governance. This principle provides a benchmark for investigators when exercising their authority not to infringe upon others. The unwritten provisions in the principle of good governance in the full investigation provide uncertainty of interpretation or unmeasurable discretion. This study seeks to provide a legal void in the setting of full investigation techniques such as controlled buy and controlled delivery to safeguard the professionalism of law enforcement of narcotic crimes which do not violate human rights.

2.2 *Protection of the rights of suspects in the implementation of controlled buy and controlled delivery*

Indonesian National Police (INP) as an independent institution has a very big role in the life of the community or in the context of the state. With a wide range of tasks within the police body, there needs to be a clear division of tasks. Thus the existence of several INP authorities to prevent, fully investigate and combat drug problems, more specifically the narcotics problem. The authority of INP to prevent and overcome narcotics problem is contained in Article 15 paragraph (1) letter c, in which INP is authorized to prevent and overcome the incidence of community illness, the aforementioned community illness, in this case, is the misuse of narcotics.

The Criminal Procedure Code (KUHAP) provides separation between Preliminary Investigator and Investigator, but their duties and obligations are almost the same. It can be said that the duties and obligations of the preliminary investigator must be the duties and obligations of the investigator as well, but the duties and obligations of the investigator are not all the duties and obligations of the preliminary investigator. Therefore, an investigator must also be a preliminary investigator, but a preliminary investigator is not necessarily an investigator. In essence, all preliminary investigators are INP's officials.

The Criminal Procedure Code (KUHAP) is a regulation for the event process, in particular, criminal proceedings for law enforcement. All forms of action of law enforcement officers, especially investigators, are regulated in KUHAP. Such arrangements are certainly to ensure legal certainty and justice against someone who is undergoing criminal proceedings (Anwar and Adang, 2011). We all certainly understand that KUHAP is essentially established to protect the legal interests of a person who is reported, allegedly, indicted and convicted. Such protection is the protection granted by the State against the arbitrariness of law enforcement officials in the event that they carry out forced efforts.

Based on the Palermo Convention of 2000, criminal narcotics abuse perpetrated by organized crime is one type of "extraordinary crime." This extraordinary crime predicate is attributed to crimes that have heavy criminal penalties and require extra effort in eradicating and preventing them (Lily, 2015).

Narcotics crime in Indonesia is regulated by Law No. 35 of 2009 on Narcotics. This law is a 3rd law product issued by the Republic of Indonesia aimed at preventing and combating drug abuse. In every revision of the narcotics regulations from 1976 to 2009 provides different policies to resolve the narcotics prevention and enforcement issues to get a deterrent effect for the offender.

The concept of controlled buy and controlled delivery is one of the forcible attempts provided by law to facilitate the enforcement of the law on narcotics misuse. These two concepts of investigator's authority may also be referred to as a special narcotics investigator's authority when compared to the ordinary criminal investigator's authority.

The objective of criminal procedure law is to seek material truth from the allegations of criminal acts that occurred in order to be able to find the perpetrators of criminal acts. The purpose of criminal procedure law is idealism in the eradication and enforcement of criminal law. The purpose of criminal procedural law is seeking material truth confirms that all provisions of this criminal procedure law mandate to uncover the cause, motive, element, and purpose of the criminal act committed by the perpetrator.

The authority of the investigator in seeking evidence is often tangent to human rights. The overly broad authority of the investigator has the potential to be misused and violate human rights. Narcotics crime, although regulated in a special law, does not mean there are human rights violations in the conduct of restrictions and prevention.

The crime of narcotics misuse includes "extraordinary" crimes or a special crime. The specificity of criminal offences for law enforcement requires a special eradication effort. One of the specificities of these law enforcement efforts is the controlled buy and controlled delivery. These two enforcement attempts are granted by law to facilitate investigators in the eradication of narcotics addicts to be easy to eradicate. The forced attempts of this narcotics full investigator make it easy to uncover the misuse of narcotics or use criminal networks that are difficult to reveal or map.

The provision of forcible attempts that grants more authority to narcotics investigators also raises potential problems in the issue of the protection of suspects' rights. The most common potential violation is the unlawful collection of evidence by virtue of the forced controlled delivery and controlled buy. Two potential violations require a rule of thumb or an objective standard to avoid misuse of authority by the investigator. One of the standards used is to evaluate any discretion of the forced attempts of covert purchases and supervised submissions. It also needs to be strengthened by pre-trial institutions that provide additional absolute competence to evaluate controlled buy and controlled delivery in order to objectively force efforts implemented by narcotics investigators.

One way to prevent the misuse of the authority of a narcotic crime investigator is to apply the principles of good governance strictly and transparently. Officials or investigators who will apply forcible efforts of controlled buy and they should have an understanding of the authority of this enforced effort. The application of controlled buy and/or controlled delivery by narcotics investigators may be free to apply because it includes the subjective authority of the investigator. Subjectivity in the application of the two authorities is based on the assessment of the investigator (Discretion) (Hadjon, 1997). Measures in assessing free authority within the domain of administrative law can be tested on the basis of good governance (AUPB) or good governance principles.

The free discretion or authority to be implemented by narcotics criminal investigators shall be based on the law on police, Law No. 2 of 2002 which provides guidance to police officers to prevent discretionary expulsion violating the following matters:

1. not contrary to a rule of law
2. in harmony with the legal obligations that require such action to take place
3. should be reasonable and include the office environment
4. appropriate consideration based on circumstances that compel, and
5. respect for human rights

The five principles of discretionary attempts to forcibly perform covert purchase and supervise submissions as described above can be used as guidelines for investigators of narcotic crime. The nature of 5 principles of investigative discretion is cumulative. All principles must be fulfilled by investigators when applying forcible measures.

3 CONCLUSION

It is becoming a priority to encourage the eradication of narcotics abuse. The increasingly expanding mode of drug abuse and the use of crime networks require extraordinary strategies and efforts in order to be eradicated. Narcotics crimes perpetrated by organized criminals require special strategies and legal efforts undertaken by investigators of narcotic criminal offences.

The authority of law enforcement of narcotic criminal acts needs to be expanded with the aim of achieving a narrow reduction in the number of drug addicts in Indonesia. As a major investigative authority, INP investigators and the BNN in applying covert purchases authority and supervised submission must be cautious and prevent any misconduct or misuse of authority. INP and BNN investigators can apply good governance principles in measuring and implementing the authority of the investigation.

REFERENCES

Anwar, Yesmil. (2011). Criminal Justice System Concepts, Components, and Implementation In Law Enforcement in Indonesia. Bandung: Widya Padjajaran.
Atmasasmita, Romli. (2010). Globalization and Business Crime. Jakarta: Kencana Prenata Media.
Atmasasmita, Romli, (2010). The Contemporary Criminal Justice System. Jakarta: Kencana Prenada Media.
Chazawi, Adami. (2011). Laws and Regulations on Proofing the Corruption. Jakarta: Bayu Media Publishing.
Lily, Robert. (2015). The Context of Criminology Theory and Its Consequences. Jakarta: Kencana Prenada Media.
Hadjon, Philipus Mandiri. (1997). Introduction to Indonesian Administrative Law. Yogyakarta: Gadjah Mada University Press.
Harahap, Yahya. (2005). Criminal Procedure Code: Investigation, Prosecution and Judicial Review. Jakarta: Sinar Grafika.
Mahmud, Peter. (2010). Legal Research. Jakarta: Kencana Prenada Media.
Remmelink, Jan. (2003). Criminal law. Jakarta: Gramedia Pustaka Utama.
Hari Sasangka., & Lili Rosita, (2003). The Proof in Criminal Case, Mandar Maju.

Advancing Rule of Law in a Global Context – Susetyo, Rinwigati Waagstein & Budi Cahyono (eds)
© 2020 Taylor & Francis Group, London, ISBN 978-1-138-32782-5

Deferred Prosecution Agreement (DPA): Can it be a solution for corporate criminal liability in Indonesia?

Arija Br Ginting & Topo Santoso
Universitas Indonesia, Indonesia

ABSTRACT: Corporate crime is ambiguous, highly complex, diffusion of responsibility and victimization, and difficult to detect and to prosecute. Although since 1955 corporations have been introduced, until now, corporate crime has rarely empowered criminal law. PERMA (Supreme Court Decree) no. 13 of 2016 on the Procedures for Handling Criminal Cases by Corporation is expected to be the solution. However, learning from the experience of the United States and Great Britain, excessive corporate prosecution had a negative impact, especially for the third parties (employees and shareholders), the economy in general, and the corporation itself. Both the United States and Great Britain then apply the Deferred Prosecution Agreement (DPA) according to the characteristics of each country. The advantages of DPA are preventive (such as compliance programs) and retributive (such as financial penalties, compensation for victims) functions for the corporation. This means the drugs offered to the company fit the company's disease approach. Thus, Indonesia is expected to formulate the Indonesian DPAs to address issues of corporate criminal liability

Keywords: Corporation, Criminal Liability, DPA

1   INTRODUCTION

Corporate criminal liability is theoretically influenced by the characteristics of corporate crimes that are different from human crimes. It is... *ambiguous, highly complex, diffusion of responsibility and victimization, and difficult to detect and to prosecute* (Ali, 2013, pp.13-14), which results in a variety of mentions for corporate crime, such as white collar crime, transnational crime, and others. The criminal law then finds it urgent to declare corporations as subjects of criminal offenses

In practice, the phenomenon of corporate crime is also felt by the community and law enforcement. *The Global Economic Crime Survey* 2014 shows the range of crimes involving corporations between 2011 and 2014 (PwC's 2014 Global Economic Crime Survey, 2014). Some corporate criminal acts have become serious problems in Indonesia, namely criminal acts that are in the top five percentages, such as bribery and corruption (Widjojanto, 2017). Corporations also have a large role in forestry crimes, such as illegal practices, conversion of forest areas, oil palm plantations, and even corporate crimes that involve conflict with the local communities (Tarigan, 2013). The number of *lex specialis* regulations governing corporate criminal liability (regardless of uniformity of formulation) has not been able to accommodate the technical rules of criminal law enforcement. The Supreme Court then initiated the formation of PERMA No.13 of 2016 as a technical reference of prosecution against the corporations. From this, it can be seen that various reforms have always been done in order to maximize the work of criminal law in tackling corporate crime.

The latest, attention-grabbing issue illustrating a pattern of crime and the mechanism of corporate criminal liability was a Rolls-Royce company dealing with PT Garuda Indonesia. They were involved in a corruption case, namely of the procurement of Jet Airbus 330 engines.

The bribery was done by Rolls, a British company, through a beneficial ownership Mugi Rekso Abadi or MRA group to a former CEO of Garuda Indonesia. The British Serious Fraud Office (SFO) has identified Rolls, which committed transnational corruption because it involved several countries, not only Indonesia. In March 2017, with a cooperative attitude towards law enforcement, Rolls Royce eventually entered the UK's Deferred Prosecution Agreement (DPA) scheme. SFO and Rolls-Royce reached an agreement "... the DPA involves payments of £497,252,645 (comprising disgorgement of profits of £258,170,000 and a financial penalty of £239,082,645) plus interest. Rolls-Royce is also reimbursing the SFO's costs in full (c£13m)" ((SFO), 2017).

The Indonesian Corruption Eradication Commission (KPK) based on preliminary evidence has declared the former CEO of Garuda Indonesia and the director of the MRA as suspects. KPK cannot be authorized to prosecute Rolls-Royce, which is domiciled in the United Kingdom. KPK can only use the information and data from SFO and CPIB (Corrupt Practices Investigation Bureau) of Singapore (DetikNews, 2017). Let us say Rolls-Royce contributes to corruption, which is causing the loss of the Indonesian state, and Indonesian authorities cannot do anything to make Rolls-Royce liable. So, what we have to think about related to the idea of Indonesian Corruption Act is to recover the state loss, including asking the corporation to be responsible for recovering state losses.

The mechanism of international corruption cases settlement like this, DPA, is known in the United States and the United Kingdom. Although KPK does not have the authority to explore the Rolls-Royce case further, in similar circumstances, the United States and Britain can do so. The introduction of DPA in the United Kingdom and the United States intended to avoid the negative impact of corporate prosecution. As Ben Morgan puts it "... the scheme was introduced to limit the negative effects of corporate criminal behavior on innocent people – employees, pensioners, and others reliant on the future of a company – suppliers, manufacturers, or customers" (Morgan, 2017). DPA indirectly serves to make corporations and law enforcement as partners to combat crime.

The problem that then arises is whether DPA can be a solution to corporate criminal liability that is like the statement mentioned earlier? To answer it comprehensively is of course not easy. Therefore, the analyzed aspect is limited to the context of corporate criminal liability, namely prevention and retribution contained within the DPA.

2 METHOD

The research method used is normative juridical by prioritizing secondary data search, in the form of both primary law material and secondary law material. The approach used is a multi-approach which includes a micro-comparison approach and a historical approach. To support the secondary data, interviews were also conducted with several sources such as law enforcement officers and academicians. All information was then analyzed qualitatively.

3 DISCUSSION

3.1 *Development of corporate criminal liability objectives*

Corporate criminal liability, recently known in the 20th century, actually has a long history. Albert W. Alschuler states that the purpose of corporal criminal liability was influenced by two forms of punishment in the ancient time, namely deodant (the punishment of animals and objects that have produced harm) and frankpledge (the punishment of all members of a group when one member of the group has avoided apprehension for a crime) (Alschuler, 2009). He then concludes that the most relevant to corporate criminal liability is the frankpledge set for instrumental reasons as the justification for the punishment of group members when one member is eluded from the punishment for his crimes (Alschuler, 2009). This further affects the attitudes of states to decide whether to recognize a corporate criminal liability, accept on the condition, or refuse at all corporate liability.

Indonesia itself decided to formulate corporate criminal liability within the *lex specialis* rule with more than 100 laws. Related to Albert W. Alschuler, Muladi and Dwidja Priyatno stated that J.E. Sahetapy once researched and stated that in some areas of Indonesia "... it often happens that the village of the villain or the village where a murder or theft of a foreigner has taken place is liable to pay a fine or loss to the family of the murdered or theft" (Priyatno M. a., 2012, p. 42). This can be said to be the forerunner of the collectivity in Indonesia. The reasons for the punishment are then underpinned by various theories of elaborated corporate criminal liability.

Much of the literature in the Indonesian language that addresses the objectives of corporate criminal liability has linked it to conventional criminal purposes, namely retaliation, prevention, and coordination. The following two expert opinions represent the literature. Mahrus Ali states that the purpose of criminal prosecution relevant to corporate crime is related to deterrence and rehabilitation theories (Ali, 2013, p. 264). This is because a corporation bases its crime on economic motive or profit and loss orientation. Theoretically, the basic assumption of the prevention theory is that a corporation when doing a certain activity thinks rationally with the main purpose to maximize the expected utility (maximizing the expected utility). This means that there is already a consideration of profit or loss requiring the corporation's economic rationality. This rationality is the basis for the imposition of criminal sanctions for corporations (Ali, 2013).

Another opinion is by Dwidja Priyatno, who views that corporative punishment is integrative. First is prevention (general and specific) in which the corporation will not commit any further criminal offenses, and other corporations are prevented from committing a criminal offense, with the aim of ensuring the protection of the community; secondly, the protection of society means that with corporate crime, one is no longer capable of committing a crime; third is to maintain community solidarity due to the enforcement of local customs and to prevent private revenge or unofficial retaliation; and fourth, the purpose of criminal punishment is that there is a balance between the punishment and criminal liability offender (Priyatno, 2004).

These latter two opinions essentially cover two important dimensions, namely prevention and (equal) retaliation. For example, rehabilitation (Ali, 2013) and proper retention and prevention of unauthorized revenge (Priyatno, 2004) concern the punishment of corporations with a dimension of retaliation. Therefore, this article focuses on the prevention and retaliation aspects contained in the mechanism of the Deferred Prosecution Agreement (DPAs). There is no statement that specifically analyzes the objective of sentencing contained in the DPA. This concerns DPA, which is a new concept and does not exist in the Indonesian criminal justice system yet. However, the two most popular theories about the objective of sentencing seem to be relevant to the DPA context.

3.2 *The notion of the deferred prosecution agreement for Indonesian corporations*

3.2.1 *General description*

The UK introduced DPA with Schedule 17 of the Crime and Courts Act 2013, and the United States regulates it through the Foreign Corrupt Practice Act (FCPA). Australia is also reviewing the possibility of implementing DPA. DPA can be defined as "... an agreement reached between a prosecutor and an organization which could be prosecuted, under the supervision of a judge. The agreement allows a prosecution to be suspended for a defined period provided the organization meets certain specified conditions" (Serious Fraud Office, 2017). Both countries also practise DPA differently according to the characteristics and needs of each country. However, both countries have similar backgrounds and finally decided to adopt the DPA due to the weaknesses of previous corporate criminal liability mechanisms.

Among the weaknesses is the occurrence of collateral consequences associated with the prosecution of corporate actors. In various sources, collateral consequences refer to the corporate death penalty. It includes direct regulatory restrictions, such as loss of good standing and loss of professional licenses (e.g., banking, audit). These direct sanctions may lead to indirect third-party effects, including loss of jobs, shareholders' losses, forced mergers, and even bankruptcy (Amulic, 2017, p. 135). Those collateral consequences even occur in the post-Enron cases in the United States. Attention to collateral consequences can be regarded as a form of corporate humanization because criminal law is also viewed as an ultimum remedium. As Dominik Brodowski points out that the principle of ultima ratio (*ultimum remedium*) becomes more pressing (Brodowski, 2014).

In addition, the opaque and sophisticated corporate crime makes it difficult to detect, leading to significant challenges for law enforcement officers with traditional investigative and prosecutorial models. In addition, a high cost has to be spent by law enforcement officers (Government, 2016).

The DPA is the fruit of long experience in addressing corporate criminal liability issues. The development of corporations and efforts to tackle corporate crime have been carried out. Such conditions may be experienced by Indonesia, as a country in a development stage. It is not excessive if the experience of those countries can be used as an alarm to anticipate a similar incident to happen the at national level.

Changes in efforts to combat corporate crime in Indonesia, for example, PERMA No.13 of 2016, indicate the realization that corporate crime is already in sight (Mugopal, 2017). This is offset by significant corporate growth (in business). According to the 2006 economic census (SE06), there were 22.73 million businesses/non-agricultural companies, while the 2016 economic census (SE2016) recorded 26.71 million businesses/companies. It shows that in 10 years, the growth of the business/company will reach 17.51% (Statistics, 27 April 2017). The introduction of the DPA concept is not too early since the corporation, and its potential criminal act is a reality.

3.2.2 *DPA: To prevent or retaliate?*

The form of prevention of crime against people with corporations is not exactly the same, especially the logic of its effectiveness in practice. The general prevention of white-collar crime (corporations) is considered appropriate due to the fact that it "... *tends to be the antithesis of crimes of passion*" because the organs of corporations in general "....*pay particular attention to precedent in determining the risks and rewards of contemplated action*" (Weismann, 2008, p.1326). Then, how can DPA prevent corporate crime? One of the most important elements of the DPA mechanism is the existence of an effective compliance program, which prevents from and detects potential deterrent and mitigates enforcement consequences (insurance) (Pappas, 2012) of a corporation. According to Merriam Webster, compliance means "the act or process of complying to a desire, demand, proposal, or regimen or to coercion conformity in fulfilling official requirements" (Webster, 2017). DPA negotiations must always be related to compliance, namely at the assessment stage (usually investigation) to determine whether the corporation will comply to DPA, and at the decision stage of the DPA, which the corporation agrees to, the corporation must comply to DPA when it is successful.

At the investigation stage (feasibility assessment), the compliance program will affect "... whether a violation will be resolved by NPA and DPA, the duration of NPAs and DPAs, and the amount of any assessed penalty" (Pappas, 2012). The term NPA is known in the United States, and it is similar in shape to DPA. Often compliance programs are "condensed" when a DPA agreement is reached, which is when a corporation has acknowledged a criminal offense and requests to be forgiven based on the requirement that the corporation will follow the compliance program. For example, the corporation will restructure its financial management or its corporate organization management.

However, according to the author, a compliance program can actually be done not only when it enters the DPA stage. DPA can be a preventive tool, so the occurrence of a crime is not a requirement to assess the effectiveness of DPA. DPA can even prevent criminal acts. The seriousness of the corporation to implement an effective compliance program in its organization can be a measure of crime. Some concrete actions, which are the steps to implement the compliance program, are as follows: (Sanders, 2017).

1. Code of business conduct preparation and updates;
2. Compliance risk assessments and preparation of risk-based compliance policies and procedures;
3. Risk-based due diligence on third-parties and business partners, and M&A due diligence;
4. Preparation of compliance contract provisions, representations and warranties, and certifications;
5. Compliance counseling advice and opinions;
6. Compliance training and preparation of training materials and presentations;
7. Compliance monitoring, auditing and testing, and related remediation activities;
8. Representation before relevant federal, state, or local regulatory or enforcement agencies during investigations, and with regard to compliance procedures and activities.

The implementation of a compliance program itself does not compel corporations without positive aspects. As Martin T. Biegelman analyzes that "a strong compliance program is absolutely necessary to protect an organization both internally and externally" (Biegelman, 2008, p. 2). This program's aim does not merely require a corporation to comply with the existing laws and regulations to avoid criminal or civil violations. Compliance actually has a deeper meaning; that is "developing and sustaining a culture based on values, integrity, and accountability, and always doing the right things" (Biegelman, 2008, p. 2). This then affects the establishment of trust among the shareholders, employees, vendors, and the public. An effective compliance program must combine one's action and one's mindset and not just lip service or just writing on paper. Thus, there is a guarantee that the corporation needs the compliance program.

Based on the conceptualization of the compliance program, DPA as an umbrella can be applied to all types of criminal acts. The general nature of this can also be seen in the formulation of RKUHP 2015 discussion in 2017. RKUHP aspires to be a legal umbrella that is *lex generalis* for corporate criminal acts in various sectors, such as environment, corruption, taxation, and others. This is slightly different from that of the United Kingdom which limits the use of DPA (not just compliance) to the economic crimes based on Part 2 Schedule 17 of the Crime and Courts Act 2013.

Although Indonesia does not formulate for what crime DPA can be implemented, the adoption of DPAs value in some criminal regulation in Indonesia can be said as the effort to internalize DPA partially. Article 57A of RKUHP states that corporate criminal liability should consider the following:

a. the level of harm inflicted on society;
b. the level of corporate leadership involvement;
c. the duration of crime committed;
d. the frequency of crime by corporation;
e. the willingness to commit a crime;
f. the crime involving a public official;
g. public reaction;
h. jurisprudence;
i. the corporate track record in committing a crime;
j. potential to be improved or not; and/or
k. the corporation's cooperation in the handling of criminal acts.

Such considerations clearly tighten corporate liability. A corporation that seriously runs compliance programs is to be convicted in the event of a crime. The eleven points are measurements that tighten the penalization of good corporations, and good corporations are those that run compliance programs.

Contrary thinking, *argumentum a contrario*, can also be made to the compliance program. It is also internalized into the RKUHP and PERMA mentioned above. This means that rejection or absence of a compliance program can be a boomerang for the corporation. This is demonstrated by an example of the case presented by Martin T. Biegelman, namely Computer Associates. There were internal corporate issues, such as disagreements that resulted in clashes of employee dismissals, minimal training, and corporate policies that did not contain internal controls. In the end, corporate officials did massive manipulation and conspiracy (Biegelman, 2008). Computer Associates was then convicted on that basis. The absence of a compliance program is also a consideration in determining the occurrence of a corporation's criminal offense under Article 5 of the Crime and Courts Act Act 2013 in the United Kingdom.

Then, if it is associated with the theory of corporate criminal liability, the compliance program is the embodiment of the corporate culture model. Corporate culture model is a theory of corporate criminal liability that focuses on express and implied corporate policies that affect the way corporations conduct their business (Sjahdeini, 2006). Mahrus Ali, citing Jennifer Hill, clarifies: "An attitude, policy, rule, course of conduct, or practice existing within the body corporate generally or within the area of the body corporate in which the relevant activities take place" Ali, 2013, p. 130). However, the theory of criminal liability is not accepted by the United States and England, but it is implemented by Australia.

The rejection of the United States and England can be understood from the implementation of the DPA which conditions the double cross-check of the offender. For example is by giving an opportunity to the corporation to confirm the position of corporate management: whether the actions of a board can be said as a representation of the corporation or individual management. This attitude is a manifestation of the rejection of the absolute application of vicarious liability doctrine (Weismann, 2008).

This is as formulated in PERMA No.13 of 2016 on the Procedures of Handling Criminal Cases by Corporations. PERMA specifies the terms to determine the following corporate errors:

a. Corporations may benefit or take advantage from such offenses or the offenses are committed for the benefit of the Corporation;
b. Corporations allow criminal offenses; or
c. Corporations do not take the necessary steps to do precautions, prevent greater impact, and ensure compliance with applicable legal provisions to avoid crime.

The third point of the determination of this corporate error is the substance of the compliance program because it contains the precautionary elements as presented by Pappas. The preventive measures undertaken by the corporation are a manifestation of the compliance program. It can be said that the DPA, which requires the compliance of the program, actually contains the objective of corporal punishment, namely the appropriate prevention and retaliation.

3 Effectiveness of the conventional compliance program *and* DPA's compliance program
A compliance program or due diligence is a form of compliance already known in Indonesian law, prior to the emergence of DPA. The form follows the field of study. For example, compliance in the health sector should be a form of compliance with all health regulations.

Failure to comply with this compliance may result in civil or criminal violations. Programs are established as a precautionary measure. While the compliance within the DPA is more comprehensive, in addition to before the occurrence of criminal acts (prevention) the program can also be established when the corporation has committed a crime. The corporation pleads guilty, pays a fine, cleans up (usually to the context of environmental pollution), and has compliance programs and other matters agreed upon by the available guidelines. It shows that the compliance program's position in DPA mechanism is stronger than the conventional program.

3.2.3 *DPA: cross-jurisdictional corporate liability*

One form of DPA's implementation for a cross-border criminal act is Innospec, which brings together the United States authority with the United Kingdom's. The two countries conduct joint investigations against multi-jurisdictional cases and conduct joint negotiations with foreign prosecutors. However, to avoid double jeopardy, the two authorities, namely the Department of Justice (DOJ; the United States) and SFO (the UK), perform the division of labor.

In the case of Innospec, the first of such investigation was conducted in the United States in 2005 and was followed by SFO in 2008. Innospec pleaded guilty in the UK to a charge brought under section 1 of the Criminal Law Act 1977 of conspiring to give or agree to give corrupt payments to government officials in Indonesia as inducements to secure, or as rewards for having secured, contracts with the Indonesian Government for the supply of a lead-based anti-knock fuel additive called Tetraethyl lead (George, 2014, p. 119). Meanwhile, DOJ demanded Innospec to plead guilty in the US District of Columbia to DOJ charges of breaching UN sanctions by bribing officials in connection with the UN's Oil for Food Programme in Iraq from 2000 to 2003 (George, 2014).

Based on this, it can be said that the agreements made in parallel can be negotiated and summarized "...that they offer the prosecuting authorities a politically satisfactory settlement with regard to the division of offences" (George, 2014) as seen in the case of Innospec.

In connection with the case of Rolls-Royce and Garuda Indonesia, the state's losses due to corruption and alleged Crime of Money Laundering committed by Rolls-Royce through a third party against ES (former head of Garuda) can be overcome. The amount of bribes allegedly received by Emirsyah amounted to 1.2 million Euro and USD180,000 or equivalent worth Rp20 billion and goods in the amount of USD2 million (equivalent to Rp26 billion), which was spread across Singapore and Indonesia (Laluhu, 2017). When the DPA regulations were already known in Indonesia, just like Innospec, Rolls-Royce should be responsible for returning the Indonesian state losses. This happened because the DPA was done in the UK related to the number of fraud and bribery incidents that the company did in other countries. However, the many aspects of DPA teach us that it is a complicated concept that we have to think about when we decide to apply it comprehensively in our criminal justice system. For some instances, Australia and most recently Canada which tried to formulate their own model of DPA can also enrich our insight to think about DPA at the national level,

4 CONCLUSION

Based on the above explanation, it can be concluded that corporate criminal liability in Indonesia can actually consider the ideas contained in the DPA. This is particularly useful for cases that have global (international) dimensions involving other countries (especially those who are also familiar with DPA). A DPA that requires an effective corporate compliance program has the power to prevent and retaliate equally to the corporation.

Compliance programs can be useful to "prevent" in the pre-crime stage and "crackdown" at the time of the crime. In the event of a crime, DPA can be a solution to ensnare corporations that commit transnational crimes. Other countries with DPA mechanisms may conduct the joint investigation and join negotiation in multi-jurisdictional cases.

REFERENCES

(SFO), S. F. (2017, January 17). *Serious Fraud Office*. Retrieved June 27, 2017, from http://www.sfo.gov.uk: https://www.sfo.gov.uk/cases/rolls-royce-plc.

(SFO), S. F. (2017, March). *Serious Fraud Office*. Retrieved June 23, 2017, from http://sfo.gov.uk: https://www.sfo.gov.uk/publications/guidance-policy-and-protocols/deferred-prosecution-agreements/

Ali, M. (2013). *Asas-asas Hukum Pidana Korporasi*. Jakarta: Raja Grafindo Persada.

Alschuler, A. W. (2009). *Two Ways to Think about the Punishment of Corporations*. California: Northwestern University School of Law Scholarly Commons.

Amulic, A. (2017). Humanizing the Corporation While Dehumanizing the Individual: The Misuse of Deferred Prosecution Agreements in the United States. *Michigan Law Review*, 122-153.

Biegelman, M. T. (2008). *Building A World-Class Compliance Program: Best PRactices and Strategies for Success*. Hoboken: John Wiley & Sons, Inc.

Brodowski, D. e. (2014). *Regulating Corporate Criminal Liability*. New York and London: Springer.

Department, A. G. (2016). *Improving Enforcement Options for Serious Corporate Crime: Consideration of a Deffered Prosecution Agreements Scheme in Australia*. Australia: Australian Government adn Attorney General Departmet.

DetikNews. (2017, January 20). *Detik News*. Retrieved June 20, 2017, from http://news.detik.com: https://news.detik.com/berita/d-3401205/tak-berwenang-periksa-rolls-royce-kpk-gali-data-dari-sfo

George, S. A. (2014). Deferred Prosecution Agreement: In Jeopardy of Falling Short? *Business Law Internasional*, 115-122.

Laluhu, S. (2017, May 9). *Sndonews.com*. Retrieved August 25, 2017, from http://www.nasional.sindo news.com: https://nasional.sindonews.com/read/1203926/13/kpk-segera-rampungkan-berkas-kasus-emirsyah-satar-1494340342.

Morgan, B. (2017, March 8). *Serious Fraud Office*. Retrieved June 23, 2017, from http://www.sfo.gov.uk: https://www.sfo.gov.uk/2017/03/08/the-future-of-deferred-prosecution-agreements-after-rolls-royce/.

Mugopal, U. (2017, Juni 2). Ultimum Remedium dalam Pertanggungjawaban Pidana Korporasi. (A. Br Ginting, Interviewer).

Pappas, J. S. (2012, December 7). *American Bar Association*. Retrieved Agustus 20, 2017, from http://www.americanbar.org: https://www.americanbar.org/content/dam/aba/events/criminal_justice/Frank furt/Session1_FCPA_Compliance_Programs.authcheckdam.pdf

Priyatno, D. (2004). *Kebijakan Legislasi tentang Sistem Pertanggungjawaban Pidana Korporasi di Indonesia*. Bandung: CV.Utomo.

Priyatno, M. a. (2012). *Pertanggungjawaban Pidana Korporasi*. Jakarta: Kencana Prenada Group.

PwC's 2014 Global Economic Crime Survey. (2014). *Economic Crime: A Treat to Business*. PwC's.

Sanders, T. (2017). *Troutman Sanders*. Retrieved July 21, 2017, from http://www.troutman.com: https://www.troutman.com/corporate_compliance/

Sjahdeini, S. R. (2006). *Pertanggungjawaban Pidana Korporasi*. Jakarta: Grafiti Pers.

Statistik, B. P. (27 April 2017). *Hasil Pendaftara (Listing) Usaha/Perusahaan Sensus Ekonomi 2016 (No. 50/04/Th. XX)*. Jakarta: BPS.

Tarigan, A. (2013). Peran Korporasi dalam kejahatan Kehutanan. *Climate Change*, 13.

Webster, M. (2017). *Merriam Webster*. Retrieved September 15, 2017, from http://www.merriam-web ster.com: https://www.merriam-webster.com/dictionary/compliance

Weismann, A. (2008). A New Approach to Corporate Criminal Liability. *American Criminal Law Review, Volume 44*:1319, 1326. Widjojanto, B. (2017, March 30). *ACCH: Anti-Corruption Clearing House*. Retrieved Juni 27, 2017, from http://acch.kpk.go.id: https://acch.kpk.go.id/id/ragam/riset-publik/kajian-awal-melacak-korupsi-politik-di-korporasi.

Advancing Rule of Law in a Global Context – Susetyo, Rinwigati Waagstein & Budi Cahyono (eds)
© 2020 Taylor & Francis Group, London, ISBN 978-1-138-32782-5

Disclosure of beneficial ownership to eradicate transnational financial crime

Riza Alifianto Kurniawan & Iqbal Felisiano
Faculty of Law, Universitas Airlangga, Surabaya, Indonesia

ABSTRACT: Eradicating transnational financial crimes requires a huge effort, especially when they involve overseas jurisdictions. Most organized crime groups use corporate vehicles to camouflage and hide their illegal assets and operations from detection. Moreover,banking secrecy agreements and protection afforded to legal entities constitute obstacles to preventing and eradicating transnational financial crimes. Most legal entities who are abused by organized crimes exploit disguises in order to deceive the government from tax obligations and cover up money laundering crimes. Beneficial owners of legal entities must be declared as having ownership of the corporation in order to fulfill accountability principles and prevent exploitation. On the other hand, the obligation to reveal ownership could be opposed by the principle of banking secrecy. The contradiction between the obligation to disclose beneficial ownership and public interest in eradicating financial crimes is an issue that must be resolved.

Keywords: beneficial owner, corporate crime

1 INTRODUCTION

1.1 *Beneficial owner regulation on transnational crimes in Indonesia*

In Indonesian regulations, the beneficial owner principle has not been applied equally and only on a sectoral basis. According to one of the provisions concerning the Beneficial Owner principle under the Regulation of the Financial Services Authority Number 22/POJK.04/2014 Concerning "Know Your Customer Principles" by Provider of Financial Services in Capital Market Sector, the Beneficial Owner shall be defined as any party that either directly or indirectly through the agreement or by any means:

a. is entitled to and/or receives certain benefits relating to (1) a securities account in the Provider of Financial Services in Capital Market Sector; or (2) the business relationship with the Provider of Financial Services in the Capital Market Sector;
b. is the true owner of the fund and/or Securities in the Provider of Financial Services in the Capital Market Sector (ultimate account owner);
c. controls a Customer's transactions;
d. authorizes transactions; and/or
e. has control of a non-Individual Customer.

Based on the aforementioned definition, the Beneficial Owner concept is not against the law if there is a disclosure according to the "know your customer" system for the financial customer.

Regulation of the Financial Services Authority No. 22/POJK.04/2014 About KYC By Financial Service Provider in Capital Market Sector defines the beneficial owner by the functions and responsibilities of each party, either directly or indirectly through contracts or other means:

a. Eligible for and/or receive certain benefits related to:
   a. A securities account at a financial services provider in the capital market sector; or
   b. Business relationships with providers of financial services in the capital market sector
b. The actual owner of the funds or securities and financial services providers in the capital markets sector (ultimate owner account)
c. Controlling customer transactions
d. Authorizing the transaction and/or
e. Controlling nonindividual customers

A law specifically governing the beneficial owner separately under several different regulations, one of which is the regulation by the financial services authority, is absent. A beneficial owner that has not been regulated should be viewed as a risk for the occurrence of organized crime. The legal vacuum concerning beneficiaries is conducive to creation of fictitious companies by organized crime that can disguise the real ownership.

Determining the beneficial owner as noncriminal or criminal raises the issue of whether the profit a beneficial owner makes from a money laundering crime can qualify as a crime or not. Therefore the beneficial owner may be qualified as part of a system to detect money laundering or organized crime. The legal basis in determining or qualifying the criminal is use of the inclusion (*deelneming*) theory to determine the role of each offender and the threat of punishment. Inclusion (*deelneming*) in Indonesian criminal law is regulated in Article 55 of the Criminal Code. Articles 55 and 56 of the Criminal Code qualify the perpetrators according to several roles:

1. The direct perpetrators
2. Persons who order the activity
3. Persons who participate
4. The accomplices
5. Persons who assist[1]

Therefore, the qualification of perpetrators under Articles 55 and 56 of the Indonesia Criminal Code (KUHP) is applicable to persons (*naturlijk persoon*) as the subjects under criminal law, whereas a corporation (*rechtspersoon*) is not a subject under the Criminal Code, only in special criminal laws or special criminal acts.

Criminal operations develop very rapidly, and at times involve entire groups. Criminal offenses may occur within cross-border or transnational jurisdictions, making the composition of actors involved very expansive, even involving corporate law subjects (*rechtspersoon*) (Lilly, 2015). The transnational crime "locus delicti" occurred in several different jurisdictions, using the criminal network as a mechanism to execute the role of each actor; therefore, organized crime groups inevitably involve many actors with different roles in the commission of crimes.

The beneficial owner in the Financial Services Authority regulation does not qualify as a perpetrator. However, according to the Criminal Code, the range of so-called perpetrators is very broad, and they may wholly or partly be engaged in doing good deeds as well or may be playing an active or passive role in the occurrence of a crime. The beneficial owner may be considered as having a partial role as a perpetrator if there is an active or passive involvement in order to complete a criminal offense (Atmasasmita, 2011). Beneficiaries in transnational crimes are particularly vulnerable to exploitation as perpetrators of a crime by those who stand to benefit from the crime. Some scholars and experts support disclosure of the identity of beneficial owners as a means of preventing transnational crimes by international crime organizations.

The Organisation for Economic Co-operation and Development (OECD) Principles of Corporate Governance state that a framework for good corporate governance requires public disclosure of all company profiles as well as the company's financial situation, the performance of the company, ownership status, and corporate governance. This principle of openness is accepted in public companies because transparency provides numerous advantages in terms of providing information on good performance that enables them to earn benefits. A company that receives the title accountable has a good rating in the capital market and is expected to be able to attract capital market investment in the company's securities.

The misuse of beneficiary positions in transnational business activities has led the Indonesian government to mitigate the risk of such misuse by enacting legislation implementing regulations to identify beneficiaries. Possible risks arising from the concept of beneficiary owners are that beneficiaries may be used to disguise the proceeds of crime and store them through some fictitious corporations or other entities abroad whose ownership status remains unclear. The risks arising from withunclear corporate entities drove Indonesia to issue the Law on the Prevention and Eradication of Terrorism Funding.

According to criminal law concepts and theories, the beneficial owner is the one who assists during or after commission of the crime.. The owner of a qualified benefit of a criminal offense has the same offender qualifications as the direct actors or those who provide assistance or intellectual support in a crime. In practice, entangling the beneficial actors who are involved in transnational crimes is more just in the context of pursuing the perpetrators rather than the corporation. This is due to the need for a *naturlijk persoon* to have committed an offense so that the law enforcement process conducted by investigators, prosecutors, and judges could be more accountable in proving responsibility for the crime under the act that applied to the perpetrator. In transnational financial crime, the beneficiary in the form of a legal entity may be held criminally liable provided that the criminal law provisions used to prosecute such special offenses recognize the corporation as a subject under the criminal law and do not require the corporation fault.

1.2 *Implementing non-conviction based asset forfeiture as an alternative to restore the state financial losses due to criminal offenses related to the state economy*

Weaknesses in imposing criminal liability in corporate crimes may actually be overcome if we do not focus solely on the role of *naturlijk persoon* as the only subject in an organized crime. Concepts used in the appropriation of assets in money laundering law, as an example, had used the principle of Conviction-Based Asset Forfeiture, or confiscation of an asset connected to criminal activity by the guilty party. This concept prioritizes punishment of the offender (in persoon). The concept basically has the disadvantage that it is not easy for the prosecution to prove the relationship between the assets to be confiscated from offenders and the crime.

Indonesia is a country that ratified UNCAC under Law No. 7 of 2006 on the Ratification of United Nation Convention against Corruption. Mentioned in the UNCAC, Article 54 paragraph (1) c that "Each State Party, in order to provide mutual legal assistance pursuant to article 55 of this Convention with respect to property acquired through or involved in the commission of an offence established in accordance with this Convention, shall, in accordance with its domestic law.... such measures as may be necessary to allow confiscation of such property without a criminal conviction in cases in which the offender cannot be prosecuted by reason of death, flight or absence..." However, based on data gained by KPK (Indonesia Corruption Eradication Commission), only 30% of the UNCAC article has been implemented.

2 DISCUSSION

The corporation has been acknowledged as a legal entity in Indonesia since 1951. The relevant legislations are Article 15 paragraph (1) of Law No.7 Drt of 1955 on Economic Crime, Article 17 paragraph (1) of Law No.11 PNPS of 1963 on the Act Criminal Subversion, Article 49 of 1976 on Narcotics, Law No.9 of 1976 on Narcotics Crime, Article 1 point 13 and Article 59 of Law Number 5 of 1997 on Psychotropic, Article 1 point 19 of Law Number 22 1997 concerning Narcotics, Article 1 number 1 and Article 20 of Law Number 31 of 1999 on the Eradication of Corruption, Article 116 of Law Number 32 of 2009 on Environmental Protection and Management, Article 1 point 20 of Law Number 35 of 2009 on Narcotics, Article 1 point 10 and 14 and Article 6 of Law Number 8 of 2010 on Prevention and Eradication of Money Laundering Crime. Thus de jure, Indonesia has acknowledged corporate criminal responsibility since 1951 (Manthovani, p. 4). However, based on the data, recovery of state financial losses has for the most part not been a success story.

Many of the perpetrators cannot be prosecuted because they have fled or are not Indonesian nationals, so that actions against the proceeds of their crimes cannot be executed. A plywood company in Riau Province, for instance, buys raw wood materials from a timber company that has no license for the utilization of forest products and conducts illegal logging in Bukit Tiga Puluh National Park. The plywood company sells wood panels to its buyers in China, South Korea, and Taiwan through an Indonesian merchant company located in Hong Kong. Timber company employees, plywood companies, and marketing companies in Hong Kong realize that the wood used to make wood panels comes from illegal logging. To disguise the fact that corporate profits come from illegal activities, these three companies adopt a different strategy. Timber companies place the proceeds of the crime into the financial system by depositing them into a bank account under a fictitious name. The marketing company does layering by diverting its monetary receipts through a bank in the Cayman Islands while plywood companies integrate their profits into legal business activities by investing the money in a tourist destination area in Bali (Setiono and Barr, p. 4). In this case, the progression of the crime through many parties that are not subject to Indonesian criminal law liability may render any action against the proceeds of the crime unexecutable.

Non-conviction based (NCB) asset forfeiture is useful as a tool for recovery of the proceeds of criminal instruments. NCB asset forfeiture may anticipate situations in which criminal proceedings for an offense are instituted against an offender but not concluded because the offender cannot be brought before a court or convicted because of a lack of evidence. These situations occur when the relevant authorities, such as an investigator and a prosecutor, fail to identify an offender or fail to prove a particular person is at fault. In practice, these situations may also arise for other reasons, for example, when suspects enjoy immunity, die in the course of criminal proceedings or are declared dead, are under the age of criminal responsibility, etc. (Euro Just, 2017, p. 9).

According to Greenberg, NCB asset forfeiture may be useful in a variety of contexts, particularly when criminal forfeiture is not possible or available, as in the following examples:

a. The violator is a fugitive. A criminal conviction is not possible if the accused is a fugitive.
b. The violator is dead or dies before conviction. Death brings an end to criminal proceedings.
c. The violator is immune from criminal prosecution.
d. The violator is so powerful that a criminal investigation or prosecution is unrealistic or impossible.
e. The violator is unknown, and assets are found (for example, assets found in the hands of a courier who is not involved in the commission of the criminal offense). If the asset is derived from a crime, an owner or violator may be unwilling to defend civil recovery proceedings for fear that this would lead to criminal prosecution. This uncertainty makes a criminal prosecution of a violator very difficult, if not impossible.
f. The relevant property is held by a third party who has not been charged with a criminal offense but is aware—or is willfully blind to the fact—that the property is tainted. While criminal forfeiture may not reach the property held by bona fide third parties, NCB asset forfeiture can forfeit the property from a third party without a bona fide defense.
g. There is insufficient evidence to proceed with criminal prosecution (Greenberg et al., 2009, pp. 14–15).

The foundation of the implementation of NCB asset forfeiture, however, has been regulated in Article 16 paragraph (1) of Law No.7 Drt of 1955 on Economic Crime that matches the example provided by Greenberg. The article mentions the following:

If there is sufficient reason to suspect that a person who dies before a case has an irreversible verdict, has committed an economic crime, then the judge—at the public prosecutor's request—by a court decision has the discretion

a. To decide upon the appropriation of goods that have been seized. In that case, article 10 of this emergency law shall be in effect;
b. To decide that the act of order mentioned in article 8 sub c and d is done by incriminating it on the property of the deceased person. (Law No. 7 on Economic Crime, 1955)

From the formulation of the Article, there are several points that may enact the NCB asset forfeiture principle toward the "unknown person," related to Article 16 paragraph (6), similar to the Greenberg's condition to enact NCB asset forfeiture.

Furthermore, the NCB asset forfeiture principle is also implicitly contained in Anti-money Laundering Law (AML) under Articles 26, 65, 67 par. (2), and 71. Under those articles, NCB asset forfeiture is described as a process that stems from transactions that are reasonably suspected to originate from criminal proceeds by financial service providers, followed by the order from Indonesia Financial Intelligence Unit (PPATK) to suspend the whole or part of the transaction in the context of analysis or examination of reports and information, and cooperation between FIU and investigators to block the suspicious transactions. Then, if the perpetrator cannot be found within 30 (thirty) days, the investigator may apply to the court to decide such assets as the assets of the state or mandate their return to the rightful owners. The procedure under the AML views assets as proceeds of a criminal act positioned as legal subjects/parties so that they consist of countries represented by the investigators of AML as applicants/claimants against the assets of suspected proceeds or means of criminal acts as the respondent. This mechanism allows for the seizure of assets without having to wait for a criminal verdict containing a statement of guilt and punishment for the offender.

3 CONCLUSION

By implementing the principles of both beneficial owner regulation and NCB asset forfeiture, restoring state financial losses due to criminal offenses, related to transnational and organized crime would not be a problem, even if both of the principles are not mentioned and regulated specifically under the law. Regulation of the Financial Services Authority No. 22/POJK.04/ 2014 About KYC By Financial Service Provider in the Capital Market Sector could at least provide the Authority with the definition of the party that may be a responsible subject and burdened with criminal liability. In the meantime, NCB asset forfeiture could also be implemented by using AML in order to restore the state's financial loss due to criminal offenses related to the state economy when the perpetrator cannot be prosecuted by considering assets as proceeds of a criminal act positioned as legal subjects/parties under Article 67 of AML.

REFERENCES

Atmasasmita, Romli. (2011). *Globalization and Business Crime*. Jakarta: Kencana Pranata Media.
Euro Just. (2012). Report on non-conviction-based confiscation, General Case 751/NMSK – 2012.
Greenberg, Theodore S.., et al. (2009). *Stolen Asset Recovery; A Good Practices Guide for Non-Conviction Based Asset Forfeiture*. Washington, DC: The World Bank.
Lilly, Robert. (2015). *Teori Kriminologi Konteks dan Konsekuensi*. Jakarta: Kencana Pranata Media.
Manthovani, Reda. Prosecution of Corporations as Crime Actors in Crime in the Forestry Sector: Optimizing the Use of Money Laundering Laws in Criminal Investigation in Indonesia's Forestry Sector Conducted by the Corporations.
Moeljatno, Kitab. (2010). *Undang Undang Hukum Pidana*. Jakarta: Citra Aditya Bhakti.
Setiono, Bambang and Christopher Barr. Using Anti Money Laundering Law to Combat forest crimes in Indonesia. Centre for International Forestry Research (CIFOR).

Advancing Rule of Law in a Global Context – Susetyo, Rinwigati Waagstein & Budi Cahyono (eds)
© 2020 Taylor & Francis Group, London, ISBN 978-1-138-32782-5

Legal cultural communication system for civil society participation in eradication of corruption

Fitriati
S2 Law Program, Ekasakti University, Jalan Bandar Purus, Padang, Indonesia

ABSTRACT: Eradication of corruption cannot be done only through formal procedures but also requires the strengthening of community participation. This is regulated in Article 41 paragraph (3) of Law No. 31 of 1999 on eradicating corruption, stating that the community has a right and responsibility in preventing and eradicating corruption. This study used a sociological juridical approach, with primary and secondary data as the source. The results show legal communication as a form of influencing societal views in nonpenal crime prevention efforts. Legal communication is done in a persuasive manner with a direct application to society, taking into account the existing culture in the community. Legal communication is done so that every society can better disseminate legal knowledge about corruption crime and increase public legal awareness. In general, law enforcement will not succeed if we do not pay attention to or ignore the value of cultural and moral values.

Keywords: communication, law, participation, society

1 INTRODUCTION

Improvement of the legal substance desired for the prevention of corruption is related to the renewal of the various sets of rules and normative provisions (legal reform), patterns, and the will of community behavior existing in the legal system. The basic strategy of combating corruption should be focused on eliminating the causes and conditions of the conditions that lead to the occurrence of crime. As contained in the 6th UN Congress report of 1981 stating that (Arif, 2008): "Crime prevention strategies should be based upon the elimination of causes and conditions giving rise to crime; (that a crime prevention strategy should be based on the elimination of causes and conditions that constitute a crime)."

Law enforcement at its core lies in the attitude and behavior of society. Communities as legal subjects directly or indirectly participate in the occurrence or absence of criminal acts. Community participation is influenced by cultural conditions. Communities have engaged in various forms of participation, which can both tackle and cause corruption-related crime. Participation takes place if something is part of the structure and norms, especially customary norms. For participation to function optimally and for the prevention of corruption, it is necessary to create a system of forms of participation.

Up to now, law enforcement is conducted mostly regardless of the values and norms comprising the cultural wealth of the community. A semiotic perspective explains the role of the legal function as facilitative, repressive, and also ideological. Semiotic analysis can be interrogated in a number of perspectives to construct a more holistic approach in the sociology of law (Susanto, 2005).

Legal communication is done in a persuasive manner with a direct application to society, taking into account the existing culture. It is done so that every society can better disseminate legal knowledge. Legal communication has actually existed unnoticed in the daily life of the community but not yet organized in an effective system. It can be grown from an existing culture in society andIt developed by the community utilizing law enforcement officers and involving academics or people who have legal knowledge. Issues to be studied are how to build community participation in combating corruption and how a culture-based legal communication system can be utilized to combat corruption.

2 METHODS

This study used socio-legal empirical researchsupported by normative law research. A descriptive-analytical approach was used. Data collection techniques were observation, questionnaire, and interview. Qualitative analysis was used for a rational analysis of the data by using certain thinking patterns.

3 DISCUSSION

3.1 *Building community participation in combating corruption*

Community participation in prevention of corruption has been found in some areas. This participation is manifested primarily in the role of customary institutions, which is to supervise acts of criminal corruption. Supervision is done by each tribe incorporated in Kerapatan Adat Nagari (KAN). KAN here has conducted the settlement of corruption cases on the nagari through the nagari leaders themselves. As a case in point, nagari tigo balai reported the suspected corrupt Wali Nagari Goro badunsanak grants from the district government.

Some other forms of community participation are:

1. Supervision of customary density of nagari (KAN) to the nagari government. For example, surveillance is done in the uninhabited housing program in Sumani village. The supervision performed has a positive impact. The program runs according to the applicable rules. Here KAN makes adaptations with beneficiaries and encourages them to have the courage to monitor the help they receive. Often people quickly feel satisfied with the help they receive. Communities are given the awareness that the assistance they receive is their right. Relief recipients are prohibited from giving "aia" money to relief officers such as wali nagari. The surveillance system is a self-monitoring system. People are given the awareness not to engage in corrupt practices they may have unconsciously done. Activation of the entire system has the effect of complete oversight of all involved actors.

2. Based on the supervision conducted by KAN, it was found that there was a misappropriation by the nagari secretary by cutting BLT funds from the BBM compensation to the recipient community in Tanjung Bingkung Nagari. The problem was initially resolved by KAN and Wali Nagari. But the reprimand against the nagari secretary had no effect. Cutting was still done. The nagari secretary felt his deeds were innocent because the cut funds were not enjoyed by themselves. Based on further investigation by community leaders, the nagari guard and some other nagari government officials also enjoyed money from the cuts. Based on the decision of the Governing Body of Nagari (BPN) or the legislature in Nagari and KAN, the case was reported to the authorities. At the local level, a dismissal action is taken against the parties allegedly involved. Reporting is done on behalf of the community through BPN and KAN.

3. Corruption occurs not only among political elites. An example is the existence of a fictitious SPJ made by the nagari secretary for aid from the government in the form of an emergency housing fund for earthquake victims in the community in nagari Kapuah. This case is cited by people who feel aggrieved when they are only asked for a signature receipt, while the money is never received. The community did not immediately report the case to the authorities. Customary leaders and community leaders formed an independent settlement team, named Team Settlement, to handle missing money cases. This team consists of elders in nagari and adat leaders; they are those who have no blood relation and kinship with the nagari secretary. This independent team, choosing to confiscate the private car of the nagari secretary, with the nagari secretary's agreement, aimed to repatriate the funds he embezzled. However, the nagari secretary was unable to meet the demands

of the community. Eventually, the community took action by making arrangements with the nagari secretary. The agreement stipulates that 100 times the harvest of the rice fields owned by the nagari secretary must be submitted to nagari to be distributed in the form of revolving loans to the community.

Community groups manifest their participation by taking action. The action they take is based on the experience that when they travel the legal path, they often fail. The community attempts to resolve every corruption case it is involved in informally. If the informal path fails, then they report to the authorities. Reporting often encounters obstacles. Many cases reported by the community are then not processed for various reasons and the case disappears without any further knowledge. The population lacks public information about the criminal process of corruption that they have reported.

Action theory states that individuals perform an action based on experience, perception, understanding, and interpretation of a stimulus object or a particular situation. Individual action is a rational social action, namely to achieve target goals by the most appropriate means (Pateman, 1990). Max Weber's theory was developed by Talcott Parsons, which states that the action is not behavior. Action is a mechanical process against a stimulus, whereas behavior is an active and creative mental process.

Talcott Parsons thinks that the main element is not individual action but the social norms and values that demand and regulate that behavior. The objective conditions united with the collective commitment to value will develop a certain form of social action. Talcott Parsons also assumes that the actions of individuals and groups are influenced by the social, cultural, and personality system of each individual. Talcott Parsons also classifies the type of role in a social system called Pattern Variables, which contains an effective, self-oriented and group-oriented interaction.

In addition to the form of community participation to crack down on corruption, there is also a form of public participation in processing criminal corruption cases, including, among others

1. The existence of individual communities that submit reports to investigators and prosecutors in the form of evidence against the case in progress
2. A willingness of the community to be questioned as a witness without any call by the investigator
3. Conveying the existence of suspected fugitives directly to the investigator

Gordon Wallport, in his book *The Psychology of Participation* (Sastropoetro, 1988), states: "The person who participates is ego-involved instead of merely tasks-involved." Hence a participating person is actually experiencing his or her own involvement or ego that is more than involvement in the work or task alone. Self-involvement encompasses his thoughts and feelings. Based on this understanding, a person performs activities according to whether they agree or disagree with his thoughts and feelings. In the context of the prevention of criminal acts of corruption, the expected public participation involves the conscious involvement of the community. Community involvement as a form of participation should be supported by relevant law enforcement officers.

The need for increased participation can be addressed by involving all components and layers of the community; developing local potentials; mobilizing self-help target groups; and developing a coaching methodology for awareness raising, initiative and motivation, and human resource development. Continuous improvement of skills and programs could be used to shift attitudes and mental perspectives toward more positive and rational direction (Sumodiningrat, 1996).

Participation in tackling corruption is not only the task of institutes set up specifically to fight corruption; non-watchdog institutions have also included anti-corruption programs as part of their activities, such as environmental and human rights activists, teachers, cooperatives, religious institutions, cultural organizations, and others, although sometimes only when their interests are disrupted. This shows an increase in community participation in the prevention of corruption.

Community participation in the prevention of corruption must arise from the legal awareness of the community. Legal awareness is a consciousness that exists in every human being about what the law is or what it should be, a certain category of our psychic life by which we distinguish between law and not law (onrecht), between what should be done and not done (Scolten, 1954).

Cultural factors have a positive and negative influence on community participation. As expressed by Chiba (1989), cultures would be in the triangle of religion/ethics and morality as well as the society of the state. Culture will be between the laws of the state and society and will affect a certain level of law enforcement.

The positive influence of culture is the customary way in which Minangkabau society resolves a problem that has occurred, especially when the problem is about community life. In Minang, the term "Batanggo naik bajanjang turun" means a problem in nagari must be resolved according to a clear stage. Stages range from simple community groups to large community groups. This can be seen as a form of community participation that seeks to resolve a case that is suspected of being a criminal act of corruption at the lowest level before reporting to the authorities to be processed formally under law.

In addition, the influence of cultural factors on the participation of the community in a positive way is seen in the involvement of customary figures and community leaders in the supervision and as facilitators in development that have a point of vulnerability of corruption. The results of supervision and facilitation undertaken will be discussed periodically at customary meetings in a nagari.

Besides positive factors in Minangkabau culture, there are also negative factors that influence community participation. Habits that are considered commonplace as modest norms can actually become the seed of corruption. For example, at Minang the term for money, "panjek" (climbing) is also known as the term for cigarettes. Money is usually given to make a business dealing go smoothly or to achieve what is desired or intended. Human behavior factors must not be understood only as those perceivedwith the five senses, but also must take into account that human beings involved in the law enforcement process always interact with the environment based on the culture, given that the legal system is not a machine; it is run by human beings (Friedman, 1975).

Another negative factor is the existence of a habit that is considered a form of ethics in Minangkabau, that of giving something to achieve a goal. For example, the community gives agricultural produce to get capital assistance from the authorities. This custom is known as "ma agiah buah tangan" (memberi oleh-oleh), where bribery as a form of corruption is justified. Bribes are bad for society and cause rapid growth of corruption at the next level.

Culture serves as the normative framework in human life that determines the behavior of community members. These societal behaviors constitute a system that indirectly becomes a legal culture. The legal culture greatly affects the effectiveness and success of law enforcement. Law is the concretization of social values formed from culture. The formation of the law should be adapted to the legal culture of society. The failure of modern law is often attributable to the fact that it is not compatible with the culture of community law.

Another factor influencing public participation in the prevention of corruption is the low level of education and public legal awareness. This impacts on the widespread involvement of the community in causing corruption. The public often cannot distinguish which activities include corruption and which do not.

The most powerful factor in encouraging the emergence of community participation in the most essential eradication of corruption is to practice the values of social justice in everyday life, whether in dealing with the government, businesses, or fellow citizens. If someone wants to participate in tackling corruption then he should be able to refrain from engaging in corruption. This attitude is related to morals. Good morals can be established with good religious understanding.

Esmi Warassih states that the community empowerment process should emphasize the process of giving or transferring some of the strength, power, or ability to the community and encourage or motivate the individual so as to have the ability or empowerment to determine life choices through the process of dialogue. Community empowerment is a strength to gain access to existing resources so that there will be a fair division of powers that can raise people's awareness of the existence of the resources (Warassih, 2005).

3.2 *Cultural-based communication system for criminal corruption*

Legal communication is a form of influencing the view of society through nonpenal endeavors as a crime prevention effort. Legal communication is conducted in a persuasive manner with a direct application to society, taking into account the existing culture. Its objective is to make it possible for every society to disseminate legal knowledge better.

Communication is a tool that can connect one individual with another. It is a persuasive approach, which in relation to law enforcement in society, especially marginal society, is considered the most effective.

Legal communication can grow from the existing legal culture in society. Culture is one of the perspectives developed in law enforcement. Legal communication is a form of local cultural development in law enforcement efforts. Satjipto Rahardjo stated that the culture of local law is a culture that applies to a particular group of people. The culture of local law should pay attention to aspects of the community concerned (Rahardjo, 2006), as these are one of the bases for establishing legal communication.

The form of legal communication that has been done by the community so far can be seen by mapping community activities that have also been used for the prevention of corruption. These activities are conducted via

1. Assembly taklim
2. Teens' mosque
3. Farmers' groups
4. Women's groups
5. Nagari meetings
6. Customary meetings
7. Informal meetings
8. Persuasive approaches by law enforcement officers

Legal communication is done in a direct way to the community through the activities undertaken in the community groups. Legal communication delivers information about the law to the community directly. Information is submitted to a community group and sometimes directly to the individual. It can be transmittedthrough casual chats with the community. Legal communication also takes place at informal meetings of the community such as in a warung stalls, resting places in the middle of work, and others.

Satjipto Rahardjo argued that the law should be withdrawn from the realm of legislation. It is an idea of legal studies that also concerns the social phenomena of the workings of the law in society (Satjipto Rahardjo, 2006). There are many classes and ethnic groups with a special culture in Indonesia. Rural and urban communities have different characteristics. Legal issues in society can be solved by the traditional approach. First, the social stratification or coating society in that environment must be known. Furthermore, it is necessary to know the social institutions that are present and highly appreciated by the largest part of the community. Societal institutions have a great influence on sociocultural stability and on changes therein that are happening or will happen in the future.

Law enforcement comes from the community and aims to achieve peace within the community. Today, community involvement is very necessary in the process of law enforcement. Communities can influence law enforcement. The legal provisions for crime prevention must stem primarily from the rational efforts of the community to cope with crime, according to to what Marc Ancel formulated as "the rational organization of the control of crime by society" (Ancel, 1965).

Legal communication by existing social institutions has not been able to touch all levels of society, and there is no specific pattern in the use of such communications for law enforcement.

Legal communication involves public awareness of the importance of keeping the law. In line with the nature of the law as a compelling norm, legal communication is also coercive. Many variables affect audiences' attitudes, such as the impact of communication from other directions (other people) or differences in the frame of reference, or the background of the audience experience or of some previous legal communication recipients who are trapped in settlements of their cases requiring guarantees and material prerequisites. Many experiences of the justice seeking community, as well as violations of the law, are a consequence of not believing in the good faith of law enforcement officers. This is proof that there has been a long-standing clash of relations between law enforcement and the public in the understanding and view of habits that can melt in the rules on the ground. The importance of assertiveness and good examples that educate law enforcement must be stressed, because if not, then the law can be broken with the pressure of public opinions that may be irrelevant. Many people become judges themselves in solving crimes. In legal communications, the role of society is manifested by establishing independent institutions to serve as forums to eradicate corruption Participation in the form of supervision and facilitation requires an existing institution. There is a need for revamping institutional institutions formed by these communities.

Cultural reality is a social wealth that needs to be continuously maintained and developed to create a conducive social base for the development of national law. Because of problems that cannot be solved solely by legislation, the laws that reside in society are the supporting factors that can be utilized to streamline and develop the national law (Arif, 1996).

Herbert L. Packer proposes that criminal acts are not only immoral but are also contrary to the general view of society (the criminal sanction should ordinarily be limited to conduct that is seen, without significant social dissent, as immoral), and added to the action that harms to others (Packer, 1968). The law that resides in the community creates a legal culture. Legal culture is a human attitude to the law and its legal system, beliefs, values, thoughts, and expectations. The legal culture also includes the social mood and social forces that determine how laws are used, avoided, or abused. Without legal culture, the legal system itself will not be empowered. Legal culture is covered in the dalah of communication law done by society.

Utilization of local culture that exists in society, including with legal communication based on culture, is a powerful method for the prevention of corruption. Research shows that community participation tends to create a local wisdom within the society but not at a fixed level and rather depends on the the the act of criminal corruption encountered.

4 CONCLUSION

Strengthening community participation is an effective method in efforts to eradicate corruption. Corruption is closely related to the morals and cultures that grow in the community, so prevention also requires improvements in the use of the cultural and legal communication system in society. Community empowerment in the handling of corruption can be started by utilizing local customs and wisdom, whichcan be realized by implementing the existing forms of communication through customary institutions in the community.

REFERENCES

Abdurrahman. (1986). *Tebaran pikiran tentang studi hukum dan masyarakat*. Jakarta: Media Sarana Press.
Ancel, Marc. (1965). *Social Defence*. London: Routledge & Kegan Paul.
Arief, Barda Nawawi. (1996). *Bunga Rampai Kebijakan Hukum Pidana*. Bandung: PT. Citra Aditya Bakti.
Chiba, Masaji. (1989). *Legal Pluralism: Toward a General Theory of Japanese Legal Culture*. Tokyo: Tokai University Press.
Friedman, Lawrence. (1975). *Legal System: A Social Science Perspective*. New York: Russell Sage Foundation.

Packer, Herbert L. (1968). *The Limits of Criminal Sanction*. Stanford, CA: Stanford University Press.

Pateman, C. (1990). *Participation and Democratic Theory*. Melbourne: Cambridge University Press.

Rahardjo, Satjipto. (2006). *Hukum dalam Jagat ketertiban*. Jakarta UKI Press.

Sastropoetro, Santoso. (1988). *Partisipasi, Komunikasi, Persuasi dan Disiplin dalam pembangunan Nasional*. Bandung: Alumni.

Scolten, Paul. (1954). Algemeen/Deel, NV Uitgeversmaatschappij W.E.J. Tjeenk Willink.

Sumodiningrat, Gunawan. (1996). *Pembangunan Daerah dan Pemberdayaan Masyarakat*. Jakarta: Bina Rena Pariwara.

Susanto, Anthon Freddy. (2005). *Semiotika Hukum, dari dekonstruksi teks menuju progresivitas makna*. Bandung: Refika Aditama.

Warassih, Esmi. (2005). *Pranata Hukum sebuah telaah sosiologis*. Semarang: PT Suryandaru utama.

Advancing Rule of Law in a Global Context – Susetyo, Rinwigati Waagstein & Budi Cahyono (eds)
© 2020 Taylor & Francis Group, London, ISBN 978-1-138-32782-5

Decree of the People's Consultative Assembly as the legal basis in the lawmaking process (analysis of the Decree of the People's Consultative Assembly of the Republic of Indonesia No. XVI/MPR/1998 on the Political Economy in terms of Economic Democracy)

Kuntari
Expert Board of the House of Representative, Jakarta, Republic of Indonesia

Fitra Arsil
Constitutional Law Department, Faculty of Law, Universitas Indonesia, Indonesia

ABSTRACT: This article discusses the validity of the Decree of the People's Consultative Assembly (Tap MPR) as the legal basis in the lawmaking process. Before the amendment of the 1945 Constitution, Tap MPR bound the legislature so that it was commonly used as the legal basis in the lawmaking process. After the amendment, only Tap MPR No. XVI-MPR-1998 on Political Economy in terms of Economic Democracy was used as the legal basis. This is because Law No. 12 of 2011 does not mention its position exactly as the legal basis, even though Article 7 puts Tap MPR as a type and in hierarchy of legislation and Figure 41 of Appendix II determines that the legal basis of legislation must be at the same level or higher. Of more than 54 Tap MPRs used as the legal basis of 228 laws from 1961 until 2014, several are still valid, including Tap MPR on Economic Democracy. Tap MPR as a further elaboration of a constitutional mandate should be used as a legal basis in the lawmaking process formally as well as in direction setting for law materially so that development policies are in line with the state purpose outlined in the Constitution. This article, based on normative juridical research, focuses on the research literature to examine and analyse the principles of the law, legal systematics, and synchronization of law. Data were analysed using descriptive qualitative methods in order to obtain the certain understanding of which Tap MPR can be used as a legal basis in the lawmaking process.

Keywords: Decree of the People's Consultative Assembly, law, legal basis

1 INTRODUCTION

The Decree of the People's Consultative Assembly (hereinafter referred to as the Tap MPR) is not regulated as a legal basis in any formulation of the lawmaking process. However, Law No. 3/2014 on Industry (see the "Considering" stipulation in Law No. 3 of 2014 on Industry, namely 1. Article 5 paragraph (1), Article 20, and Article 33 of the 1945 Constitution of the Republic of Indonesia; Decision of the People's Consultative Assembly of the Republic of Indonesia Number XVI/MPR/1998 on Economic Politics in the context of Economic Democracy) and Law No. 7/2014 on Trade (see the provisions of "Remembering" in Law No. 7 of 2014 on Trade, namely: 1. Article 5 paragraph (1), Article 11, Article 20, and Article 33 of the Constitution of the Republic of Indonesia 1945; 2. Decree of the People's Consultative Assembly of the Republic of Indonesia No. XVI/MPR/1998 on Economic Politics in the context of Economic Democracy) using Tap MPR No. XVI/MPR/1998 on Economic Politics in the

Framework of Economic Democracy (Tap MPR on Economic Democracy) as the legal basis in the lawmaking process.

This means that Tap MPR not only serves as a political foundation, but also has to be a legal basis formally and materially so that all economic arrangements coincide with the material of Tap MPR, in line with the concept of Tap MPR placement as the basic norm of state or *Staatsgrundgesetz*, meaning that the 1945 Constitution, the Tap MPR, and the Constitutional Convention are classified as *Staatsgrundgesetz*. Both the 1945 Constitution and the Tap MPR provide guidance to normative legislative activities (Attamimi, 1990) for the determination of the content of a law.

The Tap MPR on Economic Democracy contains fundamental thinking about the populist economy through the empowerment of all national economic powers, especially small, medium, and cooperative entrepreneurs by developing a fair market economic system based on natural and productive resources; human resources; independent, advanced, and empowered competitiveness; environmental insight; and sustainability. The Tap MPR as one of the MPR's legal products was born in Jakarta on 13 November 1998 as a further response to the reform movement in 1998 and became one of the existing Tap MPRs as a governing provision (Asshiddiqie, 2010).

It is seen from the concept of hierarchy in Article 7 of Law No. 12/2011 on Formulation of Regulation that the Tap MPR should be the legal basis for the lawmaking process; however, it is not clear whether the Tap MPR is the basis of the authority to formulate legislation or serves as legislation that mandates the establishment of legislation. In addition, disagreement among legislators on the existence of the Tap MPR often occurs, and this discussion sometimes leads to a long debate over whether or not the inclusion of legislation becomes the basis of law. Besides, sometimes, to avoid the debate, the inclusion of legislation is done just as a formality, as it is important that the proposed inclusion of the rules of either party be accommodated.

On the other hand, the techniques of writing legislation within the legal basis of "in view of" are also quite diverse. Although there are rules about the techniques of legislative drafting, there are no specific regulations about the writing within the legal basis, as there are still many stylistic and writing concepts influenced by their respective interpretations. The legal basis is formulated by the dictum "in view of" as one part in the systematic legislation that is divided into titles, preamble, body, closing, elucidation, and attachment. Understanding what the function of inclusion of legislation in the "in view of" legal basis is very important because it reflects the legal basis of legislation. The type of legislation that can be used as a legal basis has been determined by Law No. 12/2011 in Figures 28 through 45 of Attachment II to the Drafting of Legislation.

Although the Tap MPR is now included in the legislation, the existence of the Tap MPR as one type of legislation and in the hierarchy of legislation has not reached full concordance agreement. Bagir Manan justified the MPR's Decree level of higher degree than the law (Manan, 2016). On the contrary, the Tap MPR position in the legislation hierarchy, according to Jimly Asshiddiqie, is not correct, and the placement of the Tap MPR above the law is false because the Tap MPR should be equal to the law so that it can be revoked if it is contradictory to the constitution through testing at the Constitutional Court. If it is proven, the Tap MPR is no longer a constitutional law, as only the 1945 Constitution postulated as constitutional law or the basic law for the establishment of higher hierarchical legislation from the Tap MPR (Asshiddiqie, 2010).

Constitutionally, Maria Farida Indrati states that because the people elect the president, the president is no longer the mandate of the MPR, so there is no longer a decree that mandates the president. MPR is not authorized to make provisions that are regulatory, but is limited to Tap MPR that decide (beschikking). Furthermore, Maria Farida Indrati affirmed that Tap MPR could not be substantially classified into legislation because it contains higher norms and is different from the norms contained in the law (Indrati, 2012).

A review of all the material and legal product status of the Provisional People Consultative Assembly (hereinafter MPRS) and Tap MPR that was done in 2002 and ratified through the

Fourth Amendment of the 1945 Constitution led to the unrecognized product of the Tap MPR that is regulating or "regeling" in the constitutional system (MD Mahfud, 2010). In fact, according to Jimly Asshiddiqie, as quoted by Imam Syaukani and A. Ahsin Thohari, in the future there needs to be a sunset clause arranged in stages to end the existence of MPR/S Decree as the highest form of regulation under the Constitution and above the law in the hierarchy of legislation. That means that there will be no Tap MPR/S functioning as a political law (Syaukani and Thohari, 2012).

The problem is that the use of the Tap MPR on Economic Democracy as one of the legal bases in the Law on Industry and Trade Law reinforces the legal status of the Tap MPR into one form of legislation or legal norms regulating or "regeling." It shows that there is still ambiguity related to the function of Tap MPR, whether it still deserves to be placed into a national politics law or "legal policy" that will bind both formality formation and substance of a law as a whole. Therefore, this article will discuss the urgency of the Tap MPR on Economic Democracy as the legal basis in the lawmaking process and the implications of the Tap MPR as the legal basis of the law if it is associated with the adherence to the principle of conformity between types, hierarchy, and content material in formulating good legislation.

The main issues in this study, among others are the following:

1. Why is Tap MPR on Economic Democracy used as the basis of law in the lawmaking process?
2. What are the juridical consequences of the use of Tap MPR as the legal basis in the lawmaking process on the position of Tap MPR as a type of legislations?

2 METHOD

This study used a legal research method (Soekanto, 2010) with normative juridical research on all types of legal material obtained through library study in the form of primary, secondary, and tertiary data. The typology of this research is prescriptive, seeking to provide solutions or solutions to a particular problem (Mamudji, et al., 2005). The authors will discuss the issues raised by describing the exact legal facts that occur and identifying the problem and will try to provide a solution with ideas about where Tap MPR can be used as the legal basis of the law based on Law No. 12/2011. The type of data that will be used is secondary data from related institutions.

The analysis primarily concerned the development of the concept of the legal basis in the lawmaking process and the existence of the Tap MPR in the legislation regulating the order or hierarchy of laws and regulations both before and after the amendment of the 1945 Constitution, namely Tap MPR No. XX/MPRS/1966, Tap MPR Number III/MPR/2000, Law No. 10/2004, and Law No. 12/2011. Based on the enforcement period of the legislation, 228 laws formed from 1961 to 2014 were inventoried that had used 54 Tap MPRs as the legal basis of "in view of," grouping into four tables based on laws and regulations governing the ordering or hierarchy of laws and regulations. The 54 TAP MPRs are classified according to their characteristic types (regeling or beschikking), and their current validity is based on MPR Decision No.r 1/MPR/2003, as well as their delegation pattern.

Then the existence and substance of the Tap MPR on Economic Democracy will be analysed according to the main data from the Minutes of Discussion of the Bill using historical and legislative approaches. Then how the legislation acknowledges the existence of Tap MPR, as a type and hierarchy of legislation, will affect the degree of its strength as the legal basis, and hierarchical pattern between the Tap MPR and the law within that period. The research results are presented in an analytical prescriptive format.

3 DISCUSSION

3.1 *Tap MPR as the legal basis in the lawmaking process*

The existence of written law is a hallmark of a nation of laws. In the concept of a nation of laws (rules of law), every law of conduct should be based on written law set forth in the rules. All matters must be cleared with the law as written guidelines, and using the legality principle means that all governmental actions are based on legitimate and written legislation (Asshiddiqie, 2006). A regulation must be established based on other regulations, including existing laws and regulations. Exploring the values that exist in the community and when regulation is in place will be a source of inspiration for the regulations that will be formed later, as well as legal norms that must be guided by all legal subjects. In Adolf Merkl account, the face of the two-sided legal norm (*das doppelte rechtsantlitz*) of a legal norm upon which it originates is sourced and grounded upon the norm with higher position, but of those with lower position it also becomes the basis and the source of the underlying legal norms so that the period of validity or the effectiveness of a legal norm depends on the legal norms above it (Indrati, 2012).

The loss of the MPR's authority to issue a regulatory decree does not necessarily eliminate the status and position of the previously issued Tap MPR. According to Jimly Asshiddiqie, based on the Tap MPR No. I/MPR/2003, until now only eight decrees are still valid (as regulation/regels), one of which is Tap MPR No. XVI/1998 on Economic Democracy (Asshiddiqie, 2010): Even through the Law of Article 7 No. 12/2011 and its elucidation, the position of the Tap MPR becomes clear again as one type of regulation in the hierarchy of legislative regulations. In the elucidation of Article 7 of Law No. 12/2011, there are fourteen Tap MPR legal products that are still valid as the type of legislation that covers Tap MPR in the category of Article 2 (three Tap MPRs still in effect with the provisions) and Article 4 (totalling eleven Tap MPRs) in MPR Decision No. I/MPR/2003. Hierarchy in Article 7 paragraph (2) of Law No. 12/2011 is explained as the increment of any type of legislation based on the principle that the lower legislation regulations shall not be contradictory with the higher laws and regulations.

In the period from 1961 to 2004, Tap MPRS and MPRs were used as the legal basis of the law laid out in the considerations of the "in view of." The use of Tap MPRs as the legal basis of the law also proves the consistency of the application of the hierarchy concept in Tap MPR No. XX/MPR/1966 and Tap MPR No. III/MPR/2000 on the Source of Law and Order of Legislation Regulation, where the Tap MPR is truly recognized and interpreted to be higher than the law or any other form of regulation (Asshiddiqie, 2010). At that time it was known that there was a political foundation (other than philosophical foundation and juridical base) as a political policy line that became the basis for the policy and direction of state governance for the formation of a law commonly used by the Tap MPR law by placing it in the "Considering.: It is quite possible that before the amendment of the 1945 Constitution, the People's Consultative Assembly was the highest institution constructed as the incarnation of all sovereign people and the place to which the president must account for all the implementations of his constitutional duties.

Based on the results of the research, from 1966 until the 1999, there were forty-two decrees of the MPRS/MPR that were used as the legal basis of the law. Such conditions can occur because in 1966, the beginning of the New Order Period, Tap MPRS No. XX/MPR/1966 on the source of law and order was born in which the Tap MPR is positioned as one of the legal orders in an order that is under the 1945 Constitution and above the Law/Government Regulation in Lieu of Law (Perppu). Almost all areas of governance are fundamentally governed in the Tap MPR.

Along with the rolling out of reform in 1998, in 2000 when the amendment of the 1945 Constitution entered the second year, simultaneously the People's Consultative Assembly gave birth to the Tap MPR No. III/MPR/2000 on the Source of Law and Order of Legislation Regulations that still recognize the existence of Tap MPR as one of the legislations in the same hierarchy, namely under the 1945 Constitution and above the law (the law is in a higher order than Government Regulation in Lieu of Law; this change became one of the images that began to blaze the idea of strengthening the legislative function of parliament). Understanding the position of the Tap MPR in the order of course still does not change its existence as the legal basis of the law; even within the four years from 2000 to 2004 (simultaneously with the amendment process of the 1945 Constitution where in 2002 MPR decided the final result of the change process of the 1945 Constitution) there are eleven Tap MPRs that were used as the legal basis of the law, namely Tap MPR produced by the People's Consultative Assembly in 1998, 1999, 2000, and 2001.

In 2004 was born Law No. 10/2004 on the Establishment of Laws and Regulations, which in the provisions of Article 7 of the Tap MPR are not defined as a type or in the hierarchy of laws and regulations. Such a provision greatly influences the development of the practice of forming the law. From 2004 to 2013, there was not a single product of the law, which includes the Tap MPR as the legal basis, in consideration to be recalled. By accommodating the various developments and needs of the community for better legislation, then in 2011 Law No. 10/2004 was replaced by Law No. 12/2011. The inclusion of Tap MPR into the type and hierarchy of the Laws and Regulations is one of the substantial changes in Law No. 12/2011, although in the period of 2004–2011 there was no change or shift of Indonesian state administrative system such as the revolution in 1965 that gave birth to the Tap MPR XX/1966, or Reformation in 1998 that gave birth to Tap MPR III/2000, or change of the 1945 Constitution in 1999–2002 that gave birth to Law No. 10/2004.

The existence of the Tap MPR in the hierarchy is also not based on the review in the preparation of the NA but rather is an attempt to accommodate the input of the hearing (by listening to the opinion of the legal experts: Prof. Dr. Asep Warlan Yusuf proposed the need for a special chapter on the existence of the Tap MPR to be clear and certain in the implementation; Prof. Dr. Satya Arinanto suggested that Tap MPR should remain in the hierarchy, because there is a Tap MPR that will always exist such as Tap MPR on the Vision and Mission of Indonesia 2020; and PSHK, affirming the need for regulation of the hierarchy position of Tap MPRS and Tap MPR No. 1 of 2003 the status of which lies between the 1945 Constitution and the law). Consultation meetings with the MPR represented by Taufik Kiemas (Chairman of the MPR) and Lukman Hakim Saifudin (Vice Chairman of the MPR) and working visits to East Java have been implemented by the Special Committee on the Bill on the Establishment of Laws and Regulations.

Until then, the process of deliberation of the Industry and Trade Law goes hand in hand with the intense desire of the legislators to accommodate the substance of the legal norms of the People's Consultative Assembly on Economic Democracy because it is closely related and needed as the basis for the current regulation of the industry and trade law. Although in Attachment II of Law No. 12/2011 on the Technique of Rule Preparation it is not stated literally that Tap MPR can or should be used as a legal basis, but as a form of mutual agreement between the Republic of Indonesia's House of Representatives (DPR RI) and government, the Tap MPR is consciously recognized as the legal basis and placed in the "in view of" consideration.

In the 54 Tap MPRs that have been used as the legal basis in lawmaking process of 228 laws, the following characteristics are found:

a. Tap MPR was originally established as the implementation of the Mandate of the People's Suffering contained in the Preamble of the 1945 Constitution and the provisions of Article 1 paragraph (2) of the 1945 Constitution and then developed as an important and basic policy of the MPR in the form of broad lines of state policy as the basis for the president to exercise his power in government.

b. Based on the survey and studies of the Ad Hoc Committee II of the People's Consultative Assembly against 139 MPR/MPRS Decrees which is then set forth in MPR Decision Letter No. I/MPR/2003, there are:

1) Four MPR Decrees in the category of Article 1 of Tap MPR No. I/MPR/2003;

2) One MPR Decree in category Article 2 of Tap MPR No. I/MPR/2003; namely, Tap MPR on Economic Democracy is declared to remain in effect with the provisions that the government is obliged to encourage an economic political stance that provides more

opportunities for economic support and development, with small and medium-sized enterprises (SMEs) and cooperatives as economic pillars in generating the implementation of national development in the framework of economic democracy in accordance with the essence of Article 33 of the 1945 Constitution;

3) Four MPR Decrees in the category of Article 3 of Tap MPR No. I/MPR/2003;
4) Seven MPR Decrees in the category of Article 4 Tap MPR No. I/MPR/2003; and
5) Thirty-eight MPR Decrees in the category of Article 6 of Tap MPR No. I/MPR/2003

c. Based on Tap MPR No. 1/MPR/2003 there are currently forty-six Tap MPRs that are expressly declared void. The eight Tap MPRs still valid until now are still the legal basis in the formation of law covering Tap MPR on Economic Democracy; Tap MPR No. XI/MPR/1998 on the Implementation of a State That Is Clean and Free of Corruption, Collusion and Nepotism; Tap MPR No. XV/MPR/1998 on the Implementation of Regional Autonomy; Tap MPR No. III/MPR/2000 on the Source of Law and Order of Legislation and Regulation; Tap MPR No. V/MPR/2000 on Stabilization of National Unity and Unity; Tap MPR No. VI/MPR/2000 on the Separation of TNI and POLRI; Tap MPR No. VII/MPR/2000 on the Role of TNI and the Role of POLRI; and Tap MPR No. VIII/MPR/2001 on Recommendations of the Policy Direction for the Prevention of Corruption Prevention. This should always be the basis of the regulation and adjustment for the law that uses the Tap MPR as an effort to follow up on Tap MPR No. 1/MPR/2003.

d. In Table 5, there are TAP MPRs of *regels* type such as Tap MPRS No. V of 1965, Tap MPRS VII of 1965, Tap MPRS No. 1/MPRS/1960, Tap MPR No. XI/MPR/1998, Tap MPR No. VIII/MPR/2001, and in reference to the classification by Jimly Asshiddiqie the Tap MPR No. XVI/MPR/1998, Tap MPR No. XI/MPR/1998, and Tap MPR No. VIII/MPR/2001 also included. TAP MPRs of *beschikking* type once became the legal basis of the formation of the law, namely Tap MPR RI No. III/MPR/2001, MPRS Tap No. XXXIII/MPRS/1967, and MPRS Tap No. XXXVIII/MPRS/1968.

e. The form of delegation of Tap MPR is not strictly regulated in Law No. 12/2011 as the delegation of the law to the regulations below its position. But looking at the substance, there are at least three models of the article formulation that may be referred to by the MPR as a delegation of arrangements to further regulate: direct assignment to the president as the mandate and the holder of power establishes the law, the assignment to the president and other institutions, and direct delegation to form the law.

In the drafting of the norms of the Bill on Industry and Trade in accordance with Article 33 of the 1945 Constitution of the Republic of Indonesia, the principles of economic democracy adopted the understanding that economic activity is organized on the basis of populist principles, and economic development is aimed at increasing the people's welfare.

The regulation on industry aims to develop the industrial sector by spurring the industrialization process so that it can become the prime mover of the national economy. It follows a rationale that positions SMEs and cooperatives as the main pillars of the national economy and in financial politics by providing broad access to develop those economic pillars. The existence of the Tap MPR on Economic Democracy in consideration of "in view of" is not followed by adjustment to the draft Academic Draft Bill because the early NA had indeed discussed the importance of the Tap MPR as a premise that positions SMEs as the main pillar of the national economy and in financial politics. However, the content of the law has its own consequences and influences the direction of the regulation of the Bill on Industry that is more realistic to the domestic industry in the development of the initial draft and the end of the bill.

In contrast to the draft of the Trade Bill, the initial NA (Legal Bureau of the Ministry of Trade, 2012) submitted simultaneously with the draft Bill to the House of Representatives in the first Working Meeting did not mention the Tap MPR on Economic Democracy but merely discussed the implementation of economic democracy and concepts of a populist economy. In the initial draft of the Bill on Trade Law, the legal basis of consideration consists only of Article 5 paragraph (1), Article 20, and Article 33 of the 1945 Constitution of the Republic of Indonesia. As the discussion began, the Tap MPR on Economic Democracy proposed the Democratic Party Faction to be used as the legal basis in the Law on Trade. In addition to the consideration of the law it was suggested that economic democracy become the basis for the implementation of national economic development, with the philosophical foundation of this law to be based on the concept of economic democracy.

The position of a legal norm in the hierarchy of legislation should determine and even strengthen the existence of the rule of law so that it can become a binding base for lower legal norms of law as well as show consistency of obedience to higher legal norms. The Tap MPR on Economic Democracy will remain in force with the provision that the government is obliged to encourage an economic political stance that provides more opportunity for economic support and development, with SMEs and cooperatives as economic pillars in generating the implementation of national development in the framework of economic democracy in accordance with the essence of Article 33 of the 1945 Constitution of the Republic of Indonesia.

In the context of the constitutional system, there should be a "connecting line" or continuity of policy and thought commitment in conducting economic development based on democratic economic principles of a populist economy agreed upon by the MPR as the founder of the basic rules of the Constitution and MPR with DPR (as a legislative body on Industry and Trade Law). In particular, politicians or members of DPR must automatically assume these roles. In the regulation of the field of industry and trade, there has been an attempt for the juridical basis of the arrangement sourced from the Tap MPR material on economic democracy to side with SMEs and cooperatives oriented to the people actually in accordance with the idea of Hatta. The populist economy was conceived by Hatta as the "middle way" ideology to respond to the failure of communism and liberalism that developed at the time. The concept of a populist economy in economic politics focuses development on the people. The populist economy as a concept of economic politics contains the idea of placing the people in a substantial position in the economy.

The use of the Tap MPR on Economic Democracy as the legal basis of the law has juridical implications:

a. Strengthening the position of the Tap MPR as a type of legislation
   By placing the Tap MPRs into the hierarchy of legislation, with rules below them, the formation of the law must be based on higher rules, and lower position rules should not be contrary to higher position rules, meaning that all laws and regulations below the Tap MPR must also be guided or sourced to the Tap MPR.

b. Tap MPR as a formal and material legal basis for the formation of law
   With the placement of the Tap MPR into the hierarchy of legislation, in the formation of a law closely related to the current Tap MPR, it should be possible to use the Tap MPR as a legal basis. The breakthrough is done by placing the Tap MPR on economic democracy as the legal basis in consideration of "Considering" even though it is not explicitly specified in the drafting technique of Law No. 12 of 2011. This is done because of the mandate in Article 14 ("The Government and the House of Representatives encourage and oversee the implementation of economic politics as referred to in this Decree in the context of the realization of economic justice that is perceived as beneficial and enjoyed by the masses") and Article 15 ("To assign to the President/Mandate of the People's Consultative Assembly of the Republic of Indonesia together with the House of Representatives to further regulate in various laws as the implementation of the Political Economy in the Framework of Economic Democracy as referred to in this Decree by taking into account measurable targets and time. ") addressed to the House of Representatives, as the founder of the law together with the President (Government) to follow up the basic policy of polyeconomic ticks that have been determined.

That means the inclusion of the Tap MPR as the legal basis of the Law on Industry and the Law on Trade, either formally, on the basis of orders, or in the position as the legal basis of material where the substance of the Tap MPR also influences the direction of the regulation of the Law on Industry and Trade Law. The Tap MPR can also serve as a source of material law or material of law formation. The Tap MPR can become a legal substance as well as a source of the values of justice that grow and develop in society, social and economic conditions of society, historical and cultural heritage of the nation, and others (MD Mahfud, 2010). An understanding of the economic constitution must be embodied so that it is not merely a pile of dead words, not limited to its formal text but must also consider its contextual aspects materially (Asshiddiqie, 2010).

4 CONCLUSION

1. There should be a legal basis in the lawmaking process. When tiered higher, it can serve as a legal basis in the lawmaking process of lower legislation. The legal basis for the formation of the Law on Industry and the Law on Trade was a law derived from the president, namely Article 5 paragraph (1) and Article 20 of the 1945 Constitution. The use of the Tap MPR on Economic Democracy as the legal basis for the establishment of the two laws is a breakthrough in the customary practice of drafting legislation. The Tap MPR on Economic Democracy can be used as a formal legal basis as well as hierarchical concept in the provisions of Article 7 and item 41 of Annex II of Law No. 12 of 2011, and materially in the laws and regulations under it as a synergistic regulatory bridge in the economic field between the Constitution and the law as mandated by Article 14 Tap and Article 15 of the Tap MPR on Economic Democracy, which assigns the president and the House of Representatives to regulate more in various laws.
2. The existence of the Tap MPR on Economic Democracy as the legal basis in the lawmaking process is inseparable from the provisions of Article 7 paragraph (1) of Law No. 12 of 2011 that acknowledged Tap MPR as regulating or regeling still valid as a type of legislation and in the hierarchy or the order of legislation its position is lower than the 1945 Constitution and higher than the Law/Government Regulation in Lieu of Law. The understanding of the lawmakers has evolved and experienced a shift where the Tap MPR is used not only as a political document but also as a legal document. The position of the Tap MPR becomes the political as well as the juridical foundation. The Tap MPR has been placed as the basic norm of the state (*Staatsfgrundgesetz*) that regulates and guides the determination of the content of the law. In this case, the Law on Economics shall not be contradictory to the material in the Tap MPR on Economic Democracy.

REFERENCES

Asshiddiqie, Jimly. (2006). *Konstitusi dan Konstitusialisme Indonesia*. Jakarta: Konstitusi Press.
———. (2010). *Konstitusi Ekonomi*. Jakarta: Kompas.
———. (2010). *Perihal Undang-Undang*. Jakarta: Rajawali Pers.
Attamimi, A. Hamid S. (1993). *Hukum Tentang Peraturan Perundang-Undangan dan Peraturan Kebijakan (Hukum Tata Pengaturan)*. Jakarta: Fakultas Hukum Universitas Indonesia.
———. Hamid S. (1990). "Peranan Keputusan Presiden Republik Indonesia dalam Penyelenggaraan Pemerintahan Negara; Suatu Studi Analisis Mengenai Keputusan Presiden yang Berfungsi Pengaturan dalam Kurun Waktu Pelita I–Pelita IV." Disertasi Ilmu Hukum Fakultas Pascasarjana Universitas Indonesia, Jakarta.
Indrati, S. Maria Farida. (2012). *Ilmu Perundang-Undangan 1, Jenis, Fungsi dan Materi Muatan*. Yogyakarta: Kanisius.
Lubis, Solly. (1995). *Landasan dan Teknik Perundang-Undangan*. Bandung: Mandar Maju.
Majelis Permusyawaratan Rakyat Republik Indonesia. (2009). *Majelis Permusyawaratan Rakyat Republik Indonesia: Sejarah, Realita, dan Dinamika*. Jakarta: Sekretariat Jenderal Majelis Permusyawaratan Rakyat Republik Indonesia.
Mamudji, Sri, et al. (2005). *Metode Penelitian dan Penulisan Hukum*. Jakarta: Badan Penerbit Fakultas Hukum Universitas Indonesia.
Manan, Bagir. (1992). *Dasar-Dasar Perundang-undangan Indonesia*. Jakarta: Ind. Hill. Co.

Manan, Bagir dan Kuntana Magnar. (1993). *Beberapa Masalah Hukum Tatanegara Indonesia*. Bandung: Alumni.

MD Mahfud, Moh. (2010). *Perdebatan Hukum Tata Negara Pasca Amandemen Konstitusi*, Jakarta: Rajawali Pers.

Soekanto, Soerjono. (2010). *Pengantar Penelitian Hukum*. Jakarta: Penerbit Universitas Indonesia.

Syaukani, Imam dan A. Ahsin Thohari. (2012). *Dasar-Dasar Politik Hukum*. Depok: Rajawali Pers.

Rhamadani, Yunita. "Implikasi Hukum Atas Kedudukan Hukum Ketetapan Majelis Permusyawaratan Rakyat Dalam Undang-Undang Nomor 12 Tahun 2011 Tentang Pembentukan Peraturan Perundang-Undangan." Tesis Ilmu Hukum Fakultas Hukum Program Studi Hukum Kenegaraan Pascasarjana Universitas Indonesia, Jakarta, Juni 2013.

Advancing Rule of Law in a Global Context – Susetyo, Rinwigati Waagstein & Budi Cahyono (eds)
© 2020 Taylor & Francis Group, London, ISBN 978-1-138-32782-5

Justice principles for start-up business in Indonesia

Peter M. Marzuki, Iman Prihandono, Dian Purnama Anugerah & Dewi Santoso Yuniarti
Faculty of Law, Universitas Airlangga, Surabaya, Indonesia

ABSTRACT: The ease of doing business in Indonesia has become a pivotal focus of Joko Widodo's Presidency. Reforms are undertaken in all areas to improve the ease of business indicators. Many rules regarding business establishments have been trimmed to facilitate investors. Simplification of permits has also been made to accommodate new emerging businesses. In addition, the Indonesian government also increased economic growth by strengthening the real sectors. One of the priorities in the real sector is the creative economy. On January 20, 2015, through Presidential Regulation No. 6 of 2015 on the Creative Economy Body, President Joko Widodo establishes a new non-ministerial institution called the Creative Economy Agency. This body is responsible for the development of the creative economy in Indonesia. Creative Economy Agency is in charge of assisting the president in formulating, defining, coordinating and aligning policies in the creative economy.

Since then, new businesses have begun to emerge. The new business model is known as the Start-up Business. Start-up Business is a business that is generally based on internet applications that provide real service to its customers. Start-up Business can grow rapidly in a short period of time because it is a new breakthrough that is more efficient and practical than the conventional model of business. Some of Start-up Business becomes efficient because applying the sharing economy model of business. Thus, large investors are targeting start-ups to be developed on a larger scale. On the one hand, Start-up Business has a positive impact on consumers; on the other hand, the Start-up business may threaten the sustainability of conventional businesses that exist in society. This paper aims to analyze the problems that arise because of the Start-up Business and suggest a solution in the perspective of justice principle.

Keywords: Justice, Startup, Business, Sharing Economy

1 INTRODUCTION

Under the current administration of President Joko Widodo, the ease of doing business has become a pivotal focus in fostering Indonesia's economy. Based on World Bank's 2017 Report on the Ease of Doing Business, Indonesia ranked 91 out of 190 countries surveyed. Although Indonesia sees a gradual increase - from rank 115 in 2016 - however, compared to other neighboring ASEAN countries, Indonesia still lags behind. Singapore and Malaysia, for example, rank 2nd and 23rd respectively. The indicators for these ranking include: (1) Starting Business, (2) Dealing with Permits, (3) Taxes, (4) Credit Access, (5) Enforcing Contract, (6) Trading Across Borders, (7) Protection for investors, etc. (World Bank, 2017)

The government has taken serious concern and establishes directive frameworks through the economic policy package volumes XII to XIV to facilitate and accelerate the ease of doing business. Moreover, it is recorded that over 3.143 regulations, both national and local regulation, have been abolished in order to simplify procedures and cut time in various licensing sectors, which hampers the ease of business. (Sekretariat Kabinet Republik Indonesia, 2017) Instead, the government forms new regulations, i.e. the Government Regulation No. 7 of 2016 on Changes in Authorized Capital of Limited Liability Companies, which allows Micro,

Small and Medium Enterprises (MSMEs) to establish a limited liability company with capital under 50 million rupiahs, in accordance with the agreement of the shareholders.

Starting from a vision to make Indonesia as "The Digital Energy of Asia", first initiated by President Joko Widodo in February 2016 during his visit to Silicon Valley, California, USA, the Ministry of Communication officially announces the 1000 Start-Up Movement. This movement aims to foster the creation of 1000 quality start-ups by 2020, which will reciprocally increase the value of e-commerce transaction up to US$130 billion and a continues to grow at about 50 percent per year. (Kementerian Komunikasi dan Informatika, 2016)

On January 20, 2015, through Presidential Regulation No. 6 of 2015 on the Creative Economy Agency, President Joko Widodo establishes a new non-ministerial institution called the Creative Economy Agency (Badan Ekonomi Kreatif, BEKRAF). This body is responsible for the development of the creative economy in Indonesia. The Creative Economy Agency is in charge of assisting the President in formulating, defining, coordinating and aligning policies in the creative economy. Ever since its formation, BEKRAF has given technical assistance as well as contribute to funding new creative start-ups and promoting these qualities in the Startup World Cup to gain international exposure. (Bekraf, 2016)

Despite the reforms made to ease business and foster economic growth, the government still lags in responding to the emergence of new business models. The form of modern evolving businesses is different from that of the conventional type; it is highly influenced by the rapid development of technology and is known as Start-up Business. Start-ups, in general, are businesses that combine the use of information technology to everyday services, i.e. transportation, delivery, accommodation, and retail. This type of business is designed to develop quickly in a relatively short period of time. Start-up creators, or known as entrepreneurs, usually come up with an innovative idea, which is then manifested in the form of internet-based applications acting as a platform to connect the supply and demand of certain goods and/or services. Some well-known examples of start-ups in Indonesia are, among others, Go-jek, Grab, Uber, AirBnB, Traveloka, and Tokopedia. Ever since their emergence, start-up disrupts the way people use to consume goods and access services. Hence, this business model is highly preferred over the pre-existing businesses due to the benefits consumers attained in terms of price and efficiency.

Due to start-ups' disruptive effect, pre-existing businesses slowly face a detrimental loss as they can no longer adapt to compete in the market, which now favors the innovate start-ups. Conventional retail outlets were forced to close its outlets as customers shift to online shops; meanwhile, conventional taxi companies face a significant loss in shares and regular earnings as transportation network companies (TNCs) keeps on penetrating the market. The demonstration, which leads to mutually destructive actions, happened in major big cities in Indonesia as a form of refusal from conventional taxi drivers to the online transport drivers.

Against this backdrop, this paper seeks to analyze the nature and characteristics of start-ups, the concept of sharing economy they endorsed, as well as comparing start-up regulations in other jurisdiction. Realizing the current regulatory gap and legal uncertainty, specifically relating to tax, investment and fair business competition, this paper aims to provide a thorough analysis on the theory of justice in order to foster the creation of a just regulation for start-up businesses in Indonesia.

2 METHOD

The method used in this paper is normative legal research. There are two approaches used in this paper, the statutory approach and the conceptual approach. First, the statutory approach is used to define the existence of regulatory gaps in aspects pertaining taxation, licensing, investment and fair business competition in relations to start-up businesses, as well as analyzing statutory norms in other jurisdictions. The writer also analyses the Supreme Court Judgment No. 37/P/HUM/2017, which serves as a landmark jurisprudence for start-up businesses, especially start-up TNCs. Second, the conceptual approach is used by referring to 3 theories of justice from John Rawls, Richard Posner and Amartya Sen, in order to analyze which theory befits in formulating a just regulation.

3 DISCUSSION

The emergence of the start-up is a logical consequence to the vast development of the Internet in the late 19th century, which caused the information to move rapidly beyond geographical boundaries. The term start-up business became a worldwide trend ever since the Internet Bubble phenomenon. This has led to the reform of business models, from which conventional-sectorial industry evolved into a more globalized digital start-up business (Graham, 2005).

Indonesia is a very conducive market for start-up businesses. According to the Report from the Center for Human Genetic Research (CHGR) Institute, in 2016, the number of start-ups in Indonesia reaches around 2,000 units and is the highest compared to other ASEAN countries. Start-ups in Indonesia are expected to grow up to 6.5 fold to about 13,000 by 2020.

3.1 *Characteristic of start-up business*

The term Start-up business, in general, refers to a newly discovered company and is still in an initial stage of developing its business and gaining market access. They usually start with very limited capital and then seek funding from angel investors or venture capitals when they gradually grow bigger. Start-ups generally operate in the field of technology, or digital-based; it maximizes the benefits of technology to simplify commerce. This new business model is meant to be a disruptive invention; a game changer that is able to disrupt the existing market, designed to perform better and grow faster compared to conventional businesses. (Robehmed, 2013)

3.2 *The concept of sharing economy*

Most start-ups today run their business based on the core principles of collaborative consumption, or often referred to as 'sharing economy', where technology is used to create the efficiency and trust to match the supply and demand in the market. The sharing economy, in general, is the reinvention of traditional economic model of sharing, swapping, trading, or renting products or services through technology (Boltsman, 2017). Costas Courcobetis defines collaborative consumption as "economic model based on sharing, swapping, trading, or renting products and services, which enables access over ownership" (Coubertis, 2017). This model of economy reinvents not just the way people consume, but how people consume. Instead of paying for full ownership, people find sharing economy as a breakthrough to the possibility of taking maximum advantage over the access to things they do not literally possess. This creates a sense of belonging and collective accountability, which is mutually beneficial for both the provider and its users. (Boltsman, 2010)

Rachel Botsman defines sharing economy as "an economic system based on sharing underused assets or services, for free or for a fee, directly from individuals." Some major characteristics of sharing economy include:

- The core business idea involves unlocking the value of unused or under-utilized assets ("idling capacity") whether it is for monetary or non-monetary benefits.
- The company should have a clear values-driven mission and be built on meaningful principles including transparency, humanness, and authenticity that inform short and long-term strategic decisions.
- The providers on the supply-side should be valued, respected, and empowered and the companies committed to making the lives of these providers economically and socially better.
- The customers on the demand side of the platforms should benefit from the ability to get goods and services more efficiently; in other words, they pay for access instead of ownership.
- The business should be built on distributed marketplaces or decentralized networks that create a sense of belonging, collective accountability and mutual benefit through the community they build.

Sharing economy also create value from the 'idling capacity' of an asset, whether it is empty rooms on Airbnb, second-hand goods on Tokopedia, or spare seats on ride-sharing mode such as Uberpool or Grabhitch. Through an economic lens, the idea of sharing economy creates more efficient use of and provides added value to existing underutilized assets, and empowers both providers and customers to exchange in more direct and human ways (Cho, 2012). Start-ups, which endorse the sharing economy method, act as 'markets' or media providers (platforms) to connect individual goods or service providers, referred as the sharing economy units, with their buyers or service users.

The concept of sharing economy reaps in social benefits, such as community empowerment and social participation in capital creation activities, or referred to as crowdfunding. With the concept of no barrier-to-entry, the community, despite their background or economic status, may engage in this new economy. This is especially advantageous for the marginalized populations, those who are traditionally excluded and finds it difficult to obtain employment, or those with low education. Many also make use of sharing economy to obtain additional income by making it as a side-job; this applies for those who already have a fixed job in pre-existing businesses.

3.3 *The economic influence of start-up business in Indonesia*

The rise of sharing economy start-ups is perceived as a positive thing in Indonesia. According to the 2017 report released by AlphaBeta, the presence of ride-sharing services such as Uber, Grab and Gojek in Indonesia succeeded to reduced travel or commutation expenses by 138 trillion rupiah, or equivalent to US $ 10 billion, as well as reducing 46,000 hectares of land currently used as parking lots in 33 cities in Indonesia. Furthermore, the National Statistics Agency (BPS) showed that as of August 2016, the unemployment rate in Indonesia fell by 5.61 percent. According to BPS, there are 500,000 new jobs created in the online transport sector, which is to be associated with start-ups such as Gojek, Grab, and Uber(Wicitra, 2017). In addition, ride-sharing start-ups are expected to open employment opportunities for 7 million Indonesians and save 71 million vehicles traveling by 2020. The aforementioned statistics are an evident form of efficiency, with which resolves major social issues encountered in most urban and suburban areas in Indonesia, such as unemployment, traffic congestion, and poor access to public transports. Therefore, start-ups are more than welcomed as it provides the community with something the government cannot provide.

3.4 *Regulatory gap: tax, license, investment, fair business competition*

First, in terms of taxation, the 1945 Constitution mandates that taxes are to be levied by law. To that end, the government issued the Law No. 36 Year 2008 on Income Tax, which regulates the subjects and objects of taxation in Indonesia based on the principles applicable in tax law, including justice, certainty, efficiency and convenience. In the three-way relationship of ride-sharing economy, there are two taxable actors: (1) Transportation Network Companies, as legal entity is a taxpayer who may be subject to corporate income tax (PPh Badan), and (2) Individual public transportation service providers, as an individual subject to personal income tax (PPh Perseorangan).

The law regulates that for start-ups with income lower than Rp 4.8 billion, will only be charged over Net Income (PPh Neto), which is one percent of the business' gross income. Although this percentage seems small, because it is charged from the gross income, this PPh Neto would still be a burden for most start-ups that are not performing well. It is overwhelming that the same law provides another option for start-ups, as taxable undertakings, to be charged with 25 percent income tax. Although the amount seems big, it is however taken from the company's net income, which means start-ups do not have to pay a single dime when their business suffers loss. This proves that there is no standardization or legal certainty as to which model of taxation is applicable for start-ups, thereby leaving gaps for smart-witted entrepreneurs select taxation methods which are most beneficial for them.

Second, as liaison practitioner, start-ups do not always require a business license. The Airbnb, for example, promulgates via its official webpage, series of guidance to people intending to use it as a medium to commercialize its property for short-term lodging, to first communicate with the government on its obligations, including the matter of business license and taxation.(Airbnb, 2017). Meanwhile, Go-Jek introduces themselves as a "technology company", thence they are not required to have transportation business permit(Go-Jek, 2017), thereby questioning the protection mechanism for individual sharing economy units working as their 'partner', and as well as for their consumers.

Third, for start-up companies, the capital issue is indeed a big obstacle. In response to these problems, the government began to encourage the existence of Venture Capitals, an alternative financing institution that can offer fresh financial assistance to start-up businesses through equity participation. Neil Cross defines venture capital activity as risk-based financing done in the form of equity participation to firms with high development potentials. Venture capital firms also provide added value in the form of management assistance and contribution to the overall financial management of the start-up. The legal issue in this financing scheme revolves around the possibility for the venture capital firm to take over or even liquidate the start-up that is continuously under a certain degree of performance (Supriyanto, 2016). Thence, what is the legal protection for the entrepreneurs, who have pioneered the idea and has poured all his capabilities upon the business? Up to this day, there is none.

Lastly, conflicts of online vs conventional businesses are increasingly inevitable due to the allegation of the former's unfair business competition. The fast penetration and wide approval of start-up TNCs have led the Ministry of Transportation to issue the latest Ministerial Decree No. 26 of 2017, which is designed to balance between the rights and obligation of online transports and conventional taxis and is expected to bring equal opportunity in conducting their business activities. The points include a rental obligation, quota restriction, vehicle registration, periodic test, driver certification, etc. One major concern of the National Competition Authority (KPPU) is with regards to the fixed bound tariffs, or in economic terms referred to as the unlawful minimum price fixing, as regulated under Law No. 5 Year 1999 on Prohibition of Monopolistic Practices and Unfair Business Competition. Prices of online transports, in some occasions, are incredibly cheap; thence, TNCs' faces allegation of conducting predatory pricing - setting prices below production cost for short-run profit maximization with hopes to recover losses during their future monopoly after the recoupment phase.

Shortly after the Decree's enactment, the Supreme Court annulled some of the provisions contained therein. The raison d'être of the Court's judgment is that the decree contradicts with higher regulations, i.e. the Law No. 20 of 2008 on MSMEs and Law No. 22 of 2009 on Traffic and Road Transport. Moreover, the equal treatment between TNCs and conventional taxis, as endorsed by the decree, hinders the development of start-up businesses; and, has erred the national principle of "building national economy based on fair economic democracy and the empowerment of micro, small and medium enterprises." Thence, the society is once more left with a regulatory gap, at least before a new regulation enters into force.

3.5 *Start-up business regulations in other jurisdictions*

Start-up business regulations vary from one jurisdiction to another. Some States like Singapore, Malaysia and Indonesia, welcomes start-up as a novel and innovative idea to boost their national economy, while other States refuses the idea on the grounds that it would only cause devastating loss to their pre-existing businesses, in which they perceive as something worth to protect.

In the United States, Airbnb Inc., faces considerable obstacles following the signing of the New York State Draft Bill, which prohibits citizens from illegally renting property, such as lodging houses or vacant lots, for short period of time. The regulation imposed sanctions on fines to the violating communities with nominal penalties of up to US $ 7,500 or equivalent to Rp 97.5 million (State of New York Senate Bill, 2016). Previously, Airbnb valuation was considered quite high. For the past two years, the company's valuation has jumped three times to US $

30 billion. Furthermore, Airbnb was also barred from doing its business in San Francisco after the District Court's Judgment (Airbnb Inc v. City and County of San Francisco, 2016).

In France, Uber's transport business was brought to France Court where the company, represented by their General Manager, faces up to five years of imprisonment and a fine of € 1.5 million. Uber is charged with running an illegal taxi business, commercial lies and violations of French privacy laws for illegally collecting, processing and storing personal information. (European Commission Policy Department, 2015)

The different legal scheme of start-up regulations across the globe gave rise to numerous concerns on how States are supposed to regulate their continuous existence despite the inevitable damages suffered by conventional businesses. Thence, is banning start-up from running its business can be considered as a just solution? In order to answer these concerns, one must have a thorough understanding of the theory of justice, to then be able to formulate what is justice for the start-up business.

3.6 *The theory of justice for startup business*

The status quo shows that there are still legal uncertainties revolving around start-up businesses in Indonesia. The government is still in search of what so-called as "justice" for all. Therefore, in constructing legal regulations, one must refer to the theory of justice.

John Rawls, in his book A Theory of Justice, defines justice as fairness. This foundational idea can be given shape by avoiding bias, owns respective vested interests, personal priorities, eccentricities or prejudices. This imagined state of selective ignorance is known as Rawls theory of 'veil of ignorance'. Rawls's specification of the demands of impartiality is based on his theory of 'original position', which is central to his theory of "justice as fairness". The original position is an imagined situation of primordial equality, when the parties involved have no knowledge of their personal identities or respective vested interests. Rawls puts the points in this way, in A Theory of Justice (Rawls, 1971):

"the original position is the appropriate initial status quo which ensures that the fundamental agreements reached in it are fair. This fact yields the name "justice as fairness". It is clear, then, that I want to say that one conception of justice is more reasonable than another, or justifiable with respect to it, if rational persons in the initial situation would choose those principles over those of the other for the role of justice. Conceptions of justice are to be ranked by their acceptability to persons so circumstanced."

Outflows from Rawls principle of justice as fairness are two basic principles of justice: (1) Each person has an equal right to fully adequate scheme of equal basic liberties which is compatible with a similar scheme of liberties for all, and (2) Social and economic inequalities are to satisfy two conditions. First, they must be attached to offices and positions open to all under conditions of fair equality of opportunity; and second, they must be to the greatest benefit of the least advantaged members of the society.

Richard Posner, in his book Economic Analysis of Law, believes that law should promote efficiency and social wealth maximization. Posner defines efficiency as the condition in which resources are allocated so that its value can be maximized. As a technical term, efficiency means exploiting economic resources in such a way that human satisfaction as measured by aggregated consumer willingness to pay for goods and services is maximized. Rational decision-making, which maximizes the efficient use of resources, is the stated objective, according to Posner, of economic analysis of legal problems. Efficiency becomes the standard for all proposed solutions and is measured in terms of maximized value. In his economic analysis, efficiency must become the framework of making social decisions, as well as putting the concern to the community's welfare. Thence, efficiency in Posner's view is related to the increasing one's wealth without causing harm to others. (Posner, 1992)

Amartya Sen proposes a new and more practical approach to justice by taking into account the so-called "justice" practiced in our everyday living. For Sen, justice is not a mere theory that leads to ambiguity. In his book The Idea of Justice, Sen defines justice as a condition where there is no more injustice. He criticizes his predecessor's opinion, which illustrates justice in an imagined ideal society and in the absence of pluralism; instead, he focuses on discussing ways of reducing injustice. According to Sen, justice deals with the way people live in real life, therefore efforts to realize justice cannot be separated from the concrete factual circumstances of a dynamic and pluralistic society. Therefore, justice must be approached in a practical rather than a theoretical way that encourages people to move forward by effectively promoting justice and minimize injustice. In promoting justice, it should also pay attention to customary law, local wisdom, cultural deliberations, the willingness to communicate and share ideas, cooperation, which all in all, makes a valuable contribution in reducing injustice and seeking a more just society. (Sen, 2010)

3.7 *The ideal approach of justice for start-up business in Indonesia*

Based on the aforementioned theories of justice, the writer perceives that the most ideal approach to be taken into account in formulating a just regulation for start-up business in Indonesia is Amartya Sen's practical approach to justice.

First, the Rawls theory is most likely irrelevant, as also been criticized by Sen, that the 'original position' is an imagined ideal society, which hardly exists in reality. Even if such a thing exists, the nature of start-up as a disruptive invention does not emerge together with the conventional business, but rather a bit later, in order to disrupt the existing market. If we imply justice as fairness, then treatment of start-ups and conventional businesses will most likely be made equal, by which will hinder the ease of starting such business. Second, Posner's economic point of view, which upholds efficiency and wealth maximization, must also be rendered irrelevant. This approach is much favored by the start-up companies, as their main goal is to simplify commerce with technology. However, it sacrifices the rights of conventional business players to attain just legal protection over the fierce competition with start-ups. Thence, Sen's practical approach to justice, which embraces customary law, local wisdom, and cooperation, is the better approach to be used by legislation drafters in formulating a just regulation for start-up businesses. Mere economic analysis and theories of justice will never be sufficient; unless the government is willing to engage in deliberation with the society, or representatives of both start-ups and conventional businesses, the realization of justice is a mere daydream. All in all, a just regulation takes into account conflicting interests to create equitability, thence; every party shall attain what is meant for them in equitable distribution. This will bring in peace as an outflow of the fulfillment of all parties' rights and obligations (Marzuki, 2008).

4 CONCLUSION

Start-up often faces obstacles as they were prohibited to run its business in major developed countries due to their disruptive nature, further posing a threat to pre-existing businesses and the well-established orderly-fashioned society. On the contrary, start-ups in Indonesia are warmly welcomed both by the government and society. The underlying reason is that start-up opens up new business opportunities, employment, maximization of idling capacities, and promote the development of other informal sectors, which ultimately increases the welfare of the society. Realizing how start-ups in Indonesia are rapidly disrupting the urban society, the government cannot simply sits still and let start-ups penetrate even further, while the conventional business players slowly exeunt the market. It is the government's duty to response over the problematic status quo, where regulatory gap and legal uncertainties are still major issues.

A just regulation must be made in order to accommodate the emerging start-ups. The said regulation must articulate a clear taxation and licensing mechanisms, thus start-ups are to be made in the form of Limited Liability Company. It must also balance the rights and obligations in receiving investment, as well as promote fair business competition in accordance with the existing legal norm. Amanda Sen's practical approach to justice promotes cooperation and deliberation of respective stakeholders, in this sense includes the start-us and conventional players, is the better approach to be used by legislation drafters in formulating a just regulation for start-up businesses in Indonesia.

REFERENCES

Airbnb. (2017). What legal and regulatory issues should i consider before hosting on Airbnb.
Retrieved October 18, 2017, from Airbnb: https://www.airbnb.co.id/help/article/376/what-legal-and-regu
latory-issues-should-i-consider-before-hosting-on-airbnb.

Bekraf. (2016). Dongkrak Investasi untuk Startup Indonesia BEKRAF Gelar Startup.
Pitch Day. Retrieved September 27, 2017, from Bekraf: http://www.bekraf.go.id/berita/page/10/dongk
rak-investasi-untuk-startup-indonesia-bekraf-gelar-startup-pitch-day.

Boltsman, R. (2010). What's Mine Is Yours: How Collaborative Consumption Is Changing the Way We
Live. London: Collins.

Boltsman, R. (2017, May 27). Defining The Sharing Economy: What Is Collaborative.
Consumption And What Isn't. Retrieved September 22, 2017, from Fast Company: https://www.fastcom
pany.com/3046119/defining-the-sharing-economy-what-is-collaborative-consumption-and-what-isnt

Cho, M. J. (2012). Business Model for the Sharing Economy between Enterprises. Advances in Economics, Law and Political Science, p. 181-189.

Coubertis, C. (2017). Sharing Economy in Singapore. Asian Undergraduate Summit 2017: Disruption in
Everyday Life. Singapore: National University of Singapore.

European Commission Policy Department. (2015, October). Social, Economic and Legal Consequences
of Uber and Similar Transportation Network Companies (TNCs). Retrieved September 30, 2017,
from ECPD: www.europarl.europa.eu./RegData/etudes/BRIE/2015/563398/IPOL_BRI(2015)
563398_EN.pdf.

Go-Jek. (2017). Terms of Use 1.5. Retrieved October 18, 2017, from Go-jek: https://www.go-jek.com/
kebijakan.

Graham, P. (2005, March). Paulgraham. Retrieved September 12, 2017, from How to Start a Startup:
http://www.paulgraham.com/start.html.

Katharina, G. F. (2017, June 7). Layanan Ride-Sharing Online Kurangi Beban hingga Rp 138 Triliun.
Retrieved September 27, 2017, from Industri Bisnis: http://industri.bisnis.com/read/20170607/98/
660058/layanan-ride-sharing-online-kurangi-beban-hingga-rp138-triliun.

Kementerian Komunikasi dan Informatika. (2016, June). Gerakan Nasional 1000 Start-Up Digital:
Gotong Royong Wujudkan Solusi di Era Informasi. Retrieved October 7, 2017, from Kominfo:
https://koninfo.go.id/index.php/content/detail/7684/siaran-pers-no45hmkominfo062016-tentang-gera
kan-nasional-1000-start-up-digital-gotong-royong-wujudkan-solusi-di-era-informasi/0/siaran_pers.

Marzuki, P. M. (2008). Pengantar Ilmu Hukum. Jakarta, Indonesia: Kencana, Prenada Media Group.

Posner, R. (1992). Economic Analysis of Law. Toronto, Canada: Little, Brown and Company.

Rawls, J. (1971). A Theory of Justice. Cambridge, Massachusetts: The Belknap Press of Harvard University Press.

Robehmed, N. (2013, December 16). What Is A Startup? Retrieved September 16, 2017, from Forbes:
https://www.forbes.com/sites/natalierobehmed/2013/12/16/what-is-a-startup-#544a2a9a4c63

Sekretariat Kabinet Republik Indonesia. (2017, April 10). Sekretariat Kabinet. Retrieved September 21,
2017, from http://setkab.go.id/mengaku-sudah-batalkan-sedikitnya-3-143-peraturan/

Sen, A. (2010). The Idea of Justice. London, England: Penguin Books.

Supriyanto, E. (2016, April). Modal Ventura Alternatif Pembiayaan Usaha Rintisan. Harian Jurnal
Asia.

Wicitra, J. (2017, April 13). Un-sharing Economic Model. Retrieved September 17, 2017, from Pressrea-
der: https://www.pressreader.com/indonesia/the-jakarta-post/20170413/281676844775405

World Bank. (2017). Doing Business. Retrieved September 15, 2017, from The World Bank: http://www.
doingbusiness.org/data/exploreeconomies/indonesia

Case

Airbnb Inc v. City and County of San Francisco, Case 3:16-CV-03615 (United States District Court,
Northern District of California, San Francisco, June 27, 2016).

Bill

State of New York Senate Bill. (2016, January 6). Retrieved September 25, 2017, from Senate Bill
S6340A: http://legislation.nysenate.gov/pdf/bills/2015/S6340A.

Advancing Rule of Law in a Global Context – Susetyo, Rinwigati Waagstein & Budi Cahyono (eds)
© 2020 Taylor & Francis Group, London, ISBN 978-1-138-32782-5

The criminal responsibility of corporate banking: Case study of Century Bank

Pujiyono, Supanto & Theresia Pingky
Universitas Sebelas Maret, Surakarta, Indonesia

ABSTRACT: This research reported in this article aimed to determine the form of criminal responsibility of corporate banking and its application in the case of Century Bank. Another purpose is to discern the form of legal protection used against victims of banking crime in the case of Century Bank. This study was normative doctrinal law research that is prescriptive. The author used the deductive reasoning method for analysis of the data. Based on the results of research and discussion it can be concluded that the first concept of criminal liability of the corporation has been known since the recognition of the corporation as a subject under criminal law. Criminal liability can be charged to the board, corporates, or both. However, its application is still experiencing obstacles; for example, in the case of Century Bank. This is because the subject under criminal law against the corporation has not been recognized in Law No. 10 of 1998 concerning banking. Second, the form of legal protection for victims in the case of Century Bank consists of two forms, preventive and repressive.

Keywords: Century Bank, corporate criminal responsibility, legal protection

1 INTRODUCTION

The advance of human civilization and culture, in science and technology, particularly greater sophistication in information, communication, and transportation methods, results in more diverse community needs. These needs can be fulfilled by people themselves, and in this respectthe role of corporations is desirable to society. The mutual dependence embodied in such a relation then felt not only by the community but also by state. According to Brian Roach, i"some people perceive the ascendancy of global corporations as a positive force, bringing economic growth, jobs, lower prices, and quality products to an expanding share of the world's population."

The important role and positive influence that a corporation can offer cannot always be realized; in addition, corporations also sometimes implicated in infringements against criminal law, harming not only individuals and the wide society but also a state. An Australian criminologist, John Braithwaite, defines corporate crime as "the conduct of a corporation or employees acting on behalf of a corporation, which is proscribed and punishable by law" (Braithwaite, 2008). It is noteworthy that corporate crime often contains deceit, misrepresentation, concealment of facts, manipulation, breach of trust, and subterfuge or illegal circumvention, thereby creating great harm to the wide society (Atmasasmita, 2003:13).

The recognition of the corporation as the subject under criminal law considered as capable of committing a crime that can impose criminal responsibility on the corporation is not new, resulting in many legal issues and debates among both academicians and legal practitioners. Breakthroughs related to categorization of the corporation as the perpetrator of a crime have resulted in a demand for imposing responsibility on the corporation for misdeeds and criminal practices. The importance of the imposition of responsibility against corporation has been stated as well in some countries, including France. In particular, for crimes committed by corporations, in France corporations can be held responsible as long as they fulfill the responsibility criteria included in Code Pénal Français corresponding to Article 121-2 Code Pénal Français:

> "Les personnes morales, à l'exclusion de l'Etat, sont responsables pénalement, selon les distinctions des articles 121-4 à 121-7, des infractions commises, pour leur compte, par leurs organes ou représentants......"
>
> (Legal persons, with the exception of the State, are criminally liable for the offences committed on their account by their organs or representatives, according to the distinctions set out in Articles 121-4 and 1217). However, theoretically, the corporation is exempted from the crimes committed by an individuals as human beings (Natural Person) (Chance, 2012, 10) as follows: "In theory, a corporate entity can commit any offence except for offences which, by their very nature, can only be committed by natural persons."

One example of a corporation often engaged in criminal practice is a banking institution. According to a book entitled *Business Studies* "Bank is a lawful organization, which accepts deposits that can be withdrawn on demand. It also lends money to individuals and business houses that need it. Banks also render many other useful services – like collection of bills, payment of foreign bills, safe-keeping of jewelry and other valuable items, certifying the creditworthiness of business, and so on." Meanwhile, in the same book, banking is defined "as an activity involves acceptance of deposits and lending or investment of money. It facilitates business activities by providing money and certain services that help in exchange for goods and services. Therefore, banking is an important auxiliary to trade. It not only provides money for the production of goods and services but also facilitates their exchange between the buyer and seller."

Among several cases recently attracting the public's attention is one real example of banking crime by the Century Bank (Bank Century). Bank Century was defeated in its clearing attempts on November 18, 2008 as a result of internal problems occurring in the bank management related to their clients:

1. Customer fund corruption up to IDR 2.8 trillion (IDR 1.4 trillion for Century Bank customers and 1.4 trillions for Antaboga Deltas Sekuritas Indonesia customer)
2. Fictitious security selling of Antaboga Deltas Sekuritas product, in which the product did not have license from BI and Bappepam LK

After Century Bank had lost its case in the clearing procedures, the customers of Century Bank could make banking transactions in either cash or non-cash forms. The customers of Century Bank could not withdraw cash from ATMs of Century Bank and of Bersama. The customers then came to the Century Bank office to ask for clarification from the bank officer. However, the bank officer could not guarantee that the money could bebe withdrawn through the ATM in the following days. It led the customers to take legal measures to strive for their rights. In addition to the demand for compensation, customers also demanded the imposition of responsibility to Century Bank as a corporation. The court sentenced the administrators of Century Bank to imprisonment and a fine, and Century Bank was imposed with a fine, but an indictment stating that Century Bank was guilty as a corporation was not granted. Therefore, the author wishes to discuss further the status of Century Bank as a corporation in the case in which it is involved and the form of responsibility taken in that case.

Studies on corporate responsibility have been wide ranging. Among others, is postgraduate thesis research conducted by M. Yusfidli Adhyaksa, S. H. entitled *Pertanggungjawaban Pidana Korporasi dalam Penyelesaian Kasus Bantuan Likuiditas Bank Indonesia (BLBI)* (Corporate Criminal Responsibility in Resolving Bank of Indonesia's Liquidity Grant [BLBI] Case).Others are a postgraduate thesis study conducted by Evan Elroy Situmorang, S. H. entitled *Kebijakan Formulasi Pertanggungjawaban Pidana Korporasi terhadap Korban Kejahatan Korporasi* (Corporate Crime Responsibility Formulation Policy for the Victim of Corporate Crime) and a graduate thesis study conducted by Anggi

Hardiyanti A. Makkarumpa entitled *Analisis Pertanggungjawaban Pidana Korporasi dalam Tindak Pidana Perbankan (Studi Putusan No. 404/PID.B/2011/PN.MKS)* (An analysis on Corporate Crime Responsibility in Banking Crime: A Study on Verdict No. 404/PID.B/2011/PN.MKS). Meanwhile, this current study is different from the previous study. The difference lies in a further study of the subject, as the author has chosen Century Bank as the subject of research and focuses on how a corporation can be decided as guilty and determined as the perpetrator of the crime, which forms the basis of the title of this article.

2 METHOD

This study was normative doctrinal law research that is prescriptive and applied in nature (Mahmud Marzuki, 2014: 67). This research used statute, case, and conceptual approaches (Marzuki, 2014: 133–134). This research employed primary and secondary law materials. Primary law materials used included banking law, the penal code, and corporation law. Secondary law material used included journals related to the topic of research. Deductive syllogism was the technique used for data analysis..

3 DISCUSSION

3.1 *Status of Century Bank as a corporation in the Century Bank case*

a *Chronology of the Century Bank case*
The case against Century Bank stemmed from an internal problem involving deceit by bank management in relation to their clients. The bank management deceived customers in the form of customer fund corruption (misuse) up to IDR 2.8 trillion (IDR 1.4 trillion for Century Bank customers and 1.4 trillion for Antaboga Deltas Sekuritas Indonesia customer) and fictitious security selling of Antaboga Deltas Sekuritas products. These two problems resulted in enormous losses for the customers of Century Bank. After Century Bank was defeated in the clearing procedures, the customers of Century Bank could make banking transaction in either cash or non-cash forms. The customers of Century Bank could not withdraw cash from ATMs of Century Bank and Bersama. The customers then came to the Century Bank office to ask for clarification from the bank officer. However, the bank officer could not guarantee that the money could be withdrawn through ATMs in the following days. Thus, fund withdrawals could be done only through tellers, with the amount limited up to IDR 1 million.

On November 13, 2008, the customers of Century Bank admitted that transactions in the form of foreign currency could not be processed and even clearing and transfer could not be done. The bank management permitted only the transfer of deposited funds to savings accounts. Thus, money could not leave the bank. Customers believed that Century Bank had traded an illegal investment product, based on the fact that the Antaboga investment marketed by Century Bank was not listed in Bapepam-LK. The customers strongly protested and occupied the subsidiary office of Century Bank. The customers reported the deceit to Mabes Polri (Republic of Indonesia's Office Headquarters) and DPR (Legislative Assembly) to ask them to resolve the case immediately and to return their deposit money.

b *Analysis of the status of Century Bank as a corporation in the Century Bank case*
The concept of the corporation as a subject under criminal law developed as the result of changes occurring in the way business activities were undertaken. In the process, the position of a corporation as a criminal subject was addressed. Many arguments arose regarding the scope of crimes committed by the corporation as the criminal subject. For example, Kramer in his book *Corporate Criminality: The Development of Idea,* as cited in *Principles and Theories of Criminal Liability,* chapter 3, states: "the corporate crime involves: criminal acts (of omission or commission) which are the result of deliberate decision making (or culpable

negligence) by persons who occupy structural positions, within the organization as corporate executives or manager. These decisions are organizational in that they organizationally based-made in accordance with the operative goals (primarily corporate profit), standard operating procedures, and cultural norms of the organization-and are intended to benefit the corporation itself."

Meanwhile, there are some rationales of the importance of corporate responsibility: first, the corporation is the main actor in the world economy, so that the presence of criminal law is considered as the most effective method to affect the corporation actor's rational actions (Bucy, 2007: 1288Secondthe profit obtained by a corporation and the loss suffered by society can be so large that it will be impossibly balanced when the corporation is imposed with civil sanctions only (Stephen, 2002: 46).

Theoretically, as suggested by Sam Park and John Song, there are three basic criteria that can be used to determine that a corporation is responsible for illegal acts committed by its administrators (Park and Song, 2013: 732-740). First, the corporation is responsible for the crime committed by an administrator only if the crime is still in the scope and basic characteristic of their jobs in the corporation. Second, the corporation is not criminally responsible for a crime committed by the administrators unless the crime is intended to benefit the corporation. A corporation's factual profit from the crime committed by the administrator should not be real, even if the fact that the administrators deliberately transfer the profit to the corporation is enough. Third, in stating that the corporation is responsible for the crime committed by its administrators, the court obligatorily transfers the administrators' intentionality to the corporation.

Departing from this theory, the author tries to analyze the case of Century Bank.

1. The corporation is responsible only if the crime committed by the administrators is still in the scope and the basic job description in the corporation.
   In the case of Century Bank, the crime committed by administrators was deceit and counterfeit security selling in the scope of their occupation as the administrators of Century Bank. On the part of shareholders, Robert Tantular, using his authority, compelled managers and employees of Century Bank to sell security products from Antaboga by threatening that they would be dismissed if they refused to do so or would not be promoted and not be given a salary increase. Another infringement committed is that shareholders transferred the customers' fund to their personal accounts. All of the infringements committed are still in the scope of duty and authority in their occupations as administrators of Century Bank.
2. The corporation is not criminally responsible for the crimes committed by the administrators unless the crime is intended to benefit the corporation.
   The crime committed by administrators resulted in a condition in which Century Bank was defeated in the clearing procedures, and disrupted the bank's operation. However, as a result, Century Bank received an infusion of funds again from the Indonesian government of as much as IDR 6.7 trillion to save Century Bank. It, of course, benefitted Century Bank, which lacked capital at that time; moreover, the bail-out amount given by the government was so large.
3. The court obligatorily transfers the administrators' intentionality to the corporation.
   In the case of Century Bank, the representatives of the Century Bank Customer Forum filed a lawsuit against Century Bank in the arbitrage court of the Consumer Dispute Resolving Agency (BPSK) of Yogyakarta. The verdict of the arbitrage court stated that BPSK decided that Century Bank is guilty of marketing counterfeit security products of PT Antaboga Delta Sekuritas Indonesia. It was based on Century Bank's action, through its administrators, that did not prevent the selling of counterfeit security products, despite their knowledge that the product was not listed in Bapepam-LK.

Having analyzed the case of Century Bank, it is reasonable to say that in the legal case, Century Bank as a corporation deserves to be given the status of actor and be imposed with responsibility.

The development of corporate responsibility in Indonesia so far has shown some progress. Several executive rules have been developed to facilitate case management involving the subject under corporation law. The regulation instrument is the Supreme Court's Regulation (Perma) No. 13 of 2016 on the Procedure of Handling Crime Committed by Corporation. Perma includes a number of important points: definition of corporation, scope, procedural law, and criminal punishment. Therefore, the existence of Perma facilitates law enforcers in imposing sanctions against the criminal subject of the corporation.

3.2 *The form of Century Bank's criminal responsibility as a corporation*

In relation to the acceptance of the corporation as a subject under criminal law, according to Mardjono Reksodipuro, it means that there has been an expansion of the definition of who is the crime perpetrator (*dader*). The problem arising immediately is related to criminal corporate responsibility. The basic principle of criminal responsibility is that there should be guilt (*schuld*) on the part of the perpetrator.

Because a corporation cannot commit a crime without its administrators' intermediary, based on functional agent theory and identification theory, the corporation's guilt should be decided by seeing whether or not the administrators committing the crime for and on behalf of the corporation are guilty. When they are guilty, the corporation is stated as guilty for the crime they have committed. It is mentioned as well that the guilt of corporation administrators becomes the guilt of the corporation itself.

Considering the preceding analysis, when considered according to some legal theories related to corporate criminal responsibility, the one applicable the case of Century Bank is identification theory. It states that the administrators' deed is considered the corporation's deed. Similarly, in the case of Century Bank, the administrators' deed in the form of fund misuse and counterfeit security selling resulting in customer losses led to Century Bank's required assumption of the consequences in the form of paying an amount of the fund in addition to the imposition of responsibility against its administrators. The action was recognized as made on behalf of and for Century Bank as a corporation.

The legal instrument related to corporate crime in Indonesia still has weaknesses that result in some constraints in handling cases. In the case of Century Bank, the existing legal instrument is less supportive in recognition of the corporate legal subject. However, recently *Perma* has been developed, governing the corporation crime. However, *Perma* has existed only since the occurrence of the Century Bank case so it cannot be applied therein.

4 CONCLUSION

4.1 *First conclusion*

1. In the case of Century Bank, the status of Century Bank, as a corporation, can be stated as the perpetrator of crime (*dader*), after authentication of the fact that the administrators' action was conducted in the scope of their occupation (job) in the corporation; that the administrators' action benefited Century Bank as a corporation; moreover, the arbitrage court stated that Century Bank was guilty in the case of counterfeit security selling to the customers of Century Bank.
2. In relation to the Century Bank Case, the form of criminal responsibility used was the imposition of criminal sanctions against the administrators and fines against Century Bank. The corporate responsibility doctrine applied to the case of Century Bank is identification theory because the crime not only benefited the administrators but also was committed on behalf of the corporation, so that according to identification doctrine, when the administrators undertake the corporation's authority or will, the responsibility is also imposed against the corporation.

4.2 Second conclusion

1. The recognition of the corporation as a legal subject in Indonesia is still limited to certain laws and has not been governed clearly in KUHP (penal code), so that it often becomes the reason for a corporation avoiding a legal process. Therefore,Indonesian law should be reformed to includerules related to corporation law subjects in both KUHP and relevant other laws outside of KUHP.
2. In the case of Century Bank, there should be a reformation in the role of law enforcers to see not only the legal aspectsdealing with the case. It should also consider the presence of relevant doctrines and theories of experts.

REFERENCES

Alfitra. (2014). *Modus Operandi Pidana Khusus Di luar KUHP*. Jakarta: Raih Asa Sukses (Penebar Swadaya Group).

Atmasasmita, Romli. (2003). *Pengantar Hukum Kejahatan Bisnis*. Jakarta: Prenada Media, hal. 13.

Bucy, Pamela H. (2007). *Trends in Corporate Criminal Proseutions*. American Criminal Law Review.

Chandra, Septa. (2017). Correlation Between Theory of Criminal Liability and Criminal Punishment Toward Corporation in Indonesia Criminal Justice Practice. *Jurnal Dinamika Hukum*.

Hanafi (1997). *Perkembangan Konsep Pertanggungjawaban Pidana dan Relevansinya Bagi Usaha Pembaharuan Hukum Pidana Nasional*, Tesis, Program Sarjana, Universitas Indonesia.

Huda, Chairul. (2006). *Dari Tiada Pidana tanpa Kesalahan Menuju Kepada Tiada Pertanggungjawaban Pidana tanpa Kesalahan*, Cet. Kedua. Jakarta: Kencana.

Mazuki, Peter Mahmud. (2014). *Penelitian Hukum*. Jakarta: Kencana Prenandamedia Group.

Muladi dan Barda Nawawi Arief. (1998). *Teori-Teori dan Kebijakan Pidana*. Bandung: Alumni.

Muladi dan Dwidja Priyatno. (2010). *Pertanggungjawaban Pidana Korporasi*. Jakarta: Kencana Prenada Media Group.

Nyoman Serikat Putra Jaya, Bahan Kuliah Sistem Peradilan Pidana (Criminal Justice System) (2010). Program Magister Ilmu Hukum UNDIP, Semarang, hal. 111.

Prasetyo, Rudi. (1989). *Perkembangan Korporasi dalam Proses Modernisasi dan Penyimpangan-Penyimpangannya."* Makalah Seminar Nasional Kejahatan Korporasi, FH Universitas Diponegoro, Semarang, November 23–24.

Ray WIG (2000). Corporate Law, Penerbit Megapoin, Jakarta.

Reksodiputro, Mardjono. (1994). *Kemajuan Ekonomi dan Kejahatan (Kumpulan Karangan Buku Kesatu*, Cet. Pertama, Pusat Pelayanan Keadilan dan Pengapdian Hukum d/h, Lembaga Kriminologi. Jakarta: Universitas Indonesia.

Saleh, Roeslan. (1983). *Stelsel Pidana di Indonesia*. Jakarta: Aksara Baru.

Stephens, Beth. (2002). The Amorality of Profit: Transational Corporations and Human Right. *Berkeley Journal of International Law*.

Suhartono, Slamet. (2017). Corporate Responsibility for Environmental Crime in Indonesia. *Journal of Law and Conflict Resolution*.

Advancing Rule of Law in a Global Context – Susetyo, Rinwigati Waagstein & Budi Cahyono (eds)
© 2020 Taylor & Francis Group, London, ISBN 978-1-138-32782-5

Analysis of the position and authority of the Nagari indigenous council as a traditional representative institution from a perspective of constitutional law

Ryan Muthiara Wasti & Fatmawati
Faculty of Law, Universitas Indonesia, Indonesia

ABSTRACT: Indonesia acknowledges the existence of indigenous law communities along with their traditional rights in Article 18 of the Indonesian 1945 Constitution. One of these is the representative institutions of traditional peoples who embrace traditional values that endure to the present. Law No. 6 of 2014 on Villages has not fully accommodated traditional values that exist in the respective regions, particularly the traditional representation in Nagar Minangkabau. Therefore, there are two main issues: (1) the position and authority of traditional representative institutions within the governance structure of nagari in Minangkabau and the Village Law; and (2) the ideal regulation on traditional representative institutions in Indonesia. The analysis was carried out using the theory of traditional constitutional law as it bears a close relation to Indonesia's state constitutional values followed by acceptance of the diversity of customs that arise from an amalgamation of laws that have their own characteristics. In addition, a comparison was carried out regarding regulations that govern indigenous communities in the United States, Australia, Cameroon, and PRC. The conclusion is that the nagari indigenous representative institutions are not fully accommodated in the Village Law and are thus an ideal regulatory instrument to accommodate the needs of the nagari indigenous community in Minangkabau, including, among others, membership, method of election, and the position and authority of the indigenous representative institution. Therefore, it is necessary to amend Village Law No. 6 of 2014 in regard to customary village regulations that can be compared with other countries that have more customary village regulations and treatment in their countries such as the United States, Cameroon, PRC, and Australia.

Keywords: indigenous governance, Traditional constitutional law, traditional representative institution

1 INTRODUCTION

Minangkabau is one of the regions with a unique customary government besides its kinship system, which is the largest matrilineal system in the world (Cribb, 2004). The customary government in Minangkabau is considered to have characteristics that can be a reference for the system of government inIndonesia because the form of government is based on the peculiarities of the community as well as the needs of the region. This can be seen from the division of powers in its government, where there is a special institution that sets the representative of the consciously chosen community, and the highest institution of traditional dispute resolution known as the Nagari Indigenous Council (KAN) (Agam Regency Government, 2007). A customary government has not been accommodated by existing laws, as evident from the regulation of the customary government that is handed over to each region and lack of details to guarantee the implementation of customary law in each region although it is based on customs.

No region wants the dualism of government in which both customary government and state government exist. Based on the concept of local government regulation of Governor/Head of Central Sumatera Province on January 4, 1955 No. E/G/1955, the existing custom council can decide customary matters only, while general governance issues are resolved by the Nagari House of Representatives (DPRN) (Kemal, 2009). This matter certainly explicitly demonstrates the practice of legal dualism in which the local ways and the state normative system apply in conjunction, with the consequences of conflict in which the legal status is usually considered the most valid one (Soetandyo, 2008).

Though seen from the implementation, especially in the framework of regional development, the local culture is also influential in the formation of regional legal products because the indigenous peoples are the object of the law that will prevail in the area. In some areas in Minangkabau, as an example, inconveniences result when unexpected legal products do not fit the needs of the community. This is because the practice of regional autonomy in the nagari administration does not go well; dualism still exists in institutions that do not want to be equalized although both can function harmoniously. On the other hand, the government considered the legal one by the country is not willing to make room for the customary one to provide input and aspirations. Although already accommodated through provincial and regulatory rules, there are still hurdles faced by the Provincial Regulation of West Sumatera No. 9 of 2000 concerning Government of Nagari and Regency Regulation of 50 Kota No. 1 of 2001 in the nagari government having absolutely no tangency with the inherent government system within the Minangkabau indigenous structure (Latief, 2006).

Therefore, the main issues in this research are: (1) What are the regulations governing indigenous and tribal peoples in Indonesian legislation? (2) What is the position and authority of the Nagari Customary Council as a representative customary institutionwithin the nagari government structure in Minangkabau and Law No. 6 of 2014 about the Village? (3) What is the ideal regulation of indigenous representative institutions in Indonesia?

This article describes normative juridical research that analyzed the position and authority of a customary institution (KAN) in the nagari government system in Minangkabau and shows from sources of law and the existence of a law that actually applies in society that the law cannot be separated from what is done and becomes a habit of society. KAN is a representative institution of indigenous people because the elements contained in the KAN are a representation of various circles within the nagari region. As a consequence, KAN should certainly be an institution that fulfills the aspirations of its indigenous people.

In addition, this article also describes village regulations that are considered not in line with the needs of indigenous peoples, especially Minangkabau, so the institution of KAN is still needed. The existing rules have not been able to accommodate the needs of indigenous peoples with the principle of consensus and kinship that have been held for a long time. To achieve this result, the study was directed to descriptive research to bring to light the regulation of indigenous and tribal communities to be associated with customary governance in response to the urgency of the role of customary institutions needed by the community as a form of protection of their customary values. In discussing the regulations of indigenous peoples, the general arrangements in the world are considered along with the conditions of indigenous peoples in the world.

2 DISCUSSION

2.1 *The ideal concept of indigenous representative bodies in Indonesia*

The existence of indigenous peoples has been recognized in the constitution and the legislation in Indonesia, especially in in Article 18B paragraph (2) of the 1945 Constitution and the explanation of the 1945 Constitution (Republic of Indonesia, 2002). However, recognition of indigenous peoples in Indonesia is not only stated in the constitution; some decisions of the Constitutional Court become part of the reinforcement of the constitutional arrangements. Among others are the Decision of the Constitutional Court No. 31/PUU-V/2007 and Decision of the Constitutional Court Number 6/PUU-VI/2008. The Constitutional Court's verdict reinforces Article 18B paragraph (1) of the 1945 Constitution that has not explicitly stated the traditional rights of indigenous peoples. Based on the verdict of the Constitutional Court, it can be concluded that the unity of indigenous peoples is recognized if it meets the following requirements: (1) there are people whose members have an in-group feeling; (2) there is a customary government order; (3) there is customary property; (4) instruments of customary law norms exist; and (5) there are elements of customs territory for indigenous peoples.

The regulation of indigenous peoples is certainly not only in the recognition of its existence, but of all forms of government adopted within indigenous peoples. One of the most important things in indigenous governance is the representative institution. In general, the language of representative institutions refers to institutions that have legislative and representative functions. As with the concept of power sharing embraced in Indonesia, there are three branches of power revealed by Montesqiue: executive, legislative, and judicial. The existence of legislative institutions based on national law is integrated with the checks and balances of executive power. This principle of checks and balances is important in the doctrine of separation of powers because each branch controls and balances other branches of power so that there is no abuse of authority by any branch of power (Ashshiddiqie, 2006).

In contrast to the representative institutions called legislative bodies in the national government, the indigenous peoples are not always represented by a person or institution that specifically has legislative functions only. There are, however, similarities in which the state practice of the trias politica principle, executive power can perform legislative functions. This is evident from Article 20 paragraph (2) and paragraph (4) of the 1945 Constitution specifying that the draft law must be discussed jointly and ratified by a president. In fact, in Article 4 paragraph (1) of the 1945 Constitution, it is stipulated that a president holds the power of government. Thus, the power exercised within the Indonesian statecraft also is not strictly divided between the legislature and the executive so that execution within customary institutions is also common (Sunni, 1981).

There are several models of customary representative institutions in Indonesia:

1. Tribal leaders simultaneously act as representative bodies where there is a function of legislation, absorbing and channeling the aspirations of indigenous peoples.
2. There is a special institution that functions in legislation, supervision of chiefs (executives), and budgets.
3. The Village Assembly or Village Forum has the functions of representatives, legislation, and policymakers.

It is certain that there are several variations among existing customary representatives that can be influenced by many factors. Among indigenous peoples whose tribe chiefs possess legislative and judicative roles, the chiefs have the highest position and are usually chosen by a mutually agreed consensus system so the indigenous peoples trust them as their representatives in setting rules and absorbing aspirations and as judges in resolving disputes. However, usually the tribe chiefs will be assisted by other institutions such as the customary court, ministers, or other customary instruments in order to avoid arbitrariness.

Andrew Rehfeld related the principle of representation to the presence of constituents in a democratic concept (Rehfed, 2005). Representation as an instrument for a group to use to fight the will of the majority becomes important because it speaks for its constituents. That is why it is important for the representative to be closed to its citizens. Helena Catt stated that the principle of democratic participation is evident when everyone is comfortable together, sharing knowledge, feeling equal respect, and being able to criticize each other, so there is no fear of expressing their opinions (Catt, 1999). In addition, democratic values are relevant in terms of securing the relationship between society and government. It means that even in indigenous communities convenience must exist in order to create a harmonious relationship (customary instruments) between the community and the government.

In an area that is not as large as a village, in some countries such as the United States and Canada, there is a tendency to minimize representation. This is done to give each representative more opportunity to participate directly in the decision-making process. In addition, the least number will speed up the decision-making process itself. Meanwhile, the argument for creating a representative at the medium level is that many members are needed to represent the interests and needs of the citizens (Sarundajang, 2002). This of course is the case in an area having a large size and population.

If the number of members has met the required figure and is in accordance with the needs of local citizens, then what must be considered again is how the representative is selected. According to Sarundajang, there are three groups of regional representative bodies based on the method of election of its representative members (Sarundajang, 2002): (1) The members are selected directly or indirectly; (2) some of its members are elected and the rest are appointed by higher government units; and (3) all voters participate as members, and each voter can represent himself.

However, Sarundajang argued that elections by the people are the best option that can satisfy all parties (Sarundajang, 2002). In the concept of regional government in the world, elections are identical to the determination of members, especially in democracies. Nevertheless, the democratic principle is as stated in the 1945 Constitution and depicted in terms of government of the people, by the people, and for the people (Soeprapto, 2013). This means that whatever the mechanism of the election, it is then seen as a representative form that represents the people themselves. In addition, the holders of sovereignty are always faced with restrictions by law and the constitution as a product of mutual agreement (Ashshiddiqie, 2005).

Sovereignty practices in indigenous peoples of Indonesia also vary depending on the customary policies of the indigenous peoples. In Dayak Kalimantan, for example, the tenure of the Dayak customary council incorporated in the Customary Council is not explicitly stated, only mentioning the regulation concerning the process of establishing the Dayak Customary Council and its dismissal. In general, most indigenous peoples do not mention the term of their representative office. This is because of the position of the customary representative institution, which can take the form of customary assemblies and the representative institution itself or even join an executive body only. In the case of representative institutions, it deals with the indigenous assemblies in which usually the process of elections and dismissals shall be conducted by the consesus process of the leaders as well as their indigenous peoples and not written such as in the national government setting. This is a collective agreement that will be voluntarily accepted by all parties from their own indigenous community, even by the dismissed ones.

Within the village community itself, the term of office of the village representative agency known as the Regional Consultative Body (BPD) in Article 56 (1) of Law No. 6 of 2014 concerning the Village is 6 (six) years from the date of the swearing in/taking an oath. Members of the BPD can be elected for a term of membership of not more than 3 (three) times consecutively or nonconsecutively. In the administrative structure of the sub-district with the DPRK as the representative body, the function of the council in the Public Notice No. 7/1945 is (1) to discuss the household of the regional representatives and (2) set the rules. This shows that the DPRK is a medium for villagers to develop the village, think about village problems, and find solutions to the problems (Suhartono, 2000).

From the observation of three regions, a representative institution is itself a process of deliberations with consensus. It is not to put forward a symbol that requires the name of "representative." That is why in Minangkabau the representative institution is the Nagari Indigenous Council, which becomes the place to solve nagari problems.

However, in the development of a representative there is a shift toward greater emphasis on institutions that set a legal product only, not ones devoted to customary universal values. If promoting the universal value of customs, then the representative body is an institution in which there are direct or selected representatives. Even in the Cameroon State alone, the peoples' representatives are charged only to one person, the chief of the tribe, so that his government, which includes legislation, executive, and judicial branches, is known as a chieftaincy (the governance system under the rule of a chief) (Zambrano, 2000).

Hilman Hadikusuma pointed out that customary representation is in the form of a deliberative institution because the basic principle of governance is a consensus based on popular principles or the will of the people or represented by a group of people (Hadikusuma, 1981). This shows that the form of the representative institution does not have to be its own council that functions only in terms of legislation as adopted in the present system of government, especially in Indonesia based on trias politics theory. In fact, Sutardjo Kartohadikusumo sees the village meeting as the embodiment of democracy based on the One Supreme God at the highest level according to the philosophy of "Manunggalnya Kawulo Gusti" or the unification between the human being and God (Kartohadikusumo, 1984).

This discussion refers to one of the customary governance systems in Indonesia, namely the nagari in Minangkabau. The system of nagari government can be seen in the division of powers of customary institutions, namely the guardian nagari and council of indigenous nagari. Seen from its history, it is clear that KAN existed before the arrangement of the village. The nagari government system, according to Local Regulation No. 9 of 2000, consists of the guardian nagari and parliament nagari. The wali nagari represents the nagari internally and externally and is responsible for the nagari parliament that discusses and enforces village rules and nagari budgets and controls their implementation.

The nagari parliament consists of representatives of important categories of people in the nagari, including the three classical groups of ninik mamak, religious leaders, and scholars and *bundo kadis* (indigenous women) (Huda, 2015). The wali nagari, together with the parliament, form an official government. In addition, there are two other institutions as advisers, namely religious and customary consultative councils as advisers to the wali nagari and parliament. The term KAN was not elected in the nagari governance system in Regional Regulation No. 6 of 2000 because provincial lawmakers intended to separate the new nagari government institutions from KAN considered to be a product of the New Order (Huda, 2015).

In practice, KAN still exists even though it is not included in the regulation. This is evident from the existence of KAN in some nagari in Minangkabau, where of eight regencies in West Sumatera Province, five still have KAN: Solok Regency, Tanah Datar Regency, Sawahlunto Sijunjung Regency, Agam Regency, and Pesisir Selatan Regency (Latief, 2000).

The nagari government version of West Sumatera Provincial Regulation No. 9 of 2000 is already valid, but does not have a tangent point with the structure of indigenous peoples who already have a government structure that has been run according to customary values. Thus, KAN is seen only as a tool for forming the nagari government apparatus and not the power of government itself. This is evident from the institution that was formed not known among the Nagari Minangkabau indigenous peoples. According to Ashshiddiqie, however, customary law in a country is considered important where every country has a source of state administration that cannot be separated from local values in the region (Ashiddiqie, 2016).

The regulation of the nagari administration is also included in District Regulation of 50 Kota No. 1 of 2001 on Nagari Government. In this act even the principle of democracy applied is Western democracy. In addition, there is BPAN and Deli Nagari Consensus (MAN), which is a new institution that is not known among Minangkabau indigenous peoples. Thus, it appears that the slogan "back to the nagari" from the provincial government and the district government is limited to the formality of returning to the nagari in terms of geographical area from about 3500 villages to 600 nagari. In addition to geographical terms, other aspects are similar to the old village government, differing in name only.

Based on the observations of this study, the law has not fully accommodated the interests of indigenous peoples. This is evident from the absence of a clear definition of indigenous villages. Article 1 of the Law on the Village is regulated based only on the meaning of the village equated with the customary village. The article does not explicitly distinguish between indigenous villages and villages so that the definition of the village is implicitly addressed to the customary village as well, though based on the character of the village and the customary village there should be differences, especially in the implementation of customary values by indigenous peoples. Customary villages should be interpreted as a unity of society that purely implements customary law, which cannot be integrated with national law (Dijk, 1982). On the other hand, the customary law applied is not contradictory with NKRI (The Unitary State of the Republic of Indonesia), so it does not become the trigger of the act that divides the unity of the state. Aside from the same understanding between indigenous villages and villages, the Law on Villages is also unclear on customary villages. One of the requirements for a customary law community unit to be recognized as such is that its existence has been recognized based on the law that is applicable both generally and sectorally, whereas the law that recognizes indigenous people specifically in each region does not exist until the Law on the Village is ratified. There is no guarantee that local regulations will protect their own indigenous peoples. In the academic text of the Law on the Village, there is no explanation of the difference between the village and the indigenous village.

Such concerns can be found in the nagari administration in Minangkabau, in which the West Sumatra provincial regulation does not accommodate institutions that already maintain customary values. Arrangements in Article 108 are not enforced where the establishment of a governmental system must be on the initiative of indigenous people, while in Local Regulation No. 9 of 2000, the existence of indigenous peoples is not accommodated and is more accommodating of the national government. This could occur because of the influence of Western theory that divides power over three divisions so that the institutions that do not conform to this structure are not considered to exist. The West is different from Minangkabau, whose power lies in the *penghulu* (administrator) who holds the three powers. The executive, legislative, and judicial powers are already inherent in the body of an administrator at the tribal level and *ninik mamak* (tribal leader) at the nagari level. Ninik mamak in Minangkabau customary community is only primus inter pares, meaning first among equals. It is typically used as an honorary title for those who are formally equal to other members of their group but are accorded unofficial respect, traditionally owing to their seniority in office, not as a dictator. That is why every decision must be made in a consensus according to the truth, so that the a ruler is not an individual but the truth itself.

There are four countries that also provide appropriate regulation and recognition for their customary representative institutions: PRC, United States, Cameroon, and Australia.

Of the four countries, each has its own uniqueness in the governance of the region, more specifically the government of its indigenous peoples. In general, every country has given recognition to indigenous peoples both in the constitution and in law. This recognition is accompanied by the protection of the rights of indigenous peoples such as the right to exercise self-government, the right to control customary forests, the right to be free from threats of discrimination, and others. The recognition of indigenous peoples is not the same in every country; the United States initially voted againstUNDRIP (the United Nations Declaration on the Rights of Indigenous Peoples), which in 2007 gave recognition to indigeneous peoples in the world, but in December 2010 US president Barack Obama announced US support of the Declaration.

Recognition of the rights of indigenous peoples is supported by a number of special laws such as the IRA in the United States and ALR in Australia. In the PRC, the recognition is realized by providing an opportunity for its ethnic representation within the national parliament. In addition, indigenous peoples' government relationships with the central government are almost identical in that the indigenous peoples' governments are under subnational levels of government. This indirectly has an impact on the existence of policies that are intervened from the central government. The structure of customary representative institutions in each country varies. In Cameroon the representative body is in one institution along with the chiefs and judges, who are known for their traditional chieftaincy system. In Australia, its representative agency Maralinga Tjarutja has indirect legislative powers because the legislation is held by a special representative body of the central government, the Legislative Council. However, the different structures of the four nations leave the wealth of their respective countries distinctive. The form of customary government is also influenced by the government adopted in their respective countries.

3 CONCLUSION

Based on their research and discussion the authors reached the following conclusions:

1. The Nagari Indigenous Council is a representative institution of customs in Minangkabau. Its position should be within the structure of the nagari government, but with the equalization of the legislative functioning representative body and the chief of the executive functioning committee. KAN is thus considered as not fulfilling the role so that a new body is created that is in accordance with national law, though KAN is more than an institution that serves to deliberate a customary issue.
2. Customary representative institutions in Indonesia can be divided into three important elements:
   1. Membership
   2. Method of selection
   3. Position and authority

Membership is important in maintaining the values and traditions of indigenous peoples.

Selection methods can be a tool to ensure the availability of the right members. The election method that can be used in selecting members of the peoples' representatives is direct election followed by a deliberation process; elections are through elected representatives from the lowest level (as in the Yogyakarta DPRK), elections by appointment by the state, and direct election by deliberation process.

In terms of position and authority, a representative body of people has the highest position because it conveys the aspirations of its people. Representatives of indigenous peoples having customary assemblies such as Dayak or Minangkabau are in the assembly that has the highest position. Decision-making is conducted through deliberation and accepted by all parties.

4 SUGGESTIONS

Based on what they found in this study, the authors consider it important to propose some suggestions as follows:

a. A regulation of the nagari governance system that accommodates KAN is required as a representative customary institution. This is expected to bring back the spirit to carry out the customs at their best without any applicable duality of law.
b. There is a need to amend Law No. 6 of 2014 on the Village, especially in terms of customs village management.
c. There is a need to amend the West Sumatra Provincial Regulation No. 6 of 2000 on the Basic Provisions of the Nagari Government in which excluding KAN institution which is a reflection of customs in Minangkabau.

REFERENCES

Alder, John. (2001). *General Principle of Constitutional and Administrative Law*. New York: Palgrave Macmillan.

Apeldoorn, Van. (2004). *Pengantar Ilmu Hukum*. terj. Jakarta: Pradnya Paramita.

Ashshiddiqie, Jimly. (2015). *Gagasan Konstitusi Sosial: Institusionalisasi dan Konstitusionalisasi Kehidupan Sosial Masyarakat Madani*. Jakarta: LP3ES.

Ashshiddiqie, Jimly. (2005). *Konstitusi dan Konstitusionalisme Indonesia*. Jakarta: Konstitusi Press.

Ashshiddiqie, Jimly. (2016). *Konstitusi Masyarakat Desa (Piagam Tanggungjawab Dan Hak Asasi Warga Desa)*, diunduh dari. http://www.jimly.com/makalah/namafile/176/KONSTITUSI_MASYARAKAT_DESA.pdf, February 19, 2016.

Ashshiddiqie, Jimly. (2006). *Pengantar Ilmu Hukum Tata Negara Jilid I*. Jakarta: Konstitusi Press.

Ashshiddiqie, Jimly. (1997). *Teori dan Aliran Penafsiran Hukum Tata Negara*. Jakarta: Ind Hill.

Azed Abdul Bari dan Makmur Ami. (2013). *Pemilu dan Partai Politik di Indonesia*. Depok: Pusat Studi Hukum Tata Negara Fakultas Hukum UI.

Catt, Helena. (1999). *Democracy in Practice*. London and New York: Routledge.

Cribb, Robert. (2004). *Historical Dictionary of Indonesia*, 2nd ed. Lanham, MD: Scarecrow Press.

Department of Economic and Social Affairs of United Nation. (2016). *State of The World's Indigenous Peoples*. Diunduh dari. http://www.un.org/esa/socdev/unpfii/documents/SOWIP/en/SOWIP_web.pdf, February 21, 2016.

Hasmira, Mira Hasti. (2004). *Pengaruh Budaya Daerah terhadap Organisasi Pemerintahan; (Kasus: Pemerintahan Daerah Kabupaten Solok, Sumatera Barat)*. Tesis Universitas Indonesia.

Kartohadiprodjo, Soediman. (1970). *Beberapa Pemikiran Sekitar Pantjasila*. Bandung:Alumni.

Kelsen, Hans. (1945). *General Theory of Law and State*, terj. Cambridge, MA: Harvard University Press.

Kemal, Iskandar. (2009). *Pemerintahan Nagari Di Minangkabau, Tinjauan tentang Kerapatan Adat*. Yogyakarta: Graha Ilmu.

Mamudji, Sri et al. (2000). *Metode Penelitian dan Penulisan Hukum*. Jakarta: Badan Penerbit Fakultas Hukum Universitas Indonesia.

Manan, Bagir. (1992). *Perjalanan Historis Pasal 18 UUD 1945*. Jakarta: UNSIKA.

M. Dhany. (1987). "Mencari Ilmu Hukum yang Berciri Indonesia." *Jurnal Hukum dan Pembangunan*. No. 3 Tahun ke XVII, June.

Navis, A. A. (1984). *Alam Terkembang Jadi Guru, Adat dan Kebudayaan Minangkabau*. Jakarta: Grafiti Pers.

Pemerintah Kabupaten Agam. (2007). *Peraturan Daerah Kabupaten Agamtentang Pemerintahan Nagari*, Perda Nomor 12 Tahun 2007, LD Nomor 12 Tahun 2007.

Pitkin, Hanna Fenichel. (1967). *The Concept of Representation*. London: University of California Press.

Rady, Martyn. (2015). *Customary Law in Hungary: Court, Text and Tripartitum*. New York: Oxford University Press.

Rehfeld, Andrew. (2005). *The Concept of Constituency: Political Representation, Democratic Legitimacy and Institutional Design*. New York: Cambridge University Press.

Republik Indonesia. (1948). *Undang-UndangPenetapan Aturan-Aturan Pokok Mengenai Pemerintahan Sendiri Di Daerah-Daerah Yang Berhak Mengatur Dan Mengurus Rumah Tangganya Sendiri*. UU Nomor 22 Tahun 1948.

Ruselblum Nancy L. and Robert C. Post. (2002). *Civil Society and Government*. Oxford:: Princeton University Press.

Saragih, Bintan R. (1985). *Sistem Pemerintahan dan Lembaga Perwakilan di Indonesia*. Jakarta: Perintis Press.

Soeprapto.(2013). *Pancasila*. Jakarta: Konstitusi Press.

Sunni, Ismail. (1981). *Pergeseran Kekuasaan Eksekutif*. Jakarta: Aksara Baru.

Tim Penyusun Revisi Naskah Komprehensif Perubahan Undang-Undang Dasar Republik Indonesia Tahun 1945. (2010). *Naskah Komprehensif Perubahan Undang-Undang Dasar Republik Indonesia Tahun 1945, Latar Belakang, Proses, dan Hasil Pembahasan 1999–2002, Buku II*. Jakarta: Sekretariat Jenderal Mahkamah Konstitusi Republik Indonesia.

Wignjosoebroto, Soetandyo. (2008). *Hukum dalam Masyarakat; Perkembangan dan Masalah*. Malang: Bayumedia Publishing.

Zambrano, Eduardo. (2000). *Authority, Social Theories of*, essay prepared for the International Encyclopedia of the Social and Behavioral Sciences.

Advancing Rule of Law in a Global Context – Susetyo, Rinwigati Waagstein & Budi Cahyono (eds)
© 2020 Taylor & Francis Group, London, ISBN 978-1-138-32782-5

State-Owned Enterprises (SOEs) in Indonesian competition law and practice

Ari Siswanto

Faculty of Law, Satya Wacana Christian University, Indonesia

ABSTRACT: One of the main objectives of competition law is to provide the same level play-ing field among competing companies in a particular market. The regulation of business competi-tion is intended to create fair competition among business actors while avoiding unfair competition or anti-competitive practices that may damage the business climate. For this purpose, competition law contains provisions on what should and should not be done in the context of business competition. Since competition law is intended to create fair competition, anyone com-peting for business in a particular market must be subject to the provisions of competition law.

In the competition law discourse, one of the controversial issues is the position of State-Owned Enterprises (SOEs). There are basically two main views regarding the status of SOEs in the competition law. First, SOEs should be granted privileges, even excluded from the scope of business competition law. Secondly, since SOEs are basically businesses and competi-tors to private enterprises, SOEs must also be subject to competition law.

This paper examines the status of SOEs in Indonesian competition law, both in the context of the normative framework and in the implementation of competition law provisions. For this purpose, this paper will examine the rules of competition law governing the SOEs and analyze some cases of alleged violations of competition law examined by the KPPU as the Indonesian competition authority.

Keywords: competition law, state-owned enterprises, competitive neutrality

1 INTRODUCTION

Business competition law (or simply "competition law") is one branch of law that is closely related to commercial activities. The law is designed to provide guidelines for business actors about what action is allowed and what is not when they compete with one another. In general, it can be said that competition law encourages business actors to compete with each other honestly and fairly. For this reason, competition law contains norms that prohibit anti-competitive behavior and business structure, which contain elements of unfair competition or unfairly encourage the establishment of monopoly (Siswanto, 2002).

Competition law is not designed to protect competitors. The law is created to protect and maintain a competitive environment among the competitors. Therefore, one of the important goals in competition law is to create the same level playing field among business actors. This also implies that the norms of competition law should be applied equally to the business actors who are competitors to each other in the same market.

The principle of equality of treatment to establish an equal normative platform among com-petitors is not too much of a problem if the competing business actors are private undertak-ings. However, the situation will become more complicated when there are both private business and State-Owned Enterprises or Government-Owned Enterprises that are competing in the same market. The existence of Government-Owned Enterprises which confronts private business entities in competition raises fundamental questions about whether these

Government-Owned Enterprises should be given similar treatment as private business actors or, on the contrary, should enjoy exceptions and privileges in the context of competition law.

On the one hand, the view that Government-Owned Enterprises should enjoy special treatment is quite reasonable. This view is built on the assumption that different from private business actors, Government-Owned Enterprises have a mission to provide public services or services of general public interest (Whish & Bailey, 2012). Thus, it is appropriate that the Government-Owned Enterprise be privileged in the business competition so that its role as a public service provider can be optimized. However, on the other hand, there is the view that State-Owned Enterprises are essentially business entities that are also profit-oriented, so they should be placed in a position equal to private business actors who become competitors, in order to establish a fair competition and bring economic benefits at large.

The issue of placing the Government-Owned Enterprises appropriately within the framework of competition law is constantly faced by countries that adopt competition law and at the same time operate Government-Owned Enterprises that serve the public interest through the business services provided.

This paper is intended to view and review the issue in the context of Indonesian Competition Law. Specifically, there are two things that are the focus of the study, namely (1) How the status of Government-Owned Enterprises is regulated in Indonesian competition law, and (2) How Indonesian Competition Law treats Government-Owned Enterprises in practice.

2 DISCUSSION

2.1 *Regulation of business competition in Indonesia*

Competition is essentially an antithesis of monopoly. In a competitive situation, competing for business within the relevant product market will more likely make efforts to strive for the same consumers. Competition is believed to be able to stimulate the emergence of various conditions that have a positive impact on the economy and also on consumers. (Healey, 2011: 1). The business competition will encourage business actors to strive to offer good products at the lowest possible price to get consumers. Similarly, the business competition will encourage business actors to adopt various innovations to make their products superior to their competitors' products. From the consumer's point of view, business competition puts consumers in a more protected position, because they are in a position that requires competing business actors to value them.

All the beneficial conditions described above are not generally found in monopolistic condition in which business competition is not present. In monopolistic markets, consumer's position tends to be weaker because they are not contested by business actors. Business actors are not encouraged to invent various innovations to outperform competitors that basically do not exist. Even worse, the monopolistic position allows the holder to exploit consumers through high product pricing without giving them many alternatives.

Therefore, it can be understood that in many countries, the presence of business competition is preferred over the monopolistic conditions. Currently, most countries have already enacted laws to ensure the establishment of fair business competition, known as competition law (Whish & Bailey, 2012). In addition to providing benefits to consumers, the existence of competition law in a country serves as an indicator for a robust business climate, which in turn will be an attraction for investors to conduct business activities. Furthermore, increased business activity will encourage economic growth. Those are the reasons why competition law is adopted in many countries.

Indonesia also considers that the existence of business competition law will bring many benefits to the country's economy. For this reason, in conjunction with the law reform conducted since the late 1990s, Indonesia also promulgated a law that regulates business competition comprehensively, namely the Law Number 5 of 1999 on Prohibition of Monopoly and Unfair Business Competition ("Indonesian Competition Law").

Referring to the Indonesian Competition Law, it can be seen that this law has defined several objectives, namely: (a) to safeguard the public interest and improve the efficiency of the national economy as one of the efforts to improve people's welfare; (b) to create a conducive business climate through the regulation of fair competition so as to ensure the certainty of equal business opportunity for big business actors, medium business actors, and small business actors; (c) to prevent monopolistic practices and/or unfair business competition caused by business actors; and (d) to create effectiveness and efficiency in business activities.

Of the many objectives, one of them is particularly relevant to the analysis of the status of Government-Owned Enterprises in the Indonesian Competition Law. The phrase "public interest" contained in the first objective of the competition law, for instance, can be attributed to one characteristic of a Government-Owned Enterprise, that is, to be a provider of a product that is a basic public need.

To achieve these objectives, the Indonesian Competition Law adopts an important principle which is also relevant to the discussion of the topic of this paper, namely the principle of "economic democracy with due regard to the balance between the interests of business actors and the public interest." Similarly, as in the first objective of Indonesian Competition Law, the phrase "public interest" is also contained in the statement of the principles of the law. From one side, the phrase "public interest" can be interpreted as the basis for providing the privilege of treatment for Government-Owned Enterprises, which carries out the mission of providing products of public interest.

2.2 *Government-owned enterprises in Indonesia*

In the Indonesian context, the term Government-Owned Enterprises is used in this paper to refer to State-Owned Enterprises (SOEs) owned by the central government, as well as Local Government Enterprises (LGEs) owned by provincial and municipal governments. In relation to the competition law, both State-Owned Enterprises and as Local Government Enterprises are equally relevant because both have characteristics as business entities that can compete with private business actors. Nevertheless, given the scale of its business, this paper is more focussed on SOEs, rather than on LGEs.

In a normative sense, Government-Owned Enterprises are regulated in at least two regulations, namely Law Number 19 of 2003 on State-Owned Enterprises ("SOEs Law") and, for LGEs, Law Number 23 of 2014 on Local Government ("Local Government Law"). Referring to the relevant legislation, SOEs are defined as "business entities wholly or largely owned by the state through direct participation derived from separated state assets." Whereas LGEs is briefly defined as "business entities in which the local government own the capital wholly or partially." The two definitions have something in common in that they both emphasize the ownership of the enterprise by the state or government as public bodies. Furthermore, Article 9 of the SOEs Law stipulates that SOEs may take the form of a Limited Liability Company *(Persero)* and a Public Company *(Perum)*. The law defines *Persero* as "SOEs in the form of limited liability company whose capital is divided into shares, in which the whole or at least 51% (fifty-one percent) of shares are owned by the Republic of Indonesia whose main purpose is to pursue profit." Meanwhile, Perum is defined as "SOEs whose capital is not divided into shares, wholly owned by the state, which aims for general benefit in the form of providing goods and/or services of high quality and pursuing profit based on the principles of corporate management."

The further comparison will reveal that there are similarities and differences between *Persero* and *Perum*. Similarly, both Persero and Perum are SOEs whose ownership is dominated by the state. The difference is that the dominant state ownership in *Persero* is indicated by the ownership of capital (in the form of shares) which is determined to reach at least 51%, whereas in *Perum* the state completely owns all of its capital. Another difference that exists between *Perum* and *Persero* is in terms of its capital form. *Persero's* capital must be divided into shares, while the *Perum's* capital is not. The provision that the *Persero* capital should be divided into shares is a consequence of the *Persero* categorization as a special form of Limited Liability Company. However, there is a more substantial difference between *Persero* and *Perum*. Referring to the definitions given by the law, *Persero's* main objective is to pursue profits, whereas *Perum* is explicitly founded to carry out the function of serving public interest.

The dichotomy of SOEs to *Persero* and *Perum* is also followed by arrangements for LGEs at regional levels. The SOEs at the regional level consists of two different types, namely Regional Limited Liability Company *(Perseroda)* and Regional Public Company *(Perumda)*.

Article 339 of the Local Government Law defines *Perseroda* as "LGEs in the form of a Limited Liability Company whose capital is divided into shares with one local government holds a whole or at least 51% (fifty-one percent) of its shares." Meanwhile, *Perumda* is defined as "LGE whose capital is owned by one local government and not divided into shares".

From the above description, it can be asserted that Government-Owned Enterprises, either owned by the central government (SOEs) or local government (LGEs) have two dualistic characteristics. On the one hand, Government-Owned Enterprises have a role as the representative of the state to fulfill its obligation to provide goods and services that constitute public interest. At the same time, however, the Government-Owned Enterprises also possess the characteristic of a business entity that aims to make a profit.

As of 2017, there are totally 141 SOEs owned by the Indonesian government. Of these, 13 SOEs (9%) are *Perum,* and 138 (91%) are *Persero.*

2.3 *The status of government-owned enterprises in competition law*

Government-Owned Enterprises are also common in any other countries. The dualistic character of Government-Owned Enterprises as business entities and at the same time as government entities have ultimately raised issues in the field of business competition. The central competition law issue in relation to the existence of a Government-Owned Enterprise is whether a Government-Owned Enterprise should be placed on the same regulatory platform as private enterprises or should be excluded from the application of competition law as practiced among others by the United Arab Emirates. (Fox and Healey, 2013: 11).

There are basically two opposing opinions related to the issue of whether or not Government-Owned Enterprises should be given privileged treatment by competition law. On the one hand, there is a view that Government-Owned Enterprises should enjoy special treatment, but on the other hand, there is also the opinion that Government-Owned Enterprises should be subject to the business competition law norms applicable to all business, both private and government-controlled. The notion that Government-Owned Enterprises should not be treated equally with private entrepreneurs in the context of business competition is based on the main argument that Government-Owned Enterprises prioritize the provision of public services, so it should enjoy privileges and conveniences not given to private and purely profit-seeking business actors. From this point of view, placing Government-Owned Enterprises under the general competition law norms are considered unproductive for the state's efforts to meet the basic needs of its citizens through the activities of Government-Owned Enterprises. From a theoretical perspective, this position is also reinforced by the existence of a doctrine called the Doctrine of State Action that developed in the United States. This doctrine emerged in a lawsuit related to the US antitrust law, in particular, the Sherman Act. The case, known as *Parker v. Brown*, was examined by the US Supreme Court in 1943. In this case, the regulation issued by the State of California governing the production and price of raisin products was challenged and brought before the court for being considered contrary to the principles of business competition contained in the Sherman Act. The Government of California argued that the regulation was necessary to stabilize the selling price of the State's prime product, raisin. In that case, the US Supreme Court ultimately ruled that the issuance of regulations was within the domain of state authority and thus legitimate. However, it was also acknowledged that the regulation affected the performance of business competition amongst the producers of grapes, raisin's raw materials. In the relevant section, the US Supreme Court ruling in the case affirms:

Such regulations by the state ... are to be upheld because, upon consideration of all the relevant facts and circumstances, it appears that the matter is one which may appropriately be regulated in the interest of the safety, health and wellbeing of local communities...

In conformity with that consideration, it is also asserted that:

...the adoption of legislative measures to prevent the demoralization of the industry by stabilizing the marketing of the raisin crop is a matter of state, as well as national, concern, ... In the exercise of its power, the state has adopted a measure appropriate to the end sought.

From the ruling above, it is quite clear that the actions of the government in regulating certain lines of business may be justified if done properly for the public interest concerning the security, health and public welfare.

For those who support the exclusion of Government-Owned Enterprises from the provisions of competition law, the doctrine is used to construct the argument that the establishment of a Government-Owned Enterprise is a manifestation of action within the scope of state authority, and thus should be excluded from competition law norms that bind the private business. Referring to *Parker v. Brown*, it can be said that at the beginning of its development the US antitrust law does not include Government-Owned Enterprises, before the different preemptive approach was adopted. (Fox & Healy, 2013: 51).

On the other hand, the opinion that requires a Government-Owned Enterprise to comply with the norms of competition law affirms that Government-Owned Enterprise is a business entity that carries on business activities in the same market as private business actors (Sappington & Sidak, 2003). Based on the premise, it is argued that Government-Owned Enterprises should also be regulated by the same legal norms of business competition in order to realize a business environment conducive to the state economy. In other words, government ownership of a business entity is not a proper basis for giving special treatment that benefits the enterprises. This concept is also known as "competitive neutrality" which the OECD defines as "a regulatory framework (i) within which public and private enterprises face the same set of rules and (ii) where no contact with the state brings competitive advantage to any market participant." (Aproskie, Hendriksz & Kolobe, 2014: 5).

Between these two opposing views, there is also an eclectic position viewing that Government-Owned Enterprises are essentially subject to competition law rules as private business are, but under certain conditions it is also possible that some Government-Owned Enterprises, especially those that are vital to the fulfillment of basic needs of the society, be exempted from the enforcement of business competition law. The main problem of this view is certainly to define the concept of "vital to the fulfillment of the basic needs of society."

2.4 *The regulation of government-owned enterprises in Indonesian competition law*

The Indonesian Competition Law does not mention much about the status of the Government- Owned Enterprises in its provisions. In general, it is implied that the Competition Law applies to every entity categorized as a business actor. This is evident from the stipulation of Article 1(e), which provides the definition of business actors as:

"any individual or business entity, whether in the form of legal entity or non-legal entity, established and domiciled or conducting activities within the jurisdiction of the Republic of Indonesia, who is conducting various business activities in the field of economy, either alone or jointly through agreements."

Based on such definition, both private and public business actors, including Government-Owned Enterprises, fall within the category of business actors who are subjected to the Indonesian Competition Law regulations.

The other provision that is also relevant to understand the position of Government-Owned Enterprises in the Indonesian Competition Law is Article 51, which reads:

"Monopoly and/or concentration of activities related to the production and/or distribution of goods and/or services affecting the livelihood of the public and important production branches for the state shall be regulated by law and shall be carried out by a State-Owned Enterprise and/or a body or institution which is formed or appointed by the Government."

In addition, the other relevant provision is Article 50 which govern the exclusion of the Indonesian Competition Law which read, "…excluded from the provisions of this law are: a. actions taken and or agreements made as implementation of applicable laws and regulations …"

Referring to the provisions of Article 1(e) of the Indonesian Competition Law, it could be inferred that entities categorized as business actors are determined to have the following qualifications:

a) any individual or business entity;
b) in the form of legal entity or non-legal entity;
c) established within the jurisdiction of the Republic of Indonesia;
d) domiciled or conducting activities within the jurisdiction of the Republic of Indonesia;
e) alone or jointly through agreements;
f) conducting business activities in the field of economy.

Relying on such a definition, it may be briefly stated that a Government-Owned Enterprise meets the qualifications contained in Article 1(e). State-Owned Enterprises (both SOEs and LGEs) are business as well as legal entities, established and domiciled in the jurisdiction of Indonesia, and conducting business activities in the economic field. Therefore, under this provision, it is clear that a Government-Owned Enterprise is a business actor as defined in the Indonesian Competition Law. As a consequence, in principle, Government-Owned Enterprises are also subject to the provisions of the Indonesian Competition Law.

Nevertheless, Article 51 indicates the possibility of special treatment for certain business sectors, namely the production and distribution of goods and/or services that affect the livelihood of the people and the production branches that are important to the state. Article 51, implies that the monopoly and concentration of activities related to the production and/or distribution of goods and or services for the above business sectors shall be regulated separately by law. This means that the monopoly and concentration of production and distribution activities of goods and or services for the business sector above, - a condition that is *prima facie* contrary to the principle of fair competition - are excluded from the scope of the Indonesian Competition Law. The important phrase of that article is "held by a State-Owned Enterprise" which indicates that SOEs can become one of the parties authorized to manage the production and distribution of products deemed important to the public and the state, that can be exempted from the Indonesian Competition Law rules.

Article 50 of the Indonesian Competition Law which has been quoted above may also have close links with Government-Owned Enterprises. Any action or agreement made by a Government-Owned Enterprise that is basically in contravention of the legal norms of competition law shall be permitted if it is performed or concluded to implement statutory regulations.

2.5 *The practice of Indonesian competition law concerning government-owned enterprises*

In practice, the Business Competition Supervisory Commission (KPPU) as the authority responsible for enforcing the provisions of Indonesian Competition Law tends to regard the Government-Owned Enterprises as business actors that are subject to the provisions of competition law. This can be seen from several cases examined by KPPU regarding the alleged violation of the Indonesian Competition Law conducted by Government-Owned Enterprises. For example, in Case No. 04/KPPU-L/2012 regarding alleged violation of Article 22 of the Indonesian Competition Law concerning the conspiracy involving 2 SOEs as the respondents, namely PT Waskita Karya (Persero) and PT Adhi Karya (Persero) Tbk, the KPPU clearly qualifies both SOEs as business actors as referred to in the Indonesian Competition Law. In this case, both SOEs are found to violate Article 22 and are required to pay fines.

A similar principle can also be found in Case No.10/KPPU-L/2001 with PT Bank Negara Indonesia (Persero) Tbk as the respondent. The KPPU expressly qualifies PT Bank Negara

Indonesia (Persero) Tbk as a business actor bound by obligation to comply with the provisions of the Indonesian Competition Law. Qualification of Government-Owned Enterprises as business actors as referred to in the Indonesian Competition Law can also be found in Case No. 08/KPPU-I/2005 involving PT Surveyor Indonesia (Persero) and PT Superintending Company of Indonesia (Persero) which later were proven guilty of violating competition law and obliged to pay fines.

Close examination on competition cases involving Government-Owned Enterprises shows that there is no elaborated defense from the respondents developed upon the argument that the respondent is a Government-Owned Enterprise (particularly SOEs) that should not be bound by the provisions of the Indonesian Competition Law. The absence of a defensive argument that accentuates the position of the Government-Owned Enterprises shows that in practice, there is no doubt that the Government-Owned Enterprises are essentially covered by the definition of business actors as referred to in Article 1(e) of the Indonesian Competition Law.

So far, in practice, the treatment of Government-Owned Enterprises in competition cases litigation is in line with the prevailing regulations. It shows that the idea of "competitive neutrality" is also adopted by Indonesian Competition Law. The practice of KPPU shows that basically Government-Owned Enterprises do not enjoy preferential treatment different from private business actors.

For the present conditions, the normative principle which may be used as the basis for excluding of Government-Owned Enterprises from the scope of Indonesian Competition Law is contained in Article 50 and Article 51 of the Indonesian Competition Law. The basic principle that can be drawn from the two articles is that the Government-Owned Enterprises have the opportunity to be exempted from the provisions of competition law if there are laws and regulations that exclude such Government-Owned Enterprises because their business activities are related to the production and distribution of certain products or because they perform actions or conclude agreements as the implementation of law.

From these provisions, it can be inferred that the regulation of Government-Owned Enterprises and the possibility of those entities to obtain preferential treatment tend to emphasize formal aspects. As long as there is a legal basis, there is an opportunity to justify the actions taken by a Government-Owned Enterprise, even if it substantially violates the norms of fair competition. It should be remembered, however, that most Government-Owned Enterprises are in a competitive relationship with private business actors, and they also have the potential to engage in unfair competition. Substantially, the unfair competition practices perpetrated by any business actor remain damaging to fair competition, even if they are legitimized by legislation.

Therefore, the legislator should consider reforming Indonesian Competition Law to provide clearer guidance on the extent to which and in what matters the conduct of Government-Owned Enterprises may be exempted from the application of the competition law provisions. This criterion becomes particularly important given that the monopolistic position granted by the legislation can be implemented in such a way that it has a negative impact on business competition. A good example of this situation can be seen from Case No.8/KPPU-L/2016 in which PT Angkasa Pura Logistik, a subsidiary of PT Angkasa Pura I (Persero), a State-Owned Enterprise, obtains exclusive rights in the provision of terminal facilities for cargo and postal transport services as well as inspection services and controlling cargo and postal security at several airports in Indonesia. This exclusive right may be legitimate from the legal point of view as it originates from the legislation. However, the problem becomes more complicated because this enterprise also manages business units that compete with private business in the field of Aircraft Flight Expedition. Competition in the field of Aircraft Flight Expedition becomes unfair because PT Angkasa Pura Logistik as a legal monopolist in cargo and postal freight services has the privilege that can be enjoyed also by its own business units, including the business unit in the field of Aircraft Flight Expedition.

From this case, it is evident that the monopolistic legal position held by the Government-Owned Enterprises can also open up opportunities for unfair competition in related business fields, in which Government-Owned Enterprises competes with private business actors. In order to develop fair competition practices, these conditions need to be addressed in more detail.

3 CONCLUSION

Based on what has been described above, the following conclusions can be drawn:

a) Indonesian Business Competition Law basically applies the principle of competitive neutrality, which assumes that Government-Owned Enterprises should not enjoy a different treatment from private business actors. This principle is contained in the provisions of Article 1(e) of the Indonesian Competition Law. However, Indonesian Competition Law also provides provisions that make it possible for Government-Owned Enterprises to be exempted from competition law rules. Unfortunately, there has not been any elaborated provisions concerning the possibility of a Government-Owned Enterprise to abuse its legal position and authority which is formally justified.

b) As for the practice, so far, the KPPU has adopted the "competitive neutrality" principle in a consistent manner, which is in line with the provisions of the Indonesian Competition Law. Nevertheless, the recent case shows that a Government-Owned Enterprise that holds justified monopoly may use that monopolistic position to support its affiliating business units in their competition against private players. Therefore, the Indonesian Competition Law needs to address the issue of competition law that involves Government-Owned Enterprises more comprehensively in order to clarify the position of those enterprises in competition law.

REFERENCES

Aproskie, Jason, Hendriksz, Morné and Kolobe, Tshekishi, (2014). *State-owned enterprises and competition: exception to the rule?.s.l.*

Fox, E. M. and Healey, D. (2013). When the State harms competition – The role for competition law. *New York University Law and Economics Working Papers*, Paper 336.

Healey, Deborah, "Application of Competition Laws to Government in Asia: The Singapore Story", *ASLI Working Paper*, No. 025, March 2011.

Sappington, David E.M. & Sidak, Gregory J. (2003). Competition Law for State-Owned Enterprises. *Antitrust Law Journal*, 71(2),479-253.

Siswanto, Arie. (2002). *Hukum Persaingan Usaha*, Bogor: PT Ghalia Indonesia.

Wish, Richard & Bailey, David. (2012). *Competiton Law*. Oxford, United Kingdom: Oxford University Press.

Advancing Rule of Law in a Global Context – Susetyo, Rinwigati Waagstein & Budi Cahyono (eds)
© 2020 Taylor & Francis Group, London, ISBN 978-1-138-32782-5

Indonesian competition law reform in anticipating the single market under the ASEAN Economic Community (AEC)

Ari Siswanto & Marihot J. Hutajulu
Faculty of Law, Wacana Christian University, Salatiga, Indonesia

ABSTRACT: One aspect that is important to be assessed in line with the implementation of the ASEAN Economic Community (AEC) is the regulation of business competition, which in Indonesia is set forth in Law No. 5 of 1999 concerning Prohibition of Monopolistic Practices and Unfair Business Competition. This aspect is important, considering that one of the goals of the AEC is the establishment of an integrated and competitive ASEAN single market. In addition, the implementation of AEC will undoubtedly make the business condition in Indonesia marked by competition among domestic as well as foreign business actors, particularly from ASEAN member countries which can easily establish the business in Indonesia by utilizing the AEC. Consequently, there is an urgent need for Indonesia to ensure that Indonesian competition law meets the following qualifications: (a) contains provisions capable of effectively anticipating and regulating business competition involving foreign business actors (mainly from the ASEAN member countries), so that the ideal competitive environment can be realized in line with the commitments of the AEC; And (b) contains provisions that are compatible with the general principles of ASEAN business competition policy and law as outlined in the ASEAN Regional Guidelines on Competition Policy. This paper is intended to provide input for reform of Law No. 5 of 1999 by identifying the weaknesses vested in the current law in anticipating the AEC.

Keywords: competition law, ASEAN Economic Community, competition policy

1   INTRODUCTION

In 2015 ASEAN member countries formally implemented the agreement establishing the ASEAN Economic Community (AEC). The idea of establishing AEC is, in fact, not new. It can not be separated from the efforts to create ASEAN as a liberalized region through several regional economic and trade agreements back in the 1990s.

To form the AEC, ASEAN countries are committed to leverage existing economic agreements within ASEAN, such as the Agreement on the Common Effective Preferential Tariff Scheme for ASEAN Free Trade Area (CEPT-AFTA), the ASEAN Trade in Goods Agreement ATIGA), ASEAN Framework Agreement on Services (AFAS) and ASEAN Comprehensive Investment Agreement (ACIA). The CEPT-AFTA and ATIGA Agreements primarily govern the creation of an ASEAN free goods trade zone, while AFAS regulates the free trade in services. Freedom of investment flows in the ASEAN region is facilitated by AIA as a framework for cooperation.

The liberalization established by the AEC will subsequently make ASEAN region a free trade area, which among others is marked by the increasingly dynamic mobility of business actors across national borders. The ease in investment activities that follows will enable foreign business actors to engage in commercial interaction with domestic business actors more intensely.

Under such circumstances, legal instruments, especially in the field of business competition, are required to respond to the possibility of emerging business competition issues that confront domestic with foreign business actors. In addition, strong and harmonized business law and policy among ASEAN member countries are also needed to encourage the acceleration of ASEAN into a regional single market and production base. The ASEAN member countries are fully aware of this requirements so that this regional organization has also issued guidance in the field of business competition in the form of a document entitled ASEAN Regional Guidelines on Competition Policy (ASEAN Competition Guidelines). Although not legally binding, the principles of competition policy contained in ASEAN Competition Guidelines are expected to be adopted by ASEAN member countries, so that harmonization of competition law among ASEAN countries can be done better in order to realize the ASEAN competitive single market. Based on the description, this paper will focus on the following two issues:

1) Does Indonesia's competition law already contain adequate provisions to anticipate and regulate foreign business actors, especially from ASEAN countries, who step into the domain of Indonesian business competition in the era of AEC?
2) Is the Indonesian competition law already in conformity with the ASEAN regional business competition principles set forth in the ASEAN Competition Guidelines?

Before arriving at a specific discussion of the above two issues, this paper will begin with a general description of the AEC and its characteristics as well as the ASEAN principles of competition policy as contained in the ASEAN Competition Guidelines. Afterwards, the first issue will be discussed by considering the Indonesian business competition law, the regulation of foreign business actors and extra-territorial competition law enforcement. The discussion of the second issue will then follow with the focus on the compatibility of Indonesian competition law with the ASEAN principles of competition policy as contained in the ASEAN Competition Guidelines.

2 DISCUSSION

2.1 *ASEAN Economic Community and its characteristics*

The idea of establishing the ASEAN Economic Community (AEC) has developed as part of the desire to form the ASEAN Community which was formally initiated through the Declaration of ASEAN Concord II in Bali on October 7, 2003. According to the declaration, ASEAN countries envision that the ASEAN Community will be realized by 2020. In 2007 at the 12th ASEAN Summit in Cebu, Philippines, ASEAN countries agreed on a commitment to accelerate the establishment of ASEAN Community, including AEC, which come into force in 2015.

On November 20, 2007, the Heads of State/Government of ASEAN member countries declared the AEC Blueprint. As evidenced by the document's consideration, the AEC Blueprint is intended to minimize development gaps among ASEAN countries and to accelerate the implementation of AEC by 2015 through a system that heavily relies on regulations (rule-based system).

Referring to the document, there are 4 (four) main characteristics in the formation of AEC, namely:

1) a single market and production base;
2) a highly competitive economic region;
3) a region of equitable economic development;
4) a region fully integrated into the global economy.

These four main characteristics are then broken down into more specific elements. The characteristics of the AEC as a very competitive economic area are to be realized through enhancing the regional economic competitiveness. Increased competitiveness in the context of the AEC is further pursued through the aspects of business competition policy, consumer welfare (Whish & Bailey, 2012), Intellectual Property Rights, infrastructure development, taxation and e-commerce.

Among these aspects, the aspect that is particularly relevant to this paper is competition policy. The main objective of a business competition policy is to create a vigorous competition culture, such as the one developed in China (Wang & Su in Zimmer, 2012). For that purpose, ASEAN countries are expected to have adequate competition policy (including law), institutions and arrangements. In order to establish the competition policy, institutions and arrangements that support the AEC, ASEAN countries agreed to encourage cooperation among ASEAN competition authorities, as well as to develop regional guidance on competition policy based on international best practices.

Existing data show that the involvement of business actors from ASEAN countries in other ASEAN countries in the form of foreign direct investment (FDI) shows a substantial increase after 2010 (US $ 16,306 million) until 2016 (US $ 24,000 million) (ASEAN, 2017). These data suggest that cross-border business activities among ASEAN countries have been increasing even before the AEC is formally enacted by the end of 2015. This has been the case because the AEC also relies on existing ASEAN agreement instruments enacted even before the AEC was formally introduced, in particular, the CEPT -AFTA. The growth trend of cross-border business activities among ASEAN countries indicates that the volume of interaction among business actors of ASEAN countries is increasing, and so is the interaction between local and foreign business. Increasing interaction among business actors will in turn also increase competition among local business actors with foreign business actors in the ASEAN region. This is what makes the business competition issue highly relevant in the context of the AEC.

2.2 *Principles of ASEAN competition policy*

ASEAN member countries are fully aware that the AEC will fundamentally bring changes to their business competition characteristics. Therefore, ASEAN member countries formulate an action plan in the field of business competition, titled the ASEAN Competition Action Plan 2016-2025 (ACAP 2016-2025), which contains the objectives and direction of business competition policy in ASEAN within ten years.

The document clearly defines that there are five strategic goals to be achieved, which are (a) effective competition regimes are established in all ASEAN Member States; (b) the capacities of competition-related agencies in AMS are strengthened to effectively implement competition policy and law; (c) regional cooperation arrangements on competition policy and law are in place; (d) fostering a competition-aware ASEAN region; and (e) moving towards greater harmonization of competition policy and law in ASEAN.

Previously, in 2010, ASEAN also issued a guideline of business competition policy for its member countries in a document titled ASEAN Regional Guidelines on Competition Policy (ASEAN Competition Guidelines). The guidelines are introduced to anticipate the establishment of a single ASEAN market, which is expected to be highly competitive and integrated with the global economy. The guidelines are compiled by the ASEAN Expert Group on Competition (AEGC), which is later also given the responsibility of observing the implementation of ACAP 2016-2025). However, the ASEAN Competition Guidelines also asserted that the document is only a guideline and does not have binding power over ASEAN member countries (ASEAN, 2010).

ASEAN Competition Guidelines contains important aspects that should be considered as guidance by ASEAN member countries to implement business competition policy in the ASEAN region. There are several aspects related to business competition contained in the ASEAN Competition Guidelines ranging from legal and business competition law significance, the scope of business competition law, business competition authority, business competition law, to inter-state cooperation in the field of business competition.

Although it lacks legally binding character, ASEAN Competition Guidelines is important since it contains the general principles that if adopted by ASEAN countries the realization of harmonization of business competition law will be accelerated. In turn, the characteristics of ASEAN as a highly competitive market will also be strengthened. Therefore, in the context of this paper, some of the principles contained in the ASEAN Competition Guidelines will be presented as a benchmark to see how far Indonesia's competition law is in line with the principles concerned.

The relevant principles can be put forward as follows:

1. Competition policy (including competition law) should be a common instrument governing all business sectors and all business actors involved in commercial economic activities.
2. Competition policy should also apply to Government-Owned Enterprises, as long as there is no explicit exception by law.
3. Competition policy and law should be applied to legal entities, and its scope should be extended to include individuals authorizing, engaging or facilitating practices that are prohibited by competition policy and law.
4. Competition policy shall prohibit vertical as well as horizontal agreements between business actors that are preventive, distorting or limiting business competition.
5. Competition policy should make space for the implementation of "hardcore restriction" for actions that must have a negative impact on business competition.
6. Competition policy should also enable the adoption of a "rule of reason" approach to assessing actions that do not necessarily be categorized as actions that negatively affect business competition.
7. Prohibition of anticompetitive agreements by competition law and policy must also include trade decisions made by business associations.
8. Competition policy and law should prohibit concerted action, which covers any action taken under implicit coordination or understanding among business actors, but not yet at the stage where an agreement is explicitly made.
9. Abuse of dominant position must be prohibited.
10. Merger of companies that brings about reduced competition substantially or interfere with competition, are prohibited.
11. Competition policy and law need to be supplemented by procedure enabling competition authority to evaluate mergers so that they do not violate competition rules.
12. The implementation of competition policy and law should not prevent the state from realizing other justifiable objectives that may deviate from the principles of business competition.

As mentioned previously, these principles will serve as benchmarks to review the contents of Indonesian competition law in order to determine the compatibility of Indonesian competition law with the principles.

2.3 Indonesian competition law, foreign business actors and extra-territorial enforcement of competition law

Significantly, Indonesian competition law was established just in 1999 when the Law no. 5 of 1999 concerning Prohibition of Monopoly and Unfair Business Competition (Indonesian Competition Law) was promulgated. The issuance of the Indonesian Competition Law has particular significance since previously the legal norms of Indonesian business competition are not comprehensive and are still fragmented in various laws (Arie Siswanto, 2002).

The Indonesian Competition Law is established based on several reasons as reflected in the consideration of the law. Just like any other countries, the main reason behind the formulation of the Indonesian Business Competition Law is the interest to establish a robust and fair competition climate, while preventing the concentration of economic power on certain business actor. In addition, it can be mentioned that another reason for the establishment of the Indonesian Competition Law is to develop an economic democracy in which everyone has equal opportunity in the production and distribution of goods and services. With the support of a robust business climate, the situation is expected to boost economic growth through the efficient market economy.

As previously mentioned, in the context of establishing a single ASEAN market, there are two interesting aspects to be discussed. First, the aspect relating to the preparedness of the Indonesian Competition Law in anticipating the emergence of foreign business actors from ASEAN countries within the scope of business competition in Indonesia. Second, aspects relating to the compatibility of Indonesian competition law with the principles of ASEAN business competition as contained in the ASEAN Competition Guidelines.

The preparedness of Indonesian Competition Law in anticipating the emergence foreign business actors within the scope of Indonesian business competition can be best seen from the rules that define the possibility of Indonesian Competition Law to be applied to foreign business actors. For that reason, it is necessary to see the provisions regulating who is the subject of the regulation of the Indonesian Competition Law.

Careful observation would likely reveal that the norms of business competition law in Indonesia, whether based on behavioral or structural approach, are basically targeting the so-called "business actors" who in Article 1(e) of the Indonesian Competition Law is defined as "any individual or business entity, whether in the form of legal entity or non-legal entity, established and domiciled or conducting activities within the jurisdiction of the Republic of Indonesia, who is conducting various business activities in the field of economy, either alone or jointly through agreements." Based on these definitions, it can be concluded that the business actor who became the subjects of the Indonesian Competition Law could be an individual or business entity, in the form of a legal entity or not a legal entity, and that is established and domiciled or conducting activities within the territory of the Republic of Indonesia.

The phrase "established and domiciled or engaged in activities within the territory of the Republic of Indonesia" brings important implications for the applicability of Indonesian competition law, including within the context of the AEC which is the subject of this paper. The words "established and domiciled" indicate that any business actor to whom the rules of the Indonesian Competition Law should be the legal subject according to Indonesian law. Another part of the phrase, "conducting activities within the territory of the Republic of Indonesia" also has significance for the applicability of the Indonesian Competition Law, as it becomes the jurisdictional basis for applying the Indonesian Competition Law based on the principle of territoriality. On the basis of this principle, it is stipulated that the Indonesian Competition Law can be applied to any business actor, including foreign business actor, as long as the business actor is engaged in business activities in the territory of Indonesia.

In addition to the applicability basis that is inferred from the definition of business actors, the enforcement of the Indonesian Competition Law against foreign business actors also has another basis, namely Article 16 of the Indonesian Competition Law which reads "Business actors are prohibited to enter into agreements with other parties abroad that contain provisions that may result in the practice monopoly and/or unfair business competition." This provision allows the norms of the Indonesian Competition Law to prohibit business actors from entering into agreements with foreign parties abroad. Indirectly, this provision is clearly limiting the opportunity of business actors outside Indonesia to enter into an agreement with a business actor in Indonesia if the substance of the agreement may result in monopolistic practices and or unfair business competition.

From this interpretation it can be seen quite clearly that territoriality is a major factor in determining the normative applicability of Indonesian Competition Law, as it provides an explanation that Indonesian Competition Law applies to persons or entities established and domiciled in Indonesia as well as to persons or business entities that conduct activities within the territory of the state of Indonesia. Both situations use the territory as the starting point of the applicability of Indonesian Competition Law. In the ASEAN single market era, this provision can still be used to apply legal norms of business competition on foreign business actors from ASEAN countries as long as they run business activities in Indonesia.

However, the more complicated issue will likely arise in situations involving extra-territorial dimensions, for example when there are some foreign business actors operating outside Indonesia who agree on a cartel agreement among themselves, and the cartel agreement held by those foreign business actors abroad impacts on the Indonesian market as their export destination. The actual damage can be experienced by business actors in Indonesia, but there is no clarity as to whether the foreign business actors who entered into cartel agreements outside the territory of Indonesia can be reached by Indonesian competition law.

Competition situations that involve extra-territorial dimensions such as shown by the above example are likely to become more frequent as the global economy develops (Dabbah, 2003). The actions of business actors in a country may have serious impacts elsewhere. Under these conditions, effective regulation can not be realized simply by relying on the rigid application of the territorial principle.

2.4 *Compatibility of Indonesian competition law with the principles of ASEAN competition guidelines*

As noted earlier, the establishment of the AEC will, in turn, bring impacts on the realization of the ASEAN single market which will be characterized by an increasingly competitive phenomenon among business actors from different countries within the ASEAN region. In such a framework, it is important to examine whether the Indonesian Competition Law is in compliance with the principles of competition policy contained in the ASEAN Competition Guidelines. This section specifically contains the results of the study of the issue, by adopting the ASEAN business competition principles as the benchmark.

When confronted with the principles withdrawn from the ASEAN Competition Guidelines, it can be argued that in general, the Indonesian Competition Law has a substantial degree of conformity. As required by the ASEAN Competition Guidelines, the Indonesian Competition Law is in principle a general instrument governing all business sectors, other than those explicitly excluded by laws and regulations. The existence of this principle in the Indonesian Competition Law is implicitly evident from the provisions of Article 51 stating that "the production and or distribution of goods and or services that affect the livelihood of the public and the production branches that are important to the state shall be regulated by law." Implicitly this provision suggests that the production and distribution of goods and services not included in the category of Article 52 are otherwise subject to the Indonesian Competition Law.

According to the Indonesian Competition Law, State-Owned Enterprises are also included in the category of business actors that should be subject to the provisions of the competition law. This principle is evident from the definition of business actors in Article 1(e) which indicates that State-Owned Enterprises are included as business actors defined by the Indonesian Competition Law. This principle is also in conformity with the ASEAN competition principles.

Similarly, horizontal or vertical agreements that prevent, distort or restrict business competition are regulated by considering the variations of agreements that may be agreed upon by business actors. The regulation of prohibited agreements is provided in Chapter III (Article 4 through Article 16) of the Indonesian Competition Law.

The prohibition of actions that undoubtedly bring negative impacts on business competition, which in the ASEAN Competition Guidelines is called "hardcore restriction", is also contained in various articles of the Indonesian Competition Law, as in Article 5 paragraph (1) which prohibits price fixing, Article 6 which prohibits price discrimination, Article 10 which prohibits boycott, Article 15 which prohibits exclusive dealing and conditional exclusive dealing, Article 25 which prohibits abuse of dominant position and Article 27 on the prohibition of cross-ownership of undertakings.

Agreements or actions which do not necessarily have anticompetitive or unfair nature are treated with the standard "rule of reason" approach as defined in the ASEAN Competition Guidelines. Unlike prohibited actions or agreements that fall under the "hardcore restriction" category, there are agreements or actions that do not necessarily prevent competition (Broder, 2010). Against such agreements or actions, the Indonesian Competition Law implements cautious treatment. Agreements or actions suspected to have a negative impact on competition will be further assessed, and will only be established violating the law once it is determined that there is a negative impact on business competition. This is what in business competition law is known as "rule of reason approach". An example of an agreement that is treated with a "rule of reason approach" is a merger between two or more companies. The merger itself in many ways can bring a positive impact on business actors, so it is not necessarily prohibited. A merger will be regarded as violating the business competition provisions when the action has a negative impact on business competition. This approach is also one of the principles contained in the ASEAN Competition Guidelines and has been accommodated in the Indonesia Competition Law.

Mergers seem to be receiving great attention in the ASEAN Competition Guidelines, which contains the principle that a review mechanism should be provided so that the prearranged merger does not violate the provisions of competition law. This principle shows that competition law is not intended to prevent mergers, but to ensure that they do not violate the competition law. Indonesian Competition Law is also familiar with the procedure of pre-merger notification as referred to in the ASEAN Competition Guidelines.

In addition to mergers, actions or agreements that are also governed by the "rule of reason approach" in the Indonesian Competition Law include below-market price-fixing agreements (Article 7), regional distribution agreements (Article 9), cartels (Article 11), trust (Article 12), monopolies (Article 17), market control (Article 19), unfair pricing (Article 21) and conspiracy (Article 22).

However, while most of the business competition principles contained in the ASEAN Competition Guidelines are already accommodated in Indonesian Competition Law, there are in fact three principles that have not been clearly reflected by the Indonesian Competition Law, namely the principle that extend the application of competition law to include individuals with certain qualifications, the principle concerning the positions of business association decisions in the context of business competition law and the principle relating to concerted action.

Concerning the extension of the business competition law, the ASEAN Competition Guidelines requires that the scope of business competition law be extended to include individuals authorizing, engaging or facilitating practices that are prohibited by the competition policy. This principle is similar to the concept of "accomplices" in criminal law which is intended to provide a legal basis for applying the law to those who accompany or assist the violation of the law.

In the current Indonesian Competition Law, there is no provision applicable to individuals who authorizes, engages or facilitates the violation of competition law. Under Indonesian Competition Law, individuals are referred to in the category of business actors who can violate business competition law as well as business entities, as stipulated in Article 1(e). This makes the parties that can be covered by competition law is limited to business actors, either individual or other business entity, which is directly involved as a violator of competition law. In fact, in practice, there are situations where violations of competition law involve the role of non-business actors. Such situations can, for example, be found in a bid-rigging that may involve individual auction organizers as a party that makes bid-rigging possible. So far, Indonesian Competition Law does not yet have a clear norm to regulate individuals who participate in violation of competition law.

The second matter that has not been clearly regulated is the pronouncement in the competition law that decisions made by trade associations must also be in harmony with the legal norms of business competition. From a legal perspective, business association decisions can hardly be categorized as agreements, and when those decisions are contrary to the norms of business competition, it is also difficult to categorize them as prohibited agreements. Therefore, business association decisions are often located in gray areas that have not been explicitly covered by the provisions of competition law.

A similar problem exists in the third matter which has not been explicitly covered by Indonesian Competition Law, namely concerted action. In practice, it is quite possible that some business actors take certain actions in such a way as to achieve certain objectives that may be contrary to the competition law. These actions can be based on obvious agreements, but can also be done by individual business actors by implicit understanding. Corresponding acts among some business actors based on this implicit understanding are commonly referred to as concerted action. The position of concerted action, though often mentioned in recent cartel cases, has not been explicitly regulated in Indonesian Competition Law.

3 CONCLUSION

Based on the discussion on the two issues mentioned at the beginning of this paper, it can be concluded that Indonesian Competition Law still relies heavily on the principle of territoriality in the application of its norms. Under this principle, the main criteria for the application of the law are the place where the business actor is established and located or the territory in which the undertaking operates. The establishment of ASEAN single market through AEC will make Indonesia exposed to the possibility of interaction among foreign business actors outside Indonesia, which negatively impact the condition of business competition in Indonesian domestic market. In these extra-territorial situations, the application of territorial principles is not sufficient. Therefore, the Indonesian Competition Law must be considered to extend the range of its applicability through the extra-territorial principle based on the "effects doctrine" justifying the application of national law to any business actor that causes the disruption of fair business competition conditions at the domestic level (Terhechte, 2011).

In addition, although Indonesia's business competition law has a substantial degree of compatibility with the ASEAN competition policy principles, there are three principles that have not been adequately addressed in Indonesian Competition Law; namely the principle related to the extension of the competition law to include individuals who indirectly take part in the violation of competition law, the principle of extending the Indonesian competition law norms to include business association decisions that have the potential to violate competition law and the principle that concerted action that may disrupt competition be more strictly regulated.

Based on what has been mentioned above, it appears that reform of Indonesian Competition Law should be performed with due regard to the above-mentioned shortcomings so that the Indonesian Competition Law is able to accurately respond to the inevitably more dynamic business competition emerging as a consequence of the establishment of AEC.

REFERENCES

ASEAN. (2017). *Celebrating ASEAN: 50 years of evolution and progress.* Jakarta: ASEAN Secretariat.
ASEAN. (2010). *ASEAN Regional Guidelines on Competition Policy.* Jakarta: ASEAN Secretariat.
Broder, Douglas. (2010). *US Antitrust Law and Enforcement.* New York: Oxford University Press, Inc.
Dabbah, Maher M. (2003). *The Internationalisation of Antitrust Policy.* Cambridge, United Kingdom: Cambridge University Press.
Siswanto, Arie. (2002). *Hukum Persaingan Usaha,* Bogor: PT Ghalia Indonesia.
Terhechte, Jörg Philipp. (2011). *International Competition Enforcement Law Between Cooperation and Convergence,* Heidelberg: Springer.
Wish, Richard & Bailey, David. (2012). *Competiton Law.* Oxford, United Kingdom: Oxford University Press.
Zimmer, Daniel. (2012). *The Goals of Competition Law.* Cheltenham, United Kingdom: Edward Elgar Publishing Limited.

Authorities

Law Number 5 of 1999 concerning Prohibition of Monopolistic Practices and Unfair Business Competition.

Advancing Rule of Law in a Global Context – Susetyo, Rinwigati Waagstein & Budi Cahyono (eds)
© 2020 Taylor & Francis Group, London, ISBN 978-1-138-32782-5

The role of ulama, adat and government institutions as the stability factors of the Minangkabau system of government

Fitra Arsil
Constitutional Law Department, Faculty of Law, Universitas Indonesia, Indonesia

Ryan Muthiara Wasti
Center of Constitutional Law Studies FHUI, Indonesia

ABSTRACT: Each system of government has its own instability potential. The relation between distinct power institutions in the concept of separation of powers does not always result in a harmony of relationships. Instead, it often results in a confrontational way. One approach to overcome the instability of government is to create a conducive institutional formulation of the state in the formation of government and decision-making process. This research sees the phenomenon occurring in Minangkabau adat administration. The concept of separation of powers within Minangkabau adat administration can be traced from the existence of adat institutions, government and ulama in the government structure and the decision-making mechanisms ranging from king level to subordinate government at nagari and tribal levels. The relations of these three institutions formed a distinctive system of government that proved to produce stability at every level of government. This study compares the reality of adat rules in Minangkabau with the theory of separation of powers and the concept of stability contained in various government systems. This paper answers these relationships through normative legal research with historical and comparative approaches

Keywords: system of government, adat, ulama, government

1 INTRODUCTION

All systems of government aim for political and governmental stability. However, the fact remains that in practice, each system of government has its own problems (Arsil, 2017). Based on the findings of Robert Elgie's research, the problems faced by the government system are influenced by the institutional context and other administrative systems prevailing in the application of the system of government (Elgie, 2005). Despite the fact that the democratic values are universal and have become a global system, the mechanism of applying them in one country is adapted in accordance with the culture and characters of the said country. Democracy is conducted using various appropriate mechanisms. According to Held, each country in which the supreme power is vested on the people translates the procedures of running the democracy in their own way (Held, 1997). In practice, the system of government will be influenced by many other elements so that its success will also depend on the implementation of other elements. A system which seem strong in concept needs to be proven empirically as well.

Minangkabau, which is known for its matrilineal system, is a very unique culture (Kato, 1978). Its system of government is considered cosmopolitan because the prevailing system of government in Minangkabau is the result of a compromise between a strong indigenous culture, a strong belief system and a simultaneously recognized government. More so, in practice, these three institutions; adat (tradition), belief (religion) and government become the ways of life adapted as a whole by Minangkabau society. The existence of religion in Minangkabau was preceded by a belief in Hinduism. However, in the 7th century AD, Minangkabau beliefs began to be influenced by Islam brought by Arab and Parsee traders (Piliang, 2015). The high spirits of Islamic spread started by Syekh Burhanudin from Pariaman had made Islam easily accepted and quickly grown in Minangkabau society.

The influence of Islam is not only felt from the change of the form of Minangkabau community worship, but it also covers all aspects of life. In fact, the philosophy of adat is also juxtaposed with Islam that reads; *Adat Basandi Syarak, Syarak Basandi Kitabullah*. The saying means that adat founded upon Islamic Law, and Islamic Law founded upon the Qur'an (C.N. Latief, Dt. Bandaro, 2004). Nevertheless, the existence of Islam does not negate the adat and national government but strengthens them.

The influence of Islam is also prevalent in the Minangkabau government which is based on the spirit of the Minangkabau adat life which recognizes the ultimate authority of truth (Sjafnir, 2006). This can be seen from the existence of the leader of the people who is only *ditinggikan seranting, didahulukan selangkah* (elevated for a branch, headed for a step). It means that that the leader is not in absolute power; he only comes first in certain things which must refer to reasons and faith (Latief, 2004). In addition, in the composition of the government, there is also a philosophical formation because it is made with a very thorough thought which is reflected in the government of the *Nagari*.

Another example is in the implementation of democratic principles that only existed after the entry of Islam. Even in Islam, it is compulsory to obey leaders as long as it is not in the purpose of opposing Allah SWT. That is the reason that in its systems of government, both at the kingdom and at the nagari levels, adat leaders are always juxtaposed with religious leaders. The unique culture of Minangkabau that combines adat, religion and government is what ultimately stabilizes the government in Minangkabau.

Therefore, there are two issues that will be discussed:

a. How does the culture become the factor of stability in other countries?
b. What is the role of ulama, adat and government in shaping the stability of the adat government system in Minangkabau?

2 DISCUSSION

2.1 *Culture as the factor of stability in the system of government in some countries*

The constitution is the legal source of the state practice in the countries of the world. John Alder, a British law scholar known for the unwritten constitution, advocates the term *source of Constitution*. Alder mentions in his *General Principle of Constitutional and Administrative Law* that the sources of constitution are written and unwritten constitution and convention (Alder, 2001). According to Alder, unwritten constitution is a constitution derived from the practices, attitudes and culture of the dominant elements of the community (Alder, 2001).

Jimly points out the importance of the custom state of administration in a country. Each country has a source of state administration which is inexorable from its local values (Ashshiddiqie, 2008). Thus the constitution has its roots and is truly a part of the living system of a society, practiced and developed along with the development of that society (the living constitution) (Ashshiddiqie, 2008).

In constitutional states, the country's constitution provides a place for textually religious cultural expression primarily concerning about the reference of the country as a divine state, including the constitutions which call a particular religion as the official religion of the state. The mention of God or another religious expression in the constitution proves to be very influential in the government of a country. Hirschl puts it in his article entitled *"Comparative Constitutional Law and Religion"* that countries classified as 'Weak Religious Establishment' still have considerable juridical and constitutional implications for having a religious expression in their constitution (Hirscl, 2011).

Norway, for example, in Article 2 of its Constitution states that there is a guarantee of religious freedom. Nevertheless, it also mentions that Evangelical Lutheranism is the official religion of the State (Hirscl, 2011). The implications of this mention can be seen from other provisions of the Norwegian constitution which, among other things, dictate that the head of state must be of the Evangelical Lutheranism religion and that he is also the leader of the church. Article 12 of the Norwegian Constitution states that the King establishes *the Council of State* as the executor of the executive power, which consists of a Prime Minister and at least seven other members. More than half of those members shall be Evangelical Lutheranism adherents (Arsil, 2017a).

Such arrangements can be found in countries that mention *Evangelical Lutheran Church* as the 'state church' or official religion of the state; namely Norway, Denmark, Finland and Iceland. As is the case with Greece and Cyprus which formally state the Greek Orthodox Church their state church (Hirscl, 2011). Even in various Islamic countries, the recognition of Islam is also accompanied by the title of Islam as "the source of legislation" which means that laws implemented in those countries are formed under the official religious law according to their constitutions.

In Ireland, the original text of its constitution (1937) gave a special place to the Catholic Church. However, in 1973, through the fifth amendment of the Irish constitution, it was no longer included the text. The Catholic influence was initially seen dominant in its constitution, such as the mention of family as the basic unit in society as well as the prohibition for a divorce (Article 41). In 1983, there was also an amendment which guaranteed that a fetus has the same right to live as its mother. However, in the fifteenth (1995) amendment, divorce was allowed, and by 2015 there was a referendum on a constitutional amendment to include the marriage equality clause (Arsil, 2017a).

In Belgium, the social composition of the society determines the coalition of the governmental cabinet to be formed. This is contained in the Belgian Constitution which requires that the cabinet formed to reflect the main compositions of Belgian society, French-speaking and Dutch-speaking communities. (Arsil, 2015). In Article 104 of the Belgian Constitution, it is determined that a cabinet is made up of 15 ministers. The ministers in the cabinet, aside from the prime minister, must be evenly distributed to ministers from the Flemish-speaking community and the Wallon-speaking community (Belgian House of Representatives, 2012).

The arrangements of the constitutions of those countries show that each country has its own way of looking for ways to stabilize its government. One of them is by making religion or cultural aspects as the stability factor. These countries apply cultural injections to their constitutional systems such as governmental systems, parliaments, separation of powers and other administrative systems.

In addition, the countries that make religion as a factor that stabilizes their government are not only the so-called religious states (e.g. the Islamic state), but also the countries categorized as Weak Religion Establishment, as described by Hirscl. This shows that religion and culture can be some of the factors that make up stability in the national governments of each country.

2.2 *The role of ulama, adat and government as the stability factors in the Minangkabau government system*

In the middle of the 7th century, Minangkabau was influenced by Islam after this religion entered its vast territory carried by Arab and Parsee traders. The entry of Islam made a big change for the Minangkabau community because they were originally Hindus and Buddhists. The influence was reinforced by the persistence of Islamic propagators such as Syech Burhanudin who was a scholar of Pariaman and studied with Syech Abdur Rauf in Aceh. The well reception of the Minangkabau community of Islam brought about the Kingdom of Koto Batu, located in Pariangan in the 9th century with its king named Datuk Suri Diraja II. Koto Batu is the kingdom eventually known as the Minangkabau Kingdom based in Pariangan (Navis, 1984).

Datuk Suri Diraja practiced Islam well in his government. He is the *mamak* (uncle) of Datuk Ketemanggungan who later became the leader of the Minangkabau Kingdom. Before Datuk Suri Diraja ruled, the applicable laws in Minangkabau were (Piliang, 2005):

1. The *Simumbang Jatuh* Act (regulating the absolute power of the king, should not be opposed);
2. The *Silamo-lamo* Act (governing the enforcement of the "law of the jungle", those who are strong, rule);
3. *Sigamak-gamak* Act (regulating the free economy, economic actors may freely take economic action)

These three laws made the people miserable and were unable to live as human beings who have rights. At that time, not knowing the term democracy, the King acted at will, so there was no religious institution such as ulama in his government. What happened with the absence of other institutions that perform the check and balance functions was the dictatorship of the king and the instability of the government and the economy of the kingdom of Minangkabau.

Seeing these conditions, Datuk Suri Diraja II replaced the old law with the Law of Retort. It was the implementation of the law of Qisas in which evil deeds will be rewarded in accordance with the act. This law was considered more equitable to the public. In addition, during the time of Datuk Suri Diraja II, the public was free to criticize leaders and state officials when they made mistakes. With the fairer rule, the people became more peaceful, and the violations and crimes were reduced by themselves so that the condition was more conducive (Piliang, 2005). This shows that Islam has brought the positive side with the transformation of an authoritarian system into a democracy.

At the beginning of the reign of Sultan Sri Maharaja Diraja, the Minangkabau Kingdom still had a simple structure in which Sultan Sri Maharaja Diraja as the first king was assisted only by a statesman named Indra Jati or known as Cati Bilang Pandai who at the same time also run the government (C.N. Latief, Dt. Bandaro, 2004). In its 8th year, Minangkabau kingdom which originally stems from the Kingdom of Koto Batu divided the tasks of the kingdom to:

a. Datuk Suri Diraja, positioned as the ultimate *Penghulu,* was an intelligent mindset who could deny any challenges such as the arrival of foreign ships that repeatedly tried to outsmart the kingdom but can be won by his ingenuity.
b. Cati Bilang Pandai, the companion of the Sultan in running the government, was a formidable statesman and prioritized deliberation in any decision making so that all difficulties could be overcome.
c. Tun Tejo Gurang, was a royal architect who had high technical knowledge of his time in designing the country's development.

Of the three leaders, it can be seen that there was a power division between Datuk Suri Diraja as the head of the state and Cati Bilang Pandai as the head of the government.

From its inception, the Minangkabau Kingdom demonstrated the implementation of Islamic values. Starting from the law made by Datuk Suri Diraja II called the Law of Attraction which was very abundant with Islamic values. In addition, the entry of Islam to Minangkabau had also changed the structure of government by incorporating elements of a person with religious capacity in each unit of government. One of the examples is the birth of the concept of the king of religion at the king level. In the executive power, "*Basa Ampek Balai*" which was the council of ministers, included a *Tuan Kadhi* position in Padang Ganthing, accompanying three previous officials namely treasurer in Sungai Tarab, Indomo in Saruaso, and Makhudum in Sumanik. (Mansoer et al. 1970: 64). At the subordinate government level, in nagari and the tribe there were the "mualin" who acted as religious officials. So, a nagari institution consisted of *penghulu* who represented *ninik mamak, manti* who represented *cerdik pandai* group, *malin* who represented the religious group and *dubalang* who represented youth groups. This shows that the entry of Islam to Minangkabau was not merely in ritual sense but also thoroughly becoming a tool in the institutional adat.

The real example can also be seen from the division of power which was done in its government. In the time of King Alif in 1560, there were three power divisions in the Kingdom of Pagaruyung namely Raja Alam (the king of the world) in Pagaruyung, Raja Adat (the king of adat) in Buo and Raja Ibadat (the king of religion) in Sumpur Kudus, all of whom were often referred to as *Rajo Tigo Selo* (Kings of the Three Seats) (C.N. Latief, Dt. Bandaro, 2004). The king of the world became the king of all kings which means he became the highest command in the kingdom, while the King of Adat had the authority related to adat and the King of Ibadat had the authority related to religious and moral matters of society.

At a time when the national government had not yet acknowledged the Nagari administration as being equivalent to the village administration, there was a criticism of the implementation of adat in Minangkabau by the Padri movement where the supposed rulers should be ulama not *penghulu*. This can be seen from the agreement in the Bukit Marapalam Treaty which prioritizes the syara' (Islam). By doing so, Padri's movement had made some improvements to the implementation of that principle by inaugurating a "Tuanku Imam" and a "Tuanku Khalifah" who were tasked to enforce the Islamic Shari'a. In the Nagari that had adopted the system of "*Adat basandi syara', syara' basandi Kitabullah*" the adat leader was no longer the sole ruler; even his position was under ulama so that the Padri made the ulama as the supreme leader in the adat structure (Khadimullah, 2008). This is to confirm that Islam had been accepted in Minangkabau society.

The vast Minangkabau kingdom is divided into many small areas called nagari. This nagari has a self-governing government that abides to the King although there is no relation between one and another. Every nagari is free to make its own rules and run them. The longstanding institutions and the king play the role as the bases of the government (Navis, 1984). Overall, nagari as the lowest governmental unit in the state system has four main functions as follows (Naim, 2009):

a. Nagari as a unit of government administrative entity which has internal
b. Nagari as a unit of security union in the context of the defense of the people of the universe, in which the youth became *parik paga nagari* (the border of the state)
c. Nagari as unit of economic unity, namely Nagari is entitled and authorized to make themselves as a corporate body in the form of a corporation based on cooperatives and sharia at the same time. Thus it reserves the right to form any business entity suitable to be developed in nagari.
d. Nagari as a unit of indigenous and socio-cultural unity. *Nagari*, in this context of this custom and socio-cultural unity, in the future, will no longer separate adat and religions. They will be merged in a unified and integral whole.

Nagari is a form of village community life organization in the government system prevailing in West Sumatra. Nagari is the lowest administrative unit under the sub-district and also a unity of territory, a unity of custom as well as a unity of government administration (Sihombing, 1975 as quoted by Choiriyah, 1994). Nagari is an autonomous region within the confines of the Minangkabau kingdom and is entitled to take care of itself.

In the government of the nagari, there is a pattern similar to that of the Minangkabau Kingdom. One of the nagari devices is the duty officer in charge of developing the religious teachings to the *kamanakan* (youth) and the whole people and of taking care of the worship problem. Nagari in Minangkabau is not just a governmental area that is formally under the central government, but nagari becomes a symbol of the customs and major distinctions in the establishment of Islamic Sharia in Minangkabau. That is why a *mualin* (cleric) has a very important role in the nagari as the person who leads his people and the youth in implementing the Islamic Sharia. Another important role is in establishing a common policy with *penghulu* as the traditional leader.

The system of nagari as a form of local government in Minangkabau is a unique style which has been implemented since a long time ago. From the very beginning of the formation of nagari government, its members were elected through consensus by the representatives of their clans. In addition, each member gained trust not only because of his election but also because of his intelligence and morality which are considered good, so he deserves to get that trust (Kemal, 2009).

In addition to the *penghulu*, Minangkabau people recognize the existence of KAN (*Kerapatan Adat Nagari* – State Council) which is the highest forum in deciding cases at the nagari level.

KAN consists of representatives from each component in the community in the nagari; namely the *penghulu* (adat leader), *mualin* (religious leader), *bundo kanduang* (female representatives) and *cadiak pandai* (scholar representatives) (Agam Regency Government, 2007). In this forum, the religious leaders become one of the requirements in decision-making at the nagari level.

In general, the system of government in Minangkabau has two forms; each of which has its own peculiarities. They are Lareh Koto Piliang and Lareh Bodi Chaniago. Koto Piliang has the following characteristics (Arifin and Kayo, 1993):

1. Sovereignty is in the hands of the king, which means the nature of autocratic rule is by upholding the king's command;
2. The system of his government is known as *batanggo turun* which means the wisdom is derived from the king. This is embodied in this philosophy of harmony, namely:
   Titah datang dari ateh (the command comes from the king)
   Sambah datang dari bawah (worship comes from the people)
3. The base of its power is *titiak dari langik* (point from the sky);
4. Its stratified levels are; main *penghulu* (Dt. Ketemangungan), *penghulu pucuk* (in the nagari), *penghulu suku* (in the ethnic group), *penghulu andiko* (directly related to *mamak kamanakan* (the elders and the youth)).

Meanwhile, the characteristics of Bodi Chaniago are: (Arifin and Kayo, 1993):

1. Democratic government by upholding the results of an agreement;
2. The system of government is known as *naiak bajanjang*, the results of a discussion in one level will be discussed again in a higher level;
3. The base of the power is from below or from the a society called *mambusek dari bumi* (originated from the earth;
4. The non-stratified *penghulu* as in Laras Koto Piliang and their respective functions depends on the outcome of the discussion:

   Duduak samo randah (sitting equally low)
   Tagak samo tinggi (standing equally high)
   or
   Duduak sahamparan (sitting in one stretch)
   Tagak sapamatang (standing on one embankment)

These two systems of government look contradictory, but they actually have one common goal that benefits the Minangkabau community. The principle of democracy is also contained in both systems, whereas before the introduction of Islam, there was no democratic term in the government of Minangkabau. Both systems of governance are well run and translated into nagari devices such as the form of rumah adat (traditional house), the shape of the mosque, the arrangement of balairung and the design of the *penghulu adat*. Besides adat and religion, national government also has taken part in creating stability in Minangkabau government.

The existence of *penghulu* as a traditional leader is not sufficient in maintaining the stability of Minangkabau government. This is because the value of Islam has already ingrained in Minangkabau society. Therefore, the religious factor becomes important in their life. Thus, *penghulu* is always accompanied by ulama in running his government. This can be seen from the rule of the king in the kingdom of Minangkabau up to the nagari level. In the Kingdom of Minangkabau, the King of Adat was assisted by the King of Ibadat to keep the Islamic values growing in society by taking care of the worshiping matter as well as being an adviser to the King of Adat in making decisions. While at the nagari level, mualin becomes the support for the *penghulu* in making decisions. *Adat basandi syarak, syarak basandi kitabullah* becomes the basic foundation of the role of ulama and the adat leaders in Minangkabau.

3 CONCLUSION

Each country has a different culture and has its own characteristics in accordance with the character of the nation. Culture is for each country a feature that they are proud of and embraced as the foundation of life for the community. In fact, the culture is formally embodied into the institutions of its government that are poured in the constitution of their respective countries. Some countries have exemplified how culture can be of value in their constitutions such as Belgium, Norway and Ireland. Although considered as one form of social behavior of the society, it turns out that the culture is formalized into a norm that must be obeyed by every citizen. It even becomes a requirement in establishing the stability of government, because if it is not implemented, it will result in instability of the government itself.

The same thing happens in Minangkabau as an ethnic group known for its uniqueness because it embraces the matrilineal lineage system. Minangkabau applies the same pattern of institutionalizing ulama and adat in formal form in a norm. Indigenous leaders and religious leaders become a symbol of the establishment of democracy within Minangkabau society, not only in the scope of the kingdom, but also at the nagari level. In fact, scholars also participated in giving consideration to the king before taking policy. The role of adat leaders and ulama is what makes the government in Minangkabau adat post Islam entry becomes stable.

REFERENCES

Books

Alder, John. *General Principle of Constitutional and Administrative Law*. New York: Palgrave Macmilan. 2001.

Arsil, Fitra. *Teori Sistem Pemerintahan: Pergeseran Konsep dan Saling Kontribusi Antar Sistem Pemerintahan di Berbagai Negara*. Theory of Government System: Concept Shift and Mutual Contribution between Governance System in Various Countries Depok: Rajawali Pers. 2017.

Arifin, Bustanul dan Dt. Bandaro Kayo. *Budaya Alam Minangkabau*. Culture of Minangkabau Nature Jakarta: Art Print. 1994.

Axelrod, Robert. *The Conflict of Interest*. Chicago: Markham. 1970.

Held, David. *Models of Democracy*, 2nd edition. Cambridge: Polity Press. 1997.

Khadimullah, Tuanku Kayo. *Menuju Tegaknya Syariat Islam di Minangkabau*. Towards Enhancement of Islamic Sharia in Minangkabau. Cikarang: Andalan Umat. 2008.

Kemal, Iskandar. *Pemerintahan Nagari Di Minangkabau, Tinjauan tentang Kerapatan Adat*. Nagari Government in Minangkabau, an Overview of Adat Council Yogyakarta: Graha Ilmu. 2009.

Muhammad Yamin. *6000 Tahun Sang Merah Putih*. 6000 Years of the Red and White. Jakarta: Balai Pustaka. 1958.

Latief, CN. Dt. Bandaro, dkk. *Minangkabau yang Gelisah*. The Restless Minangkabau. Bandung: CV. Lubuk Agung, 2004.

Naim, Mochtar. *Suara Wakil Daerah 1; Kumpulan Pidato, Tulisan, dan Buah Pikiran Selaki Anggota DPD RI dari Sumatera Barat (2004-2009)*. The Voice of Regional Representatives 1; Collection of Speech, Writing, and Thoughts as an DPD Member from West Sumatra (2004-2009) Jakarta: CV Hasanah. 2009.

Navis, A.A. *Alam Terkembang Jadi Guru, Adat dan Kebudayaan Minangkabau*. The Lands are the Teachers, Adat and Culture of Minangkabau. Jakarta: Grafiti Pers. 1984.

Piliang, Edison. *Budaya dan Hukum Adat di Minangkabau*. Culture and Adat Law in Minangkabau. Bukittinggi: Kristal Multimedia. 2015.

Riker, William H. *The Theory of Political Coalitions*. Greenwood Publishing Group. 1984.

Timmermans, Arco I. *High Politics in the Low Countries: An Empirical Study of Coalition Agreements in Belgium and Netherlands*. Hants dan Burlington: Ashgate Publishing. 2003.

Journals, Dissertation and Papers

Arsil, Fitra. *Koalisi Partai Politik di Indonesia: Kajian terhadap Pengaturan dan Praktik terkait Koalisi Partai Politik di Indonesia Periode 1945-1959 dan 1998-2014*. The Coalition of Political Parties in Indonesia: Study on the Arrangements and Practices of Coalitions of Political Parties in Indonesia Period

1945-1959 and 1998-2014. Defended on the open defense in the Academic Senate of the University of Indonesia on Saturday 17 January 2015.

Arsil, Fitra. "Keterangan Ahli pada Perkara Nomor 46/PUU-XIV/2016 Perihal Pengujian Undang-Undang Nomor 1 Tahun 1946 Tentang Peraturan Hukum Pidana atau Kitab Undang-Undang Hukum Pidana Juncto Undang-Undang Nomor 73 Tahun 1958 Tentang Menyata-kan Berlakunya Undang-Undang Nomor 1 Tahun 1946 Tentang Peraturan Hukum Pidana untuk Seluruh Wilayah Republik Indonesia dan Mengubah Kitab Undang-Undang Hukum Pidana Terha-dap Undang-Undang Dasar Negara Republik Indonesia Tahun 1945" The Description of Expert on Case Number 46/PUU-XIV/2016 about the Testing of Law Number 1 Year 1946 Regarding the Rule of Criminal Law or the Criminal Code in connection to Law Number 73 Year 1958 about Stating the Entry into Law Number 1 Year 1946 Regarding Criminal Law Regulations for the Entire Territory of the Republic of Indonesia and Amending the Criminal Code Against the 1945 Constitution of the State of the Republic of Indonesia. Presented on Constitutional Council on 1 February 2017.

Ashshiddiqie, Jimly. *Konstitusi dan Hukum Tata Negara Adat*. Constitution and Adat State Law. Pre-sented as Keynote Speech at National Seminar on Islamic Sultanates Constitution in West Java and Banten. UIN Gunung Djati. Bandung. 5 April 2008.

De Winter, Lieven, Marc Swyngedouw dan Patrick Dumont. "Party System(s) and Electoral Behaviour in Belgium: From Stability to Balkanisation", in *West European Politics*. Vol. 29 (5). 2006.

Choiriyah, Sri Zul. *Sistem Pemerintahan Desa di Minangkabau dalam Kaitannya dengan Undang-Undang Nomor 5 Tahun 1979*. Village Governance System in Minangkabau in Relation to Law Number 5 Year 1979. Master Thesis. University of Indonesia. Jakarta: 1994.

Elgie, Robert. "From Linz to Tsebelis: Three Waves of Presidential/Parliamentary Studies?" *Democra-tization*, Volume 12. Number 1/February 2005.

Sihombing, Herman. *Ketatanegaraan Desa/Nagari di Sumatera Barat*. Country/Nagari State Administra-tion in West Sumatera. West Sumatra Symposium. Bukittinggi. Fakulty of Law. Universitas Andalas. 1975.

Strom, Kaare; Ian Budge; dan Michael J. Laver, "Constraints on Cabinet Formation in Parliamentary Democracies", *American Journal of Political Science*, Vol. 38, No. 2. (Mei, 1994).

Sjafnir. *Sirih Pinang Adat Minangkabau; Pengetahuan Adat Minangkabau Tematik*. Thematic Knowledge of Adat Minagkabau. Padang: Sentra Budaya. 2006.

Tsuyoshi Kato, *Change and Continuity in the Minangkabau Matrilineal System*. Jurnal Indonesia. No. 25 (Apr.1978). New York: Southeast Asia Program Publications at Cornell University.

Timmermans, Arco dan Catherine Moury. "00Coalition Governance in Belgium and The Netherlands: Rising Government Stability Against All Electoral Odds" *Acta Politica*. Vol. 41. 2006. Palgrave Mac-millan Ltd: 2006.

Articles and Websites

Konstitusi Belgia 17 Februari 1994. "The Belgian Constitution" See www.dekamer.be/kvvcr/pdf_sec tions/publications/constitution/grondwetEN.pdf. Retrieved on 7 May 2013, at 10.50 WIB

"Belgium at A Glance" http://www.belgium.pl/belgium_at_a_glance.pdf. Retrieved on 4 March 2014 at 12.47 WIB.

Article 99 of Belgium Constitution. See www.dekamer.be/kvvcr/pdf_sections/publications/constitution/ grondwetEN.pdf. Retrieved on 7 May 2013 at 10.50 WIB.

Mochtar Naim. *Membangun Nagari ke Depan*. Developing the country to the future. Downloaded from https://mochtarnaim.wordpress.com/2009/08/29/membangun-nagari- ke-depan/on 24 November 2015.

Norwegian Constitution. https://ihl-databases.icrc.org/ihl-nat/6fa4d35e5e3025394125673e 0050814/ eee956c813a2da0ec1256a870049de0c/$FILE/Constitution.pdf

http://www.monarchie.be/history/baudouin. Retrieved on 24 March 2013 at 17.15 WIB.

Pemerintah Kabupaten Agam. *Peraturan Daerah Kabupaten Agam tentang Pemerintahan Nagari*. The Regulation of Agam District on the Nagari Government. Perda Nomor 12 Tahun 2007. LD Nomor 12 Tahun 2007.

Advancing Rule of Law in a Global Context – Susetyo, Rinwigati Waagstein & Budi Cahyono (eds)
© 2020 Taylor & Francis Group, London, ISBN 978-1-138-32782-5

The authority of the Indonesian power holder related to the state of emergency in terms of the law on states of emergency

Dewo Baskoro, Fitra Arsil & Qurrata Ayuni
Faculty of Law, University of Indonesia, Indonesia

ABSTRACT: Indonesia is a country that is prone to a state of emergency. Therefore, the existing legal instruments must be sufficient to overcome this situation. However, the power granted to the executive power holder when the state is in exception is often followed by arbitrary actions. This article describes and reviews the arbitrary actions allowed the executive power holder in order to overcome the exception from time to time and classifies the State of Emergency theory adopted from each period of the arrangement. This article also compares the constitutional arrangement in Indonesia regarding the state of emergency with the constitution in other countries. Finally, this study finds that the authority of the executive power holder related to Indonesia's current state of emergency covers areas related to people, places, goods, freedom of expression, communication, transportation, and legislation. The study described in this article concluded that the authority arrangement of the executive power holder in Indonesia is currently leaning toward Carl Schmitt's State of Emergency theory, and the constitutional arrangement in Indonesia does not follow the trend of other countries.

Keywords: authority, executive power holder, Indonesia, law on state of emergency, state of emergency

1 INTRODUCTION

Indonesia is a country that is prone to a state of emergency so that the Indonesian State of Emergency Law should be set to facilitate the state's authority in overcoming the crisis. Indonesia's State Intelligence Agency predicted that from 2014 until 2019 Indonesia would receive threats and pressures, or be significantly affected in the field of defense by actors from countries who have a strategic interest in Indonesia, namely the United States, China, India, Australia, Malaysia, Singapore, and the Philippines, as well as receiving threats from nonstate actors such as the Free Aceh Movement (Gerakan Aceh Merdeka), transnational crime organizations, Free Papua Organization (Organisasi Papua Merdeka), and terrorists (Hikam, 2014). In addition to these military threats, there are also nonmilitary threats toward Indonesia. Based on the Indonesia Disaster Risk Index released by National Agency for Disaster Management, disaster comprises the following variables: hazards, vulnerabilities, losses, and environmental damage by earthquakes, tsunami, volcanic eruptions, floods, landslides, droughts, wildfire, extreme weather, and tidal wave. It was mentioned that of 33 provinces at that time, 26 were in the High-Risk class, and the remaining 7 were in the Medium Risk class; of 496 cities/regencies at that time, 322 were in the High-Risk class, and the remaining 174 were in the Medium Risk class (Badan Nasional Penanggulangan Bahaya, 2014).

Those data implied that Indonesia is vulnerable to a state of emergency. There is also an urgency to conduct in-depth research on how Indonesian law should overcome problems that may arise when the country is in a state of emergency. Moreover, Asshiddiqie postulated that there is a scarcity of literature on the State of Emergency Law in Indonesia, and the existing books are outdated and should be updated (Asshiddiqie, 2007). As of now, it has never been concluded whether or not the arrangements in the legislation are in accordance with the existing theory.

Zwitter identified that in order for the democratic system in the state of emergency to be effective and efficient, there are three possible actions to be taken, and all of them give more power to the executive institution (Zwitter, 2012). Furthermore, Vincent Iyer mentions that one of seven categories of authoritative emergency action is the granting of legislative power to the executive (Iyer, 2000). On the judicial grounds and doctrines previously mentioned, it could be concluded that in a state of emergency, power and authority need to be granted to the executive. However, current empirical research indicates that the state harms personal integrity rights because political leaders have a will to repress, and they have the opportunity to do so (Keith and Poe, 2004). The Weimar Constitution in Germany, as quoted by Schmitt, through Article 48 authorized the president, under certain circumstances (emergency), to take emergency measures without legislative consent, including amending the constitution (Schmitt, 2004). This law implies that there is a problem that may arise from implementation of a state of emergency, which is the granting of great authority to the president in an emergency situation, and it can lead to tyranny. As an attempt to avoid the Unitary State of the Republic of Indonesia from tyrannical rule, we consider it urgent to discuss the authority of the President of the Republic of Indonesia in a state of emergency so that the public acknowledges what authority is possessed by the executive power holder regarding the state of emergency.

In accordance with Article 1 paragraph (3) of the 1945 Constitution of the Republic of Indonesia, Indonesia is a legal state. Atmosudirjo defined authority as a power that has a legal basis (Admosudirjo, 1988). In other words, the authority possessed by the president of the Republic of Indonesia, either under normal circumstances, when declaring the state of emergency, or as long as the state is in a state of emergency, shall be based on law. Therefore, a legal study on the authority of the president of the Republic of Indonesia during a state of emergency is needed.

1.1 *Theoretical review*

Several theories were used in this study: the State of Exception theory by Carl Schmitt, State of Emergency theory by Hans Kelsen, and Andrej Zwitter's opinion on both theories.

Even though both Carl Schmitt and Hans Kelsen talk about the state of emergency, their theories come from different schools of thought. The first sentence of Carl Schmitt's *Political Theology* is "Sovereign is he who decides on the exception" (Schmitt, 1985). What Schmitt means by 'State of Exception' or *Ausnahmezustand* is, according to George Schwab, all kinds of economic and political disturbances that require the implementation of extraordinary measures. This situation is a sign for the constitution to provide the procedures to control crisis so that order and stability are restored (Schmitt, 1985). According to Bredekamp, as quoted by Zwitter, Schmitt sees the state of emergency as a time or moments beyond the sustainability of ordinary circumstances (Zwitter, 2012).

Then how does the sovereign, the focus of Schmitt's discussion, become relevant to this study? From the previous discussion, Schmitt's view on the state of exception is obvious but has not yet fully answered Schmitt's opinion on the relation between the state of emergency and the law. Schmitt basically agrees that the state of emergency in formal form (referring back to the discussion of formal and material state of emergency) is part of the law (Zwitter, 2012). Consequently, if there are no norms that do not prohibit the sovereign through checks and balances, he can override the law as a whole (Schmitt). Schmitt concludes this by looking at Article 48 of the Weimar Constitution, which allows the dictator to override the law in order to deal with the state of emergency (Zwitter, 2012). Hence, Schmitt suggested that when the sovereign has the authority to decide whether the state of the nation is ordinary or emergency, it becomes a political instead of a legal decision (in other words, the state of emergency goes beyond the law) (Zwitter, 2012).

Kelsen in Zwitter (2012) stated that even though the executive and other administrative organs are under unimpeded legislative restriction during a state of emergency, they are still bound to the law because their emergency authority is derived from the Constitution (Zwitter, 2012). According to Kelsen, as parts of law, emergency law is formally and materially *Lex Specialis* and the *Lex Generalis* is the law in ordinary circumstances (Zwitter, 2012). According to him, in accordance with Identity Thesis (stating that the state is the same as the order of the law) in the state of emergency, the state remains under the guidance of democratic and legal principles (Zwitter, 2012). As the order of the law, the state under any circumstances must manage the order of the law, even in a state of emergency (Zwitter, 2012).

From the discussion, it is clear that the state of emergency in Schmitt's point of view is different from Hans Kelsen's. Schmitt argued that the state of exception is outside the law but is created by law,whereas, Kelsen argues that it is in the domain of law and is guided by the law (Zwitter, 2012). Kelsen's theory has an advantage because it is in accordance with the jurisprudence of the International Court of Justice that views the law applied in the state of emergency as *lex specialis* from its *lex generalis,* the International Human Rights Law (Zwitter, 2012).

Zwitter indicated that the theory adopted by a country depends heavily on the provisions of the Constitution in the country concerned (Zwitter, 2012). There are three criteria in determining whether a country follow the State of Emergency theory of Schmitt, namely (Zwitter, 2012):

1. The executive is authorized to determine the state of emergency without guidance from the parliament.
2. The executive may impose material law in the state of emergency without temporal limitation.
3. The executive may change the formal laws related to emergency authority.

The more criteria fulfilled by the Constitution's arrangement, the more likely the country is to follow Schmitt's State of Exception theory. On the contrary, the more criteria that are not fulfilled, the more likely the country is to follow Hans Kelsen's theory (Zwitter, 2012).

2 METHOD

The study described in this article is legal research, scientifically based on method, systematics, and certain thoughts intended for studying one or several legal phenomena by analyzing them to try solving problems that arise in these phenomena (Soekanto, 1986). Normative juridical legal research was applied in this study, using both written and unwritten legal principles, legal systematics, vertical and horizontal levels of synchronization of the legislation, comparison of laws, and the history of law (Mamudji, 2005).

This is descriptive-analytical research that will be the proposal for a compulsory bachelor's thesis. The article provides a description of the research that has been conducted. It means that this research is still at the intellectual level of descriptive analysis (Mamudji, 2005).

The data used in this research are secondary data consisting of primary, secondary, and tertiary legal materials. The primary legal material for this research consists of laws and regulations and their derivative regulations, which are the 1945 Constitution of the State of the Republic of Indonesia Article 12 and Article 22 paragraph (1) along with its derivative regulations such as the *Undang-Undang Prp* (Law upgraded from Government Regulation in Lieu of Law) No. 23 of 1959. The secondary materials are books, journals, theses, dissertation, and data obtained from credible sources on the internet. The tertiary materials are law materials that provide guidance and explanations for primary and secondary materials, namely dictionaries and encyclopedias.

This research employed a statute approach, examining the arrangements in the existing legislation. This research also used a comparative approach by comparing the arrangements in other countries' constitutions as well as the historical approach by looking at previous arrangements to see the progress of Indonesian law on states of emergency.

3 DISCUSSION

The arrangement regime on the state of emergency can be divided into the period of the 1945 Constitution of the Republic of Indonesia; the period of S.O.B. (*Staat van Oorlog en Beleg—* State of War and Siege); the period of Law No. 6 of 1946; the period of Law No. 74 of 1957; the period of *Undang-Undang Prp* No. 23 of 1959 based on the 1945 Constitution of the Republic of Indonesia; and the period of *Undang-Undang Prp* No. 23 of 1959 based on the 1945 Constitution of the Republic of Indonesia.

By the time the 1945 Constitution of the Republic of Indonesia was applied on August 18, 1945, in accordance with Article IV of Transitional Rules, the president became a dictator who held the power of the People's Consultative Assembly, the House of Representatives, and the Supreme Advisory Council until such institutions were established (Suny, 1986). Therefore, in relation to the state of emergency, the executive power holder had the authority to declare a state of emergency in accordance with Article 12 of the 1945 Constitution of the Republic of Indonesia (with conditions and consequences prescribed by law), but any limitation provided by the law might be changed by the president as the temporary authority holder of the House of Representatives.

In the period of S.O.B. law, in accordance with Article 37 S.O.B. Law, the executive had the authority to apply material law in a state of emergency without temporal limitation. The executive also had the power to change formal provisions of emergency authority (Arifin, 1957). The statement of the state of emergency could be carried out by the president without asking for a recommendation from the parliament (the House of Representatives) based on Article 1 S.O.B. Law (Arifin, 1957).

IN the period of Law No. 6 of 1946, based on Article 2 of the law, the statement ofa state of emergency must be passed through the Constitution. Furthermore, the president was authorized to cancel the state of emergency, but he or she must obtain approval from the House of Representatives or the Working Committee of the Central Indonesian National Committee. Therefore, the executive power holder based on this law can only declare a state of emergency with control from the parliament (the House of Representatives or the Working Committee of Central Indonesian National Committee).

The implementer of this law under the state of emergency is the National Defense Council consisting of the prime minister, the Minister of Defense, the Minister of Home Affairs, the Minister of Finance, the Minister of Prosperity, the Minister of Transportation, the Commander-in-Chief, and three representatives of the people's organization who are chosen by the Council of Ministers (Article 3 of Law No. 6/46). This National Defense Council, in a state of emergency with a level of "attack," has the authority to enact regulation equivalent to the law, but within ten days, the president must ask the approval of the regulation from the House of Representatives (Article 7 of Law No. 6/46). According to Article 11 of this law, the regulation referred to in Article 7 may be valid only for maximum 30 days. Therefore, according to Law No. 6 of 1946, the executive power holder does not have the authority to amend the formal provisions concerning the authority in the state of emergency, and there is a temporal limitation in every material law in the form of regulation during a state of emergency.

When the Constitution of the Republic of the United States of Indonesia of 1949 came into force, there was absolutely no change in the authority of the executive power holder in a state of emergency. The 1949 Constitution of RIS (Republic of the United States of Indonesia) regulated the state of emergency in Article 184 paragraphs (1) and (2), whereby the authority to declare a state of emergency lies with the government with terms and procedure arranged by federal law. Government, as defined in this constitution, is the president and the ministers. However, since the federal law did not exist, Law No. 6 of 1946 and S.O.B. were still in force based on Article 92 paragraph (1) of the Constitution of the Republic of the United States of Indonesia of 1949, which reads:

The existing laws and administrative provisions that has existed when the Constitution comes into force, shall remain in force and remain unchanged as the rules and provisions of the United States of Indonesia itself, as long as the rules and regulations are not revoked, supplemented, or amended by law and administrative provisions of this Constitutional power.

Similarly, the Provisional Constitution of 1950 also regulated the state of emergency in Article 129 paragraph (1), which reads:

By the procedures and matters to be determined by the law, the President may state the territory of the Republic of Indonesia or parts of it in a State of Emergency, whenever he considers it necessary for the interest of internal security and security against foreigners.

As long as the new law has not been enacted under this Constitution, the old law is still in force (Article 142). Therefore, it is clear that there was no change in the authority of the executive power holder related to the state of emergency; all referred back to the S.O.B. Law and Law No. 6 of 1946 until finally revoked by Law No. 74 of 1957.

Based on this Constitution, the authority to declare a state of emergency (state of emergency or state of war) is held by the president based on the Council of Ministers' decision if security or legal order throughout the territory or in some parts of Indonesia is threatened by rebellion, riots, or consequences of natural disasters, so that it is feared these cannot be overcome by usual measures; or war or the danger of war or the rape of Indonesian territory by any means arises (Article 1). However, what exactly is meant by "the Council of Ministers' decision"? In the elucidation of Article 1 of Law No. 74 of 1957, it was made clear that the statement of state of emergency referred to by this Act is a presidential decree and the affirmation "the Council of Ministers' decision" is an extension of Article 83 paragraph 2 of the 1950 Constitution in which the ministers are responsible for the government policy (elucidation of Article 1). Since the form of the state of emergency statement under this Act is a presidential decree, based on Article 85 of the 1950 Constitution, the establishment of the presidential decree shall be subject to the approval from the ministers. Thus, there is no control from the parliament in the statement of a state of emergency based on this law.

The power in overcoming the state of emergency is not held by either the president or the ministers, but by emergency authority (Head of Region as the chairman, Chief of Police as vice chairman, and member of the Regional Government Council as members) (Article 7 paragraph (2) of the Law 74/57) and warlord (the Commander-in-Chief as Chief, Army Chief of Staff for Army, Navy Chief of Staff for Navy, and Air Force Chief of Staff for Air Force). As previously discussed, administrative and military authorities are included in the executive authority so that the emergency authority as the implementer of this law and warlord as the executive of military power are also included in the executive power holder in this research (Suny, 1985).

The warlord is authorized to set rules that deviate from the central legislation with the consent of the Council of Ministers, and without the consent of the Council of Ministers if the circumstances are very compelling (General Explanation). Therefore, the executive power holder, during a state of war, may change the formal regulations on the authority in the state of emergency.

In relation to the authority of the executive power holder in Indonesia in applying the material law without any time limitation, the law provides the limits for the enforceability of the regulations made by the emergency authority in the state of emergency for as long as two months since the abolition of the state of emergency (Article 12 paragraph (3) of Law 74/57) and the regulations made by the warlord in the state of war for as long as three months after the abolition of the state of war (Article 30 paragraph (3). However, by looking at the ability of the warlord to enact rules contrary to the central government, this law can in principle be altered by the warlord.

Therefore, based on Law No. 74 of 1957, Indonesia seems to be inclined to embrace Hans Kelsen's theory on the state of emergency. In fact, it is not due to the ability of the executive power holder to change or deviate from the Central Constitution, then Indonesia actually fulfills all three criteria of Carl Schmitt's theory.

Based on Article 12 of the 1945 Constitution of the Republic of Indonesia, the president shall be authorized to disclose the state of emergency whose consequences are regulated by the Constitution. Asshiddiqie argued that the enactment of the state of emergency must be done by the president as the sovereign head of state (Asshiddiqie, 2007). However, the form and the procedures are not explained in the Constitution and the extension of the law which is the *Undang-Undang Prp* No. 23 of 1959. From Article 1 of the *Undang-Undang Prp* No. 23 of

1959, the statement of the state of emergency is the authority of president/supreme commander of the armed forces (the use of "supreme commander of the armed forces" is based on a psychological, not legal consideration), in accordance with Article 12 of the 1945 Constitution of the State of the Republic of Indonesia. Therefore, there is no parliamentary intervention in the statement of a state of emergency based on this law.

Then, does the executive power holder have the authority to establish material law in the state of emergency without temporal limitations? This law provides restrictions on the enforceability of the regulations established by the Civil Emergency Authority until the civil state of emergency is abolished and regulations established by Regional Civil Emergency Authority can be maintained for as long as four months after the abolition of the state of emergency if deemed necessary by the regional head (Article 8 paragraphs (2) and (3) *Undang-Undang Prp* No. 23 of 1959). Also the rules established by the Military Emergency Authority shall cease to be effective when the state of military emergency is abolished, and the rules established by the Regional Military Emergency Authority may be extended by the regional head for as long as six months after the abolition of the state of military emergency (Article 22 paragraphs (2) and (3)). Similarly, the rules laid down by the warlord at the state of war are void when the state of war is abolished, and the regulation issued by the regional warlord can be maintained for as long as six months after the abolition of the state of war (Article 35 paragraphs (2) and (3). However, as there is no temporal limitation in the statement of state of emergency by the president, the executive power holder may enforce the material law in the state of emergency without temporal limitation.

Article 44 paragraph (1) of *Undang-Undang Prp* No. 23 of 1959 reads:
The Warlord has the right, by deviating from the provisions of the central constitution, regulate or take measures of any nature, other than those permitted under the terms of Chapter II, Chapter III, Chapter IV of this Regulation, if it is deemed necessary because of the circumstances that jeopardize the urgent security of the State at that time.

Based on that Article, the executive power holder has the authority to amend the formal law of his or her authority in the state of emergency through his or her authority to enact legislation contrary to the central legislation in circumstances that endanger the state's security at that time.

Based on *Undang-Undang Prp* No. 23 of 1959 derived from the 1945 Indonesian Constitution, as a result of the presidential decree of 1959, Indonesia fulfilled the three criteria of Carl Schmitt's state of emergency, so that it is totally incompatible with Hans Kelsen's theory. The *Undang-Undang Prp* No. 23 of 1959 remained in force even after the reform and amendment of the Constitution into the 1945 Constitution of the State of the Republic of Indonesia today based on Article 1 of the Transitional Constitution of 1945. Although the constitution used is still the same, the arrangement becomes entirely different because the position of the executive power holder changes within the new constitutional structure.

From a merely semantic approach, the president is entitled to declare the state of emergency with the terms and consequences prescribed by the Act (Article 12). This is reinforced in the *Undang-Undang Prp* No. 23 of 1959. An explanation of that Article 1 in the *Undang-Undang Prp* No. 23 of 1959 states that in the statement of the state of emergency by the president, supervision by the House of Representatives and by the judges is dispensed with because the president is responsible only to the People's Consultative Assembly so as to have no supervision in the statement of the state of emergency by the president. However, such an understanding is not appropriate because it has been confirmed by Attachment II of Law No. 12 of 2011 and Nos. 176 and 177 that the explanation serves only as an official interpreter of the legislator based on certain norms in the body and cannot be used as a legal basis for further regulation. Thus the supervision from the House of Representatives and judges becomes possible.

The oversight of the House of Representatives and judges is closely related to the legal form of the statement of the state of emergency. Although not governed by *Undang-Undang Prp*

No. 23 of 1959, implicitly the legal form of a state of emergency statement can be found in Article 46 paragraph (1) sub-paragraph C of Law No. 10 of 2004 which reads:

"(1) Legislation enacted in the Official Gazette of Republic of Indonesia, includes:
c. Presidential Regulation concerning:

1. the endorsement of the agreement between the Republic of Indonesia and other countries or other international bodies, and
2. the declaration of State of Emergency."

Thus based on Law No. 10 of 2004, the declaration of a state of emergency is avowed through presidential regulation. However, this law has been declared invalid by Law No. 12 of 2011, which also does not explicitly state the legal form of a state of Emergency declaration (Article 102). However, it is implicitly set out in Attachment I No. 48 Act No. 12 of 2011, which also regulates it in the form of presidential regulation. This arrangement does not rule out the possibility of a state of emergency declaration in the form of other legal products. According to Jimly Asshiddiqie, the declaration of a state of emergency as "*beschikking*," in other words containing the "concrete and individual' norm," can be done in the form of presidential decree, and if the norm is "concrete, individual and at the same time general and abstract (*regelingen*)," then the preferred form is the Government Regulation in Lieu of Law (Asshiddiqie, 2007).

From the previous discussion, we found that there are three possible declarations of a state of emergency, explicitly through presidential decree, presidential regulation, or government regulation in lieu of law. If the form used by the president is presidential decree, based on Article 31 paragraph (1), the Supreme Court has the authority to examine the legislation under the law to be used against the law. If the state of emergency is affirmed in the form of a regulatory presidential decree, the Supreme Court has the authority to test it (Article 56 of Law No. 12 of 2011). Moreover, when the state of emergency is declared by presidential decree, the contents of which are concrete and individual, the presidential decree may be disputed in the State Administrative Court (Article 4 of Law No. 5 of 1986). Then it appears that there is indirect supervision from the judges in the declaration of the state of emergency in a form of presidential decree and presidential regulation. However, the judges are not the parliament, so the executive power holder can still declare a state of emergency without control of the parliament. If the form is the government regulation in lieu of law, which is the legislation that the president sets out in strained times (Article 1 paragraph (4) of Law 12/11), then after the establishment the president must submit the government regulation in lieu of law to the House of Representatives in a form of bills on the establishment of the government regulation in lieu of law into the legislation at a subsequent session in which the House of Representatives may approve or decline and stipulate its revocation through the legislation (Article 52 of Law 12/11). Then there is a method of oversight from the parliament (the House of Representatives) in which the House of Representatives may approve or decline the state of emergency declaration. It appears that the executive power holder has no authority to declare a state of emergency with the control of the parliament.

The temporal limitation on the application of the material law in the state of emergency relies heavily on the abolition of the state of emergency (Article 8 paragraphs (2) and (3), Article 22 paragraphs (2) and (3), and Article 35 paragraphs (2) and (3) *Undang-Undang Prp* 23/59). As there is no temporal limitation concerning the duration of the state of emergency, there is also no temporal limitation concerning the application of material law in the state of emergency.

In the state of emergency, the executive power holder retains the ability to amend the formal provisions of the authority in the state of emergency. Based on Article 7 paragraph (2) of Law No. 12 of 2011, the legal strength of the legislation is in accordance with the hierarchy of the legislation. The meaning of hierarchy refers to the stage of each legislation based on the principle that the lower legislation should not be contrary to the higher legislation. The hierarchy of legislation according to Law No. 12 of 2011 is the 1945 Constitution of the Republic of Indonesia, the government legislation in lieu of law, government legislation, presidential

Regulation, and regional regulation. However, based on Article 8 of Law No. 12 of 2011, which reads:

(1) The type of law and regulation other than as meant by Article 7 paragraph (1) covers the regulation stipulated by the People's Consultative Assembly, the House of Representative, Regional House of Representative, the Supreme Court, the Constitutional Court, The Audit Board of Indonesia, Judicial Commission, Bank Indonesia, the Ministers, Boards, Institutions, or Commissions established by law or Government on the order of the Legislation, the Regional Province House of Representative, Governor, the Regency House of Representative, Regent/Mayor, Village Head or those on the same level. (2) Laws and regulations referred to in paragraph (1) shall be recognized and have binding legal force as long as it is commanded by a higher Legal Regulation or constituted by authority.

"Constituted by authority" is defined as the administration of certain governmental affairs in accordance with the legislation (elucidation of Article 8 paragraph 2). By reviewing the authority of the warlord to create regulations deviating from the central legislation (Article 44 of *Undang-Undang Prp* 23/59), the executive power holder may amend a formal arrangement of authority in the state of emergency through the authority to enact legislation deviating from such legislation.

According to *Undang-Undang Prp* No. 23 of 1959 based on the 1945 Constitution of the Republic of Indonesia (Amendment Results), Indonesia fulfills two of the three criteria of Carl Schmitt's theory. At this time, Indonesia fulfills only one criterion of Hans Kelsen's theory, so it can be concluded that Indonesia is currently leaning toward Carl Schmitt's theory of the state of exception.

4 CONCLUSION

1. The state of emergency from the perspective of the law on the state of emergency
   From the perspective of the law on the state of emergency, a country in a state of emergency is the object of study for the law on a state of emergency. The state of emergency (in terms of legal circumstances) is a condition for the enforcement of a law on a state of emergency.
2. The authority of the executive power holder in the Indonesian law on a state of emergency
   This research describes the authority of the executive power holder in the law on the state of emergency from time to time. Currently, based on *Undang-Undang Prp* No. 23 of 1959, the president (executive power holder) is generally authorized to appoint auxiliary officers in a state of emergency and also to establish the area of legal authority in the state of emergency.
   In the state of emergency, the executive power holder has the authority over legislation, individual, restriction and seizure of goods, use of place, restriction in freedom of expression, access and closing of transportation facilities, and also access and limitation of communication.
   The higher the level of a state of emergency, the greater the authority given to the president. Even in a state of emergency at a war emergency level, the president may issue rules that completely deviate from central legislation.
3. The authority of the executive power holder in Indonesian related to a state of emergency in the law on the state of emergency not in accordance with Hans Kelsen's theory of State of Emergency and the arrangement trend in several countries' constitutions.

By using the three criteria from Andrej Zwitter in determining whether a county is in accordance with Hans Kelsen's theory or Carl Schmitt's theory viewed from the authority of executive power holder in declaring a state of emergency and in the law on the state of emergency, it could be concluded from this research that in the period of the 1945 Constitution of Republic of Indonesia, the period of S.O.B., the period of Law No. 74 of 1957, and the period of *Undang-Undang Prp* No. 23 of 1959 based on the 1945 Constitution of Republic of Indonesia as a result of the 1959 presidential decree, Indonesia was totally incompatible with Hans

Kelsen's theory. During the enactment of Law No. 6 of 1946, Indonesia was fully in accordance with Hans Kelsen's theory. In the period of *Undang-Undang Prp* No. 23 of 1959 based on the 1945 Constitution of Republic of Indonesia (Amendment Results) as well as the period during which this study was conducted, Indonesia was inclined toward Carl Schmitt's theory and not Hans Kelsen's theory.

By looking at the comparative arrangement of the authority of the executive power holder in relation to the state of emergency in the law on the state of emergency in various countries' constitutions, it can be concluded that Indonesia does not follow the trend of the constitution in other countries, which at least regulate one of the three criteria formulated by Zwitter.

4.1 *Suggestion*

According to the author, by looking at the arrangement of the authority of the executive power holder in Indonesia in declaring a state of emergency and the law on a state of emergency Indonesia currently does not fulfill the criteria for Hans Kelsen's theory because it does not provide any temporal limitation in enforcing material law issued by the executive power holder in the state of emergency and provides for the discretion to amend formal regulations regarding the authority of the executive power holder in the state of emergency. The arrangement of the authority of the executive power holder in the state of emergency and law on the state of emergency must be revised so as to provide temporal limitations on imposing material law in the state of emergency, prohibit the executive power holder from amending the formal regulations with respect to his or her authority in the state of Emergency, and also not to remain in control of the parliament in declaring a state of emergency by the executive power holder. These improvements can be made by creating new legislation or an amendment to the Constitution, which respects these three things.

REFERENCES

Admosudirjo, Prajudi. (1998). *Hukum Administrasi Negara*. Jakarta: Ghalia Indonesia.
Asshiddiqie, Jimly. (2007). *Hukum Tata Negara Darurat*. Jakarta: Rajawali Pers.
Badan Nasional Penanggulangan Bahaya, Indeks Risiko Bencana Indonesia Tahun 2013. (2013). Indonesia: Direktorat Pengurangan Risiko Bencana Deputi Bidang Pencegahan dan Kesiapsiagaan.
Keith, Linda C. and Steven C. Poe. (2004). "Are Constitutional State of Emergency Clauses Effective? An Empirical Exploration." Human Rights Quarterly, *26* (4), hlm. 1075.
Mamudji, Sri, et al. (2005). *Metode Penelitian dan Penulisan Hukum*. Jakarta: Badan Penerbit Fakultas Hukum Universitas Indonesia.
Schmitt, Carl. (2004). *Legality and Legitimacy*, trans. Jeffrey Seitzer and John P. McCormick. Durham, NC: Duke University Press.
Soekanto, Soerjono. (1986). *Pengantar Penelitian Hukum*, 3rd ed. Jakarta: Universitas Indonesia Press.
Zwitter, Andrej. (2013). "The Rule of Law in Times of Crisis: A Legal Theory on the State of Emergency in the Liberal Democracy". University of Groningen Faculty of Law Research Paper No. 10/2013, the.

Advancing Rule of Law in a Global Context – Susetyo, Rinwigati Waagstein & Budi Cahyono (eds)
© 2020 Taylor & Francis Group, London, ISBN 978-1-138-32782-5

Criminal law reform through the Constitutional Court's decisions

Reza Fikri Febriansyah & Topo Santoso
Faculty of Law, University of Indonesia, Indonesia

ABSTRACT: With its premier authority, namely constitutional review, the Indonesian Constitutional Court has become an alternative and new hope for Indonesian criminal law reform. Today, the Constitutional Court has a more significant role than lawmakers in the frame of criminal law development because its several decisions become an alternative and effective solution as well an instrument to polish up most of the philosophical, juridical, and sociological values in the Indonesian Criminal Code (which is the 'legacy' of the Dutch East Indies colonial government) and another legislation regarding criminal law so that the laws remain constantly in line with the constitutional supremacy principles, although we should recognize that there are many critics and a variety of responses in society regarding those decisions. This modest research discusses the dynamic paradigm regarding criminal law reform through the Constitutional Court decisions, the empirical tendency of the Constitutional Court regarding criminal law, as well as the utilities and feasible follow-up of the decisions in the future.

Keywords: Constitution, Constitutional Court's decision, criminal law

1 INTRODUCTION

The Constitutional Court of the Republic of Indonesia (*Mahkamah Konstitusi* or MK) as the one significant result of the 3rd amendment of the Indonesian Constitution of 1945 (UUDNRI Tahun, 1945), has a premier authority, namely constitutional review. This authority afterwards brakes an old settled paradigm in Indonesian legal history that the law is impossible to be examined (*lex dura sed tamen scripta*) so we are increasingly aware that philosophically, a law made by men (*lex positivis*) is always bound by space and time according to Von Savigny's opinion that "law is never made, but grow together with the society (*Das Recht wird nicht gemacht, est ist und wird mit dem Volke*) (Shidarta, 2013).

At first, according to Article 50 Law No. 24/2003 jo. Law No. 8/2011 concerning the Constitutional Court, MK has only authorized examination of constitutional review cases limited to the laws promulgated after the amendment of the UUDNRI Tahun 1945. However, based on the Constitutional Court Decision No. 066/PUU-II/2004 afterwards, MK stipulated that Article 50 of Law No. 24/2003 regarding the Constitutional Court is unconstitutional so that MK subsequently enlarged its authority to examine constitutional review cases for all the laws that are considered against UUDNRI Tahun 1945.

Not many people seem to have realized that several of MK's decisions have become a significant factor in Indonesian criminal law reform, particularly over the three basic problems of criminal law as well as some crucial issues related to the basic principles and theories of criminal law. These research questions are: (1) To what extent is criminal law theory and reasoning used in the MK's decision? (2) What tendency occurs in the MK's decision regarding criminal law? (3) What utility can be gained as well as what feasible follow-up can be done after the MK;s decisions regarding criminal law?

2 DISCUSSION

2.1 *Basic problems of criminal law*

It is generally acknowledged that there are three basic problems in criminal law: offense, guilt, and punishment (Packer: 1968) that are also then modified in Indonesia by Barda Nawawi Arief (Arief, 1994). According to the three basic problems in criminal law, we use five MK decisions as an object of this research that related to the constitutionality of (1) the defamation against the president or vice president (Criminal Code); (2) the age limit of juvenile responsibility (Law on Juvenile Criminal System); (3) capital punishment (Narcotics Law); (4) retroactive principle in criminal law (Law on the Eradication of Criminal Acts of Terrorism and Law on Human Rights Court); and (5) a doctrine of *materiele wederrechtelijke* in a positive function (Law on the Eradication of Corruption). These five MK decisions are selected based on their significant influence on Indonesian criminal law reform as well as indicate the importance of MK's role in Indonesian criminal law development, particularly related to the three basic problems of criminal law.

2.2 *General tendencies of the Indonesian criminal law*

2.2.1 *Overcriminalization*

Overcriminalization, which is generally conceived as an excessive and unnecessary use of criminal law (Husak, 2008), in legislation, adjudication, and correctional phases, is a dangerous tendency in Indonesian criminal law development, and it also turns out to be a dangerous tendency in US criminal justice as Douglas Husak's opinion that there are two specific dangerous tendencies that occur either at the federal or state level, namely increase of criminal law usage and excessive criminal law because of a myth that increasing criminal law usage should always decrease the crime rate (Husak, 2008).

Since 1986, Barda Nawawi Arief has been anxious about overcriminalization, particularly on excessive imprisonment policy in Indonesia by uncovering data that approximately 97.96% items in the criminal code involve imprisonment as well as approximately 91.67% items of criminal law outside the criminal code (Arief, 1986). This phenomenon persistently escalated after reformation in 1998 and the process of amendment of UUDNRI Tahun 1945 four times, which is shown by so many criminal provisions produced, either in special criminal law or administrative criminal law.

Salman Luthan (2014) revealed five) tendencies of the Indonesian criminal law after reformation 1998: (1) increase in number of offenses, (2) increasing criminal punishment, (3) the establishment of a special court such as Human Rights Court and Corruption Court, (4) implementing retroactive principle, and (5) implementing shifting of burden proof (*omkering van het bewijslast*). According to Salman Luthan's opinion, those tendencies occurred for two reasons: (1) a lack of lawmakers' awareness regarding the essence of criminal law as ultimum remedium and/or (2) the intentional use of criminal law to protect a specific political interest (Salman Luthan, 2014). In addition, Mardjono Reksodiputro stated that "criminal law not necessarily tend to protect public interests even often precisely used by the powerful entity to protect their interest" (Reksodiputro, 1993).

2.2.2 *The sluggish role of the lawmakers*

Harkristuti Harkrisnowo sharply criticized the ambiguity in the criminal law paradigm in Indonesia that causes every single man who is involved in the criminal legislation process to freely innovate (Harkrisnowo, 2003). Barda Nawawi Arief also expressed his anxiety regarding the slow progress and reluctance of the current lawmakers to fulfill the society's legal needs, particularly related to the reforming process of the Bill of Indonesian Criminal Code, which is expected to become the main and basic guideline for the entire Indonesian criminal law system (Arief, 1994). The reform is needed because the philosophical, sociological, and juridical values of the old Indonesian Criminal Code are so obsolete and removed from contemporary circumstances, particularly regarding incompatibility between inner values and the norms inside.

Indeed, this anxiety is not surprising because the old Indonesian Criminal Code that was based on the concordance doctrine of 1918 by the Dutch East Indies colonial government has absolutely ignored legal values and legal culture of domestic society in Indonesia, while the general problems of special criminal law and administrative criminal law are the ambiguity of meaning and the tendency of retaliation. These violate constitutional principles, especially regarding fair and legal certainty (Article 28D (1) UUDNRI Tahun 1945) and human rights protection. Therefore, several provisions in the Indonesian Criminal Code as well as in special criminal law and administrative criminal law were decided by MK as unconstitutional (including either conditionally constitutional or conditionally unconstitutional), which then spawned a new trend that MK has a more significant role than the lawmakers in framing Indonesian criminal law reform.

2.3 *Criminal law reform and judicial activism*

Today, the role of MK's in criminal law reform has become more significant because several basic principles of criminal law have become a constitutional norm instead of merely general rules (*algemene bepalingen*) of the 1st Book of Criminal Code, so it is often used as the primary resource in some constitutional review cases. Therefore, if there are several criminal provisions considered against thoseprinciples, it is not merely disharmonization (both vertical and horizontal), but a constitutional matter. Nevertheless, there are some concerns regarding this phenomenon such as a lack of constitutional justices who possess an academic criminal law background, but in our opinion, it does not matter actually because the most important requirements to become a constitutional justices are statesmanship as well as a deepunderstanding of the constitution and state administration. In the case of MK's need to expand its considerations related to criminal law in every single case, MK has the power to present criminal law experts, either by its order or invited by the parties in order to strengthen MK's considerations (*ratio decidendi*).

Historically, we know well that Indonesian criminal law development is influenced by three main legal systems: Islamic law, civil law, and customary law. This requires an outstanding legal knowledge, wisdom, and statesmanship on the part of the constitutional Justices, who are frequently personified as "the guardian and the sole interpreter of the constitution." One reflection of these requirements is the judicial activism (Schlesinger: 1947) frequently conducted by the constitutional justices, particularly regarding constitutional reviews related to the Criminal Code as well as special criminal law and administrative criminal law.

Theoretically, judicial activism could be divided into two concepts: "interpretation" or "construction." According to Keith E. Whittington, "interpretation" is related more to a process whereby the legal meaning of a text is "extracted" from its semantic meaning, while "construction" is essentially creative, though the foundations for the ultimate structure are taken as given (Whittington, 1999). Keith E. Whittington argued that interpretation is related more to finding a linguistic meaning of the text of the constitution, while construction is related more to the court's role to give a new impactful meaning of the text of the constitution (Whittington: 1999). By the interpretation, the interpreters (the constitutional justices) are expected to be able to transform either human language to legal language or transform static law to dynamic law through the transformation of the words and the spirit contained in the norms so it could minimize the discrepancy (lsegal gap) of law in the books and law in action because we knew very well that law is always left behind the facts (*Het recht hinkt achter de feiten aan*) (Shidarta, 2011). Aharon Barak also elaborates the concepts of judicial activism (particularly of interpretation) as rational activity giving meaning to a legal text (whether it be a will, contract, statute, or constitution). According to Barak, "interpretation" obviously is both primary task and the most important tool of a court (Barak, 2006).

Based on justice life experiences, P. N. Bhagwati positively revealed that judicial activism is a basic need in state democracy, but nevertheless reminds us that the justices should not be trapped by the Anglo-Saxon paradigm: "The function of a judge is merely to find the law as it is. The lawmaking function does not belong to him, it belongs to the legislature..." (Bhagwati, 2014) as well as Subekti's opinion that the lawmakers' role and the judges' role is strictly forbidden to be interchangeable (Pangaribuan, 2016). P. N. Bhagwati expressed his anxiety that if the justices are trapped by the aforementioned Anglo-Saxon paradigm, it would obscure the essence of the fair trial process so we should frequently consider two shortcomings of judicial activism, *counter-majoritarian activism* (the court's reluctance to become a fatalistic follower of the legislators) or *jurisdictional activism* (the court's failure to consider and realize its limiting powers) (Marshall, 2002). Oliver Wendell Holmes Jr. also revealed a positive paradigm regarding judicial activism. According to him, law essentially should be contextual. Holmes argued that "a law is not a crystal, transparent and unchanged. It is the skin of a living thought and may vary greatly in color and content according to the circumstances and the time in which it is used. It is for the judge to give fmeaning to what the legislature has said and it is this process of interpretation which constitutes the most creative and thrilling function of a judge" (Bhagwati, 2014).

Judicial activism's measures among the countries are different depending on both the legal system and structures. However, exactly, when the constitutional justices engage in judicial activism, both internal and external factors should influence them. The constitutional justices as human beings are ascertained never to be independent or totally neutral in their subjectivity, so their political ideology, philosophical tendencies, even their religion are significant factors that influence their decision or at least the dissenting opinions. As John Chipman Gray stated: "All the law is judge-made-law" (Friedmann, 1990).

The aforementioned subjective factors are reflected in several decisions related to criminal law cases, such as the one of MK's considerations (*ratio decidendi*) in the MK's Decision No. 2-3/PUU-V/2007 (the constitutionality of the death penalty) which considered aspects of international politics, particularly the fact that Indonesia is the biggest Muslim country in the world and a member of Organization of Islamic Cooperation (OIC). It is therefore difficult for Indonesia to express that the death penalty (capital punishment) is unconstitutional because it potentially shifts the paradigm and values of other modern Muslim countries concerning the qishash principles that are clearly commanded by Allah in the Al-Quran (Al Baqarah, verse 178–179, Al Isra', verse 33, and Al Maidah, 45), particularly a primary principle in Al Baqarah, verse 179 that "..on qishash, there is a guarantee of life."

Actually, both of the judges and the constitutional justices, are allowed to apply judicial activism by The UUDNRI Tahun 1945 and several related laws. The phrase "fairly" before "legal certainty" in the article 28D (1) UUDNRI Tahun 1945 has a specific meaning that the implementation of legal certainty should not always be interpreted narrowly confined to written legal certainty but include fair legal certainty. If the written law is expected to be unfair, it violates the constitution. This principle is also mentioned in Article 5 of Law No, 48/2009 on Judicial Power that "The Judges and The Constitutional Justices are obliged to explore, to follow, and to have a good understanding of the sense of justice and the living law of the people."

In several constitutional review cases regarding the Indonesian Criminal Code to date, MK is expected to understand the "spirit" (*voluntas*) or "the real intent" contained in the examined norms so that MK is not stuck on semantic matters. When the spirit of the Indonesian Criminal Code is considered as violating constitutional supremacy, MK is mandated to stand by the constitutional supremacy in the frame of itsrole as the guardian and the sole interpreter of the constitution to cultivate a constitutional democracy in Indonesia.

The five MK decisions discussed in this research show that MK has applied either *interpretation* or *construction* in the frame of criminal law reform in Indonesia, which may be outlined below.

2.4.1   *Defamation of the president or vice president (Criminal Code)*

MK's Decision No. 013-022/PUU-IV/2006 revealed that Articles 134, 136bis, and 137 of the Indonesia Criminal Code were unconstitutional. MK argued that those articles violatre several constitutional principles, such as equality before the law, fair legal certainty, freedom of expression, as well as the right to obtain information based on Articles 27 (1), 28D (1), 28E (2) and (3), and 28F of UUDNRI Tahun 1945. MK also provides alternative solutions so that by law, the defamation of the president or vice president could be criminalized and penalized only based on: (a) Article 310-321 of the Indonesia Criminal Code which classifies as general defamation (*belediging*) in the case of defamation addressed to his personal matters; or (b) Article 207 of the Indonesian Criminal Code in the case of defamation addressed to his office (incumbency). Because of the difficulty of segregating presidential defamation between his personal matters and his office (incumbency), MK argued that this matter violated the constitution, particularly a fair legal certainty principle based on Article 28D (1) of the UUDNRI Tahun 1945. In this decision, there were four dissenting opinions that were expressed by Justice I. Dewa Gede Palguna, Justice Soedarsono, Justice H. A. S. Natabaya, and Justice H. Achmad Roestandi.

2.4.2   *The age limit of juvenile responsibility*

Based on MK's Decision No. 1/PUU-VIII/2010, the age limit of juvenile responsibility has to be interpreted as twelve years old. MK argued that the phrase "8 (eight) years old" based on Law No. 3/1997 regarding the juvenile system improperly as well as potentially limits children's protection rights as well as children's growth and development rights. MK has argued that the Indonesian juvenile system has to stand for the social, mental, and moral protection of children instead of merely sentencing with the commitment toward the best interest of the child. Moreover, MK also uses an Islamic law perspective that the children (who are yet not *akil baligh*) could have not criminal responsibility because they are assumed not able to understand yet either their rights or obligations. This decision is also comparative and forward-looking because MK also considers the age limit rules in several countries, UN recommendations regarding children rights, and the Bill of Criminal Code (*ius constituendum*). In this decision, there was one dissenting opinion, expressed by Justice Akil Mochtar.

2.4.3   *Death penalty/capital punishment*

MK's Decision No. 2-3/PUU-V/2007 rejected the petition on the death penalty/capital punishment in Law No. 22/1997 regarding narcotics as constitutionally fixed.

The discourse related to the constitutionality of the death penalty/capital punishment actually focused on the Constitutional justices's reading of the linkages between Article 28I (1) and Article 28J (2) of the UUDNRI Tahun 1945. By using systemic interpretation (*sistematische interpretatie*), MK argued that the concept of human rights based on Article 28A of 28I UUDNRI Tahun 1945 should comply with the limitations based on Article 28J (2) UUDNRI Tahun 1945. MK also argued that human rights protection in Indonesia is not unlimited; instead it would be restricted by laws in the spirit of TAP MPR No. XVII/MPR/1998.

In the context of the death penalty, MK also suggested that we should keep the equilibrium between "right to life" and "right of life" because the crime committed is exactly an assault on either right to life or right of life. That is why MK also argued that it is so difficult to fully ignore the retaliation principle of sentencing because in the criminal law concept, the retributive aspect is inseparable. Nevertheless, it is not a main principle but a preventive and educative measure.

In this decision, regarding the constitutionality of the death penalty/capital punishment, there were three dissenting opinions, expressed by Justice Laica Marzuki, Justice Maruarar Siahaan, and Justice Achmad Roestandi. Justice Harjono expressed his dissenting opinion only with regard to the fact that foreign citizens are assumed to have legal standing before MK in constitutional review cases.

2.4.4 *Non-retroactive principle*

There are two outstanding MK decisions regarding the constitutionality of non-retroactive principle: (1) MK Decision No. 013/PUU-I/2003 and (2) MK Decision No. 065/PUU-II/2004.

On the MK Decision No. 013/PUU-I/2003, MK argued that the retroactive principle can be used in a limited maner (for the sake the equilibrium of legal certainty and justice) with the considerations that (1) the value of justice does not derive merely from legal certainty but keeps a balance between the legal protection of victims and that of offenders; and (2) for more serious crimes, justice should move forward rather than merely looking at legal certainty.

MK Decision No. 013/PUU-I/2003 noted that Law No. 16/2003 as an *ex post facto law* is unconstitutional because it is ineligible to be qualified to overrule the non-retroactive principle. According to MK, the first Bali bombing that occurred on October 12, 2002 cannot be categorized as an *extraordinary crime,* so the law cannot be applied retroactively, instead of a cruel crime, which is still covered by the existing criminal law. MK was also concerned that the existence of an *ex post facto law* would become a bad precedent in the future, whereas we still have a sufficient existing law to address that crime. In this decision, four Constitutional justices expressed a dissenting opinion: Justice Maruarar Siahaan, Justice I. Dewa Gede Palguna, Justice H. A. S. Natabaya, and Justice Harjono.

As for the conception of retroactive principle in Law No. 26/2000 in Human Rights Court, MK decided that applying an ad hoc Human Rights Court for several incidents before the enforcement of Law No. 26/2000 on Human Rights Court is constitutionally fixed. MK argued that to interpret Article 28I (1) of the UUDNRI Tahun 1945 it should always be inseparable from Article 28J (2) of the UUDNRI Tahun 1945 so it reflects MK's consistency in how to read and interpret the linkage between Articles 28I (1) and 28J (2) of the UUDNRI Tahun 1945, as MK has previously done on Decision No. 2-3/PUU-V/2007 regarding the constitutionality of the death penalty/capital punishment.

MK warned that Article 28I (1) of the UUDNRI Tahun 1945 should not just be interpreted in textual matters, but also encompass the history of the retroactive principle comprehensively.

In Decision No. 065/PUU-II/2004, MK also argued that the equilibrium between legal certainty and justice, particularly regarding the retroactive principle, should consider three legal orientations: legal certainty (*rechtssicherkeit*), justice (*gerechtigkeit*), and usefulness (*zweckmassigkeit*). The criminal law applying to *extraordinary crimes* can thereby be enforced retroactively based on the *aut punere aut dedere* philosophy that there is no crime that cannot be prosecuted. In Decision No. 065/PUU-II/2004, there were three dissenting opinions, expressed by Justice Achmad Roestandi, Justice Laica Marzuki, and Justice Abdul Mukhtie Fajar.

2.4.5 *The doctrine of* materiele wederrechtelijke

MK Decision No. 003/PUU-IV/2006 revealed that the doctrine of unlawful (*materiele wederrechtelijke*) on the positive function based on living law measures, especially in corruption cases, is unconstitutional because there are no definitive measures for the prudential values based on the living law so it produces uncertain measures that are in exact opposition to the fair legal certainty principle in Article 28D (1) UUDNRI Tahun 1945. Moreover, MK also revealed that punishment for the attempt (*poging*) that is equal to the punishment for the offense (*voltooid*) is a matter of open legal policy. MK argued that it is a part of extraordinary measures to overcome corruption. In this decision, Justice Laica Marzuki expressed his dissenting opinion, stating that the word "may" in the law's explanation is unconstitutional.

3 CONCLUSION

Based on the five MK decisions discussed above, it seems clear that MK's role in the Indonesian criminal law reform is significant, particularly regarding the dynamic of the criminal law paradigm in Indonesia on three basic principles in criminal law as well as several basic principles in criminal law.

For the "offense" as the first basic problem in criminal law, MK tends to use legal reasoning based on a purposive (*doelmatigheid*) rather than a textual approach (*rechtmatigheid*), so MK seems to favor the quality of democracy rather than retaining the text of the criminal provision regarding defamation of the president and vice president.

For "criminal responsibility" as the second basic problem in criminal law, there is no significant MK role yet regarding some crucial issues on criminal responsibility, such as the discourse between the monistic and dualistic paradigm as well as about strict and vicarious liability. MK's role is limited to reconstructing the age of juvenile offenders from "8 (eight)" to "12 (twelve)"; however it deserves an appreciation because the judicial activism and forward-looking approach looked predominantly on this case until adoption of Law No. 11/2012 concerning the juvenile system.

For the "punishment" as the third basic problem in criminal law, MK strictly revealed that the death penalty/capital punishment in Indonesia is constitutional based on the expectation that its implementation should reflect the values contained in the Bill of Criminal Code as a middle way proposal (the death penalty/capital punishment as privileged punishment and that should be ruled alternatively) to bridge the persistent controversy regarding the death penalty/capital punishment. The other interesting matter regarding punishment is MK's consideration in a death penalty case that retribution is an inherent aspect of the punishment, particularly in capital punishment, even though it is not a primary pattern in the Indonesian criminal justice system.

MK became very conservative when deciding the doctrine of *materiele wederrechtelijke* on positive function with a consideration regarding the fair legal certainty principle based on Article 28D (1) UUDNRI Tahun 1945 and tended to obey the formalistic legality principle rather than to engage in judicial activism by considering the decisions of the supreme court related to corruption cases.

Special attention deserves to be given to the confusing policy of MK related to the constitutionality of the non-retroactive principle in criminal law. Although the Constitutional Court expressly decided that the retroactive principle may be carried out in a very limited manner (for the sake of the equilibrium of legal certainty and justice), it seems clear that the Constitutional justices' configuration is not consistent so that from the two outstanding decisions related to the retroactive issue in criminal law, there are three configurations: (1) The majority group, consisting of Chief Justice Jimly Asshiddiqie, Justice Harjono, Justice Soedarsono, Justice H. A. S. Natabaya, and Justice I. Dewa Gede Palguna, consistently revealed that retroactive criminal law can be applied restrictively (for the sake of the equilibrium of legal certainty and justice) by first distinguishing between the concepts of the enactment of the retreat law and *ex post facto law*. This group "permits" the enactment of the retreat law, but "prohibits" *ex post facto law*. (2) Some Constitutional justices (i.e., Justice Laica Marzuki, Justice Abdul Mukhtie Fajar, and Justice Achmad Roestandi) consistently revealed that the principle of non-retroactivity in criminal law is an absolute constitutional principle that should not be ruled out for any reason. (3) Justice Maruarar Siahaan consistently revealed that the retroactive principle might be applied restrictively, either in the conception of an *ex post facto law* (retroactive legislation) or enactment of the retreat law.

Eventually, the major problem regarding the constitutionality of criminal law always focused on the way we interpret the linkage between Articles 28I (1) and 28J (2) of the UUDNRI Tahun 1945. We agree with Made Darma Weda's opinion that one of the solutions is to delete the phrase "in any conditions" in Article 28I (1) so we would have a new meaning of criminal legal policy that constitutionally, retroactive criminal law can be enacted in a limited way according to Article 28J (2) of UUDNRI Tahun 1945 as well as in accordance with the majority of the Constitutional justices's opinions in several related cases.

Therefore, as a way to read and understand MK's decision as a reference for criminal lawmaking in the future (*ius constituendum*), such as the Bill of Criminal Code and several bills related to criminal law, lawmakers absolutely should read and understand the decision, including all its considerations (*ratio decidendi*) and avoid any waiver or attempt to retain the norms that violate the UUDNRI Tahun 1945.

REFERENCES

Agustina, Rosa. (2003). *Perbuatan Melawan Hukum*. Jakarta: Univeristas Indonesia.

Amsari, Feri. (2011). *Perubahan UUD 1945: Perubahan Konstritusi Negara Kesatuan Republik Indonesia Melalui Mahkamah Konstitusi*, cet. ke-2 (revised edition). Depok: PT. RajaGrafindo Persada.

Arief, Barda Nawawi. (1986). *Kebijakan Legislatif Mengenai Penetapan Pidana Penjara Dalam Rangka Usaha Penanggulangan Kejahatan*. Dissertation, Faculty of Law, Padjajaran University, Bandung.

Arief, Barda Nawawi. (1994). *Beberapa Aspek Pengembangan Ilmu Hukum Pidana: Menyongsong Generasi Baru Hukum Pidana Indonesia*, Pidato Pengukuhan Jabatan Guru Besar Ilmu Hukum FH Undip Semarang, June 25.

Barak, Aharon. (2006). *The Judge in a Democracy*. Princeton, NJ: Princeton University Press.

Faiz, Pan Mohammad. (2016). "Dimensi Judicial Activism Dalam Putusan Mahkamah Konstitusi." *Jurnal Konstitusi, 13*(2).

Friedmann, W. (1990). *Teori dan Filsafat Hukum: Telaah Kritis atas Teori-Teori Hukum (Susunan I)*, terjemahan Mohamad Arifin. Jakarta: Rajawali.

Garner, Bryan A., editor in chief. (2009). *Black's Law Dictionary*, 9th ed.Eagan, MN: Thomson Reuters, West Publishing Co.

Harkrisnowo, Harkristuti. (2003). "Rekonstruksi Konsep Pemidanaan: Suatu Gugatan Terhadap Proses Legislasi dan Pemidanaan di Indonesia." In *Orasi pada Upacara Pengukuhan Guru Besar Tetap Dalam Ilmu Hukum Pidana*. Jakarta: Universitas Indonesia.

Hosen, Nadirsyah. (2016). "The Constitutional Court and 'Islamic Judges' in Indonesia." *Australian Journal of Asian Law, 16*(2), 1–11.

Huda, Chairul. (2006). *Dari Pidana Tanpa Kesalahan Menuju Kepada Tiada Pertanggungjawaban Pidana Tanpa Kesalahan: Tinjauan Kritis Terhadap Teori Pemisahan Tindak Pidana dan Pertanggungjawaban Pidana*, Cet. Ke-1, Edisi Pertama. Jakarta: Prenada Media.

Husak, Douglas. (2008). *Overcriminalization: The Limits of the Criminal Law*. New York: Oxford University Press.

Luthan, Salman. (2014). *Kebijakan Kriminalisasi di Bidang Keuangan*. cet.Pertama. Yogyakarta: FH UII Press.

Packer, Herbert L. (1968). *The Limit of the Criminal Sanctions*. Stanford, CA: Stanford University Press.

Pangaribuan, Luhut M. P. (2016). *Catatan Hukum Luhut Pangaribuan: Pengadilan, Hakim, dan Advokat* (Hendrata Yudha, ed.), cet. Pertama. Jakarta: Pustaka Kemang.

Prodjodikoro, Wirjono. (1993). *Perbuatan Melawan Hukum*. Bandung: Sumur.

Reksodiputro, Mardjono. (1993). "Sistem Peradilan PidanaIndonesia: Melihat Kepada Kejahatan dan Penegakan Hukum Dalam Batas-Batas Toleransi." *Pidato Pengukuhan Guru Besar Tetap*. Jakarta: Universitas Indonesia.

Sapardjaja, Komariah Emong. (2002). *Ajaran Sifat Melawan Hukum Materiel Dalam Hukum Pidana Indonesia: Studi Kasus tentang Penerapan & Perkembangannya Dalam Yurisprudensi*, Edisi Pertama, Cet. ke-1. Bandung: Penerbit Alumni.

Senoadji, Indriyanto. (2006). *Korupsi, Kebijakan Aparatur Negara & Hukum Pidana*. Jakarta: CV. Diadit Media.

Shidarta. (2011). *Penemuan Hukum Melalui Putusan Hakim*, Makalah yang disampaikan pada Pemerkuatan Pemahaman Hak Asasi Manusia Untuk Hakim Seluruh Indonesia, Medan, May 2–5.

Shidarta. (2013). *Hukum Penalaran dan Penalaran Hukum, Buku I (Akar Filosofis)*, cet. I, Yogyakarta: Genta Publishing.

Weda, Made Darma. (2006). "Pemberlakuan Hukum Pidana Secara Retroaktif di Indonesia." Dissertation, Universitas Indonesia, Depok.

Advancing Rule of Law in a Global Context – Susetyo, Rinwigati Waagstein & Budi Cahyono (eds)
© 2020 Taylor & Francis Group, London, ISBN 978-1-138-32782-5

Exploring initiatives for a human rights-centered approach in post-disaster housing reconstruction in Tacloban city, Philippines

Dakila Kim P. Yee
University of the Philippines, Visayas Tacloban College, Philippines

1 INTRODUCTION

The right to housing after natural disasters is an issue that is starting to gain traction in development discourse in other parts of the world (Carver 2011). Development experts and scholars are now exploring potential channels to realize this right in the aftermath of natural disasters. However, the concept of the right to housing in post-disaster context has not yet been mainstreamed in the Philippines, even as it is one of the earliest countries to create national laws that respect the right to adequate housing (Carver 2011). The marginalization of the right to housing creates potential complications in the provision of housing solutions in the aftermath of natural disasters. When Typhoon Haiyan hit the Philippines on November 8, 2013, it left behind a significant trail of death and destruction. The typhoon has killed at least 6,300 people and left behind 1,000 people missing. The typhoon also left behind substantial challenges to post-disaster reconstruction and recovery, particularly in the housing sector as at least 1,084,762 houses were damaged by Haiyan. It is important that in the aftermath of the disaster, duty bearers such as the state and non-government organizations are aware of the components of the right to adequate housing. They should also be aware that the disaster event does not suspend the right of the people to adequate housing, rather, those who are rendered homeless and displaced after the disaster continue to be entitled to this particular right (Carver 2011). It is important that the human rights of survivors, including the right to housing, be observed. Otherwise, this may lead to an environment where the disaster survivors are rendered helpless through a series of human rights violations (Gould 2009).

With the end of the relief phase, the reconstruction and recovery phase, particularly in the housing sector has been beset by issues and concerns which can be addressed by exploring options for human rights-based approaches to housing reconstruction in Tacloban City. The city absorbed the heaviest casualties brought about by Haiyan. According to a report by the National Disaster Risk Reduction and Management Council, at least 2,671 unidentified deaths were recorded in the city (National Disaster Risk Reduction and Management Council 2014). As part of the disaster risk reduction efforts undertaken by the local government unit of Tacloban, 14, 433 families living within danger zones of the city are currently targeted for relocation to resettlement sites located in the northern section of the city. As of 2016, only 13per cent of the families targeted for relocation has been successfully transferred into the resettlement sites.

The post-Haiyan resettlement process is part of the local government unit's efforts in reducing the future risks of storm surge and tsunami in the city. However, the resettlement process is marred by the overall slow pace of construction of housing units. In resettlement sites constructed by the government, housing beneficiaries complain about the inadequate infrastructure for basic social services and the lack of a clear policy framework for security of tenure. This paper argues that the resettlement process by the government has not integrated human-rights based principles in implementing resettlement housing construction in the city. The lack of a human rights-based approach has complicated the recovery of Haiyan survivors, compounding their vulnerability as they deal with everyday problems in resettlement sites.

2   METHOD

For this research, I examined three sites to gauge the integration of human-rights based approaches in post-disaster housing reconstruction. These three sites were selected according to the lead implementing agency in the construction of the houses – government, private sector and non-government organizations. I conducted semi-structured interviews with key stakeholders in the post-disaster housing reconstruction in Tacloban City. Interview participants include government officials from the resettlement and housing cluster, NGO officials and workers engaged in housing interventions in Tacloban City, local resettlement site leaders and officials, and beneficiaries of the housing sites. The fieldwork lasted for two months in 3 different housing sites in Tacloban City.

3   DISCUSSION

3.1   *Challenges to post-disaster resettlement in tacloban city after typhoon haiyan*

Arguably, the biggest challenge to beset post-Haiyan reconstruction is the immense magnitude of the damage to housing units. In a report prepared by the Presidential Commission for the Urban Poor, the resettlement process has been characterized by slow construction and low occupancy rates in Tacloban City. The National Housing Authority (NHA) has 18 resettlement sites as part of the Yolanda Permanent Housing Project for Region VIII (where Tacloban City is located). According to the report by the PCUP, a total of 3, 366 housing units were completed, with only 47% or 1,569 being occupied. The remaining 1,797 targeted beneficiaries. Officials interviewed for this project have blamed land acquisition issues (such as titling and conversion of land from agricultural to residential use) have contributed to the slow pace of construction of permanent houses (Presidential Commission for the Urban Poor 2016).

In most of the sites where I conducted fieldwork, residents of the permanent resettlement sites constructed by the NHA note that there was a lack of participatory processes in their transfer to permanent housing sites. Residents note a confluence of top-down programming wherein government officials already devolved the housing design to contractors, leaving no room for beneficiaries to provide input regarding the selection of sites for permanent housing as well as the design of the houses where they will eventually be transferred. Residents also note that there was a confluence of intimidation and deceit in their eventual relocation to the housing sites. Some of the residents note that they were intimidated in transferring to resettlement sites, noting how the barangay officials and local housing personnel warned them that they ouldnot receive any aid if they do not avail of permanent housing. Other residents note that they were promises an array of services and aid such as cash assistance and a steady stream of relief if they transferred to resettlement sites. Thus, there is an absence of informed decision making at the end of beneficiaries. Not surprisingly, this created a bevvy of problems and concerns that compound the vulnerability of the people living in the area. Based on the interviews and focus group discussions that I conducted in a resettlement site under the National Housing Authority, the following concerns emerged:

1. Spaces of the residential units are too small
2. The substandard structure indicated by cracks in the housing unit's walls and clogged toilets as reported by the beneficiaries
3. A limited supply of water; no access to drinking water except buying bottled water
4. Lack of security as indicated by lack of policemen/security personnel in the site
5. Unequal distribution of relief and livelihood assistance to the beneficiaries of housing units
6. Lack of sustainable livelihood opportunities in the site; residents complain that not everyone is given livelihood assistance
7. Lack of a coherent structure for resettlement governance (non-functioning homeowner's association)
8. Lack of communication and consultation process within the residential site

These challenges compound the vulnerability of people who have survived the onslaught of Haiyan. While they were relocated from coastal areas where they were exposed to risks associated from typhoons, storm surges and tsunamis, they are now exposed to risks associated with economic and political marginalization. Resettlement beneficiaries are also excluded from formal participation and decision making regarding the design of the housing structure where they were transferred. Thus, the needs of the beneficiaries were not considered. Some houses have undergone building modifications even if these are not formally allowed by authorities to expand the living space, provide additional space for livelihood opportunities such as selling items and expanding the residential area of the entire housing units. Some residents also claim that there was no formal consultation that tackled the concerns of those who were targeted for resettlement; residents inform that they were merely told that they ouldbe removed from their houses in coastal communities and be transferred. The process of resettlement after typhoon Haiyan has significantly affected the capacities and capabilities of the beneficiaries to engage in fruitful forms of employment and the enjoyment of freedoms from scarcity and want. While they are grateful that they are removed from an area that was generally exposed to coastal risks, they are now exposed to new forms of risks associated with eviction and displacement.

I argue that these problems stem from a lack of appreciation for a human-rights based approach that places a strong emphasis on participatory approaches from the beneficiaries. This is shaped by two conditions. First, the aid and relief processes that emerged after Haiyan created a steady stream of dependency on relief and aid actors, especially international aid workers, to address the malaise that the survivors have encountered. These processes inculcated a sense of gratitude so strong that the beneficiaries are "obliged to be grateful" (Ong, Flores and Combinido 2015). Second, there is a trend towards the logic of security as the dominant framework of urban governance (Hodson and Marvin 2009). This logic of security renders life as permanently at risk for being harmed; at a larger scale, this logic aims to protect populations of cities from threat to life (Hodson and Marvin 2009; Zeiderman 2013). As Zeiderman notes in the case of earthquake-prone zones in Colombia, the logic of security is in constant tension with the recognition of citizenship to the population of a territory, and consequently a tension with the recognition of rights (Zeiderman 2013). Such practices prevent a full appreciation of the bundle of rights that people are entitled to in the aftermath of a disaster, including the right to housing (Carver 2011). However, since the framework of interventions is generally designed to be implemented in a top-down flow, the formulation of the design of the project comes from the experts, neglecting and ignoring potentially fruitful and productive input from people and communities on the ground. An approach sensitive to the right to participation of the people, as well as one that recognizes the centrality of the state as duty-bearer is important in addressing problems in housing provision after Haiyan.

3.2 *Integrating a human-rights based approach in post-disaster housing reconstruction*

An alternative direction that can address the problems that stem from the top-down, contractor-driven process of resettlement and relocation is the integration of a human-rights based approach to post-disaster housing reconstruction. The emergence of the human-rights based approach in development is a response to the failure of top-down strategies of development observing the human rights of people who are targeted for development (citation). The focus on a human rights-based approach is to highlight how disenfranchised groups advance their claims on the duty-bearers (ex. the state) (Ensor et. 2015). Drawing from the principles of integrating human-rights based approaches to development, this approach highlights the following components (van der Ploeg and Vanclay 2017):

1. To advance the realization of human rights;
2. To integrate human rights standards and principles into all activities through focusing on processes and outcomes;
3. To contribute to the development of the capacities of duty bearers (both state and non-state actors) to meet their obligations;

4. To contribute to the development of the capacities of rights-holders so that they are empowered to claim and exercise their rights;

This approach aims to enhance developmental outcomes by integrating the observance of human rights principles in implementing projects. I seek to document existing practices in post-disaster housing reconstruction and resettlement that aims to advance this objective.

3.3   *The policy environment*

The Philippines is a signatory to various international human rights agreements and instruments that observe the right to housing. These instruments include the following:

1. Article 25 of the Universal Declaration of Human Rights which states that "Everyone has the right to a standard of living adequate for the health and well-being of himself and of his family, including food, clothing, *housing* and medical care and necessary social services, and the right to security in the event of unemployment, sickness, disability, widowhood, old age or other lack of livelihood in circumstances beyond his control." (Emphasis mine)
2. Article 11 of the International Covenant on Economic, Social and Cultural Rights (ICESCR) which states that "The States Parties to the present Covenant recognize the right of *everyone to an adequate standard of living for himself and his family, including adequate food, clothing and housing*, and to the continuous improvement of living conditions." (Emphasis mine)

At the national level, the Philippines is also one of several countries that vehave enshrined the principles of the right to housing in the national constitution (Leckie 1989). Article XIII, Section 10 of the Philippine Constitution states that the urban poor and informal settlers shall not be evicted from their homes unless it is the law and is done in a just and humane manner. The section also stipulates that no resettlement shall be undertaken without adequate consultation with those that will be evicted. Republic Act No. 7279 or the Urban Development and Housing Act serves as the main legal framework for incorporating housing issues and concerns, particularly those related to the urban poor in the Philippines (Simon 2016). Section 2 (e) of the law mandates that the law aims to "encourage more effective participation in the urban development process." The law also contains Section 23 "Participation of Beneficiaries" which aims to incorporate greater participation from program beneficiaries regarding decision-making processes.

In 2012, following the onslaught of the aftermath of Typhoon Washi (local name: Sendong), the Commission on Human Rights (CHR) issued a human rights advisory that defined the human rights principles that need to be observed in the post-disaster environment (Simon 2016). This includes the following:

- Non-discrimination and equitable assistance in interventions to housing, land and property rights
- Observing human rights standards in the relocation of populations affected by the disaster and implementing eviction only as a measure of last resort
- Ensuring the consultation of all affected populations, including providing them with information that will enable them to make informed choices
- Provision of meaningful mechanism for freedom of expression and redress of grievances as mandated by the Constitution

While the policy framework for ensuring the observance of the right to adequate housing, there is a need for pressure from communities to activate the capacities of the local and national state as duty-holders to ensure the realization of this right. Towards this end, various non-government organizations and civil society organizations, as well as their partners in Haiyan, affected communities have mobilized to assert their right to housing, especially as they are recovering from disasters. Mobilization is an important component that empowers local citizens and activates the state. A Philippine-based NGO, the Philippine Legislator's Committee on Population and Development (PLCPD) have engaged in campaigns, in partnership with Haiyan disaster survivor groups to lobby for the implementation of a law that will promote a resilient housing framework that will complement national housing laws to address disaster vulnerabilities. Community leaders from resettlement areas in Tacloban coordinate with PLCPD in numerous campaigns on ensuring resilient housing in the Philippines. Some of these community leaders provide input for the crafting of the Resilient Housing and Human Settlements bill, although NGO representatives note that this bill is not a priority in the current Congress.

Some NGOs have also engaged in participatory housing reconstruction to realize their right to adequate housing. The Pope Francis Village, a resettlement site constructed under the auspices of FRANCESCO, a consortium of NGOs, religious organizations/associations and people's organizations that aim to realize a people-driven reconstruction effort in the aftermath of Typhoon Haiyan. Community organizers from one of the member NGOs of the consortium, the Urban Poor Associates (UPA), stated that they want to empower the survivors of Haiyan through implementing a people-driven reconstruction process that prioritizes the input from the beneficiaries that is harnessed through a participatory process. All aspects of the project implementation are guided by the principles of people's participation, such as:

- Selection of the resettlement site
- Conceptualization of the housing design
- Selection and screening of beneficiaries
- Construction of the houses (counterpart labour from beneficiaries)
- Implementation of community savings schemes and other livelihood support projects

While the housing project is still undergoing construction, with a targeted date of completion set later in 2017, the houses are larger than the houses constructed by contractors in already-finished sites that are under the authority of the state. The site of the resettlement is also near the city centre, thus minimizing the effects of dislocation that characterizes massive resettlement and eviction schemes implemented by the government.

Implementation of a people-driven scheme also enhances the capacities of the beneficiaries in realizing their rights and in demanding the protection of these rights from the state. First, community members mobilized to protest the immediate eviction in the coastal communities covered by the UPA to give way to a tide-embankment project that is supposed to provide infrastructural resilience to the city. Community member argues that the plan was hasty, that there were no ready relocation sites at that time, and that there were no consultations held, which goes against the provisions set in the Philippine Constitution and the UDHA. Second, beneficiaries of Pope Francis Village, including those organized by the UPA has engaged the state in fulfilling their obligations as duty holders, thereby enabling the state to commit to its role in upholding the right to housing. For example, community members have also mobilized to demand additional state support in the funding of resettlement schemes, such as the augmentation of resettlement fund from the Department of Social Work and Development which committed 70,000 pesos to the total housing cost per unit. According to an organizer from Pope Francis Village, this was done through multiple consultations and dialogues with the DSWD to secure their commitment. Another example is how the consortium that handles Pope Francis Village engaged the military as well as the National Housing Authority in getting their commitment to helping in the site development. Thus, through the process of community empowerment and mobilization, state agencies can provide commitment that can help realize the right to housing.

4 CONCLUSION

Implementing a human-rights based approach to post-disaster housing reconstruction remains a daunting task. The demand for immediate results leads to the prioritization of top-down programs in reconstruction that overlooks the needs and concerns of targeted beneficiaries. Despite the presence of relevant policy instruments that promotes the human rights of people that are targeted for eviction and displacement, these are overlooked by implementers of reconstruction programs. However, local civil society organizations, NGOs and people's organizations are paving the way for integrating elements of HRBAs into the reconstruction process. This provides the public with a framework for addressing future problems stemming from disaster-induced displacement.

ACKNOWLEDGEMENT

This paper has been supported by the Strengthening Human Rights and Peace Education in Southeast Asia/ASEAN (SHAPE-SEA) Research Grants Programme 2017 funded by SIDA. The content expressed in the paper is of the author alone and do not reflect the official policy of SHAPE-SEA and SIDA.

REFERENCES

Carver, Richard. 2011. "Is There a Human Right to Shelter after Disaster?" *Environmental Hazards* 10 (3-4): 232–247.

Ensor, J.E., S.E. Park, E.T. Hoddy and B.D. Ratner. "A Rights-Based Perspective on Adaptive Capacity." *Global Environmental Change* 31: 38–49.

Leckie, Scott. 1989. "The UN Committee on Economic, Social and Cultural Rights and the Right to Adequate Housing: Towards an Appropriate Approach." *Human Rights Quarterly* 11 (4): 522–560.

Gould, Charles W., 2009. "The Right to Housing Recovery after Natural Disasters." *Harvard Human Rights Journal* 22: 169–204.

Hodson, Mike and Simon Marvin. 2009. "'Urban Ecological Security': A New Urban Paradigm?" *International Journal of Urban and Regional Research* 33 (1): 193–215.

National Disaster Risk Reduction and Management Council. 2014. "NDRRMC Update: SitRep No. 108 Effects of Typhoon Yolanda." Retrieved May 1, 2015 (http://www.ndrrmc.gov.ph/attachments/article/1329/Effects_of_Typhoon_YOLANDA_(HAIYAN)_SitRep_No_108_03APR2014.pdf)

Ong, Jonathan C., Jaime Manuel Flores and Pamela Combined. 2015. *Obliged to be Grateful: How Local Communities Experienced Humanitarian Actors in the Haiyan Response*. Woking, United Kingdom: Plan International. Retrieved October 5, 2017 (https://lra.le.ac.uk/bitstream/2381/33421/2/Obliged%20to%20Be%20Grateful%20-%20final.pdf)

Presidential Commission for the Urban Poor. 2016. "Tacloban Yolanda Survivors' Forum Proceedings." Prepared by the Relocation and Resettlement Monitoring Division.

Simon, Floreen M., 2016. "Making Communities Safe and Resilient: A Primer on Resilient Human Settlement in the Philippines." Quezon City, Philippines: Philippine Legislators' Committee on Population and Development Foundation, Inc.

Van der Ploeg, Lidewij and Franc Vanclay. 2017. "A Human Rights-Based Approach to Project Induced Displacement and Resettlement." *Impact Assessment and Project Appraisal* 35 (1): 34–52.

Zeiderman, Austin. 2013. "Living Dangerously: Biopolitics and Urban Citizenship in Bógota, Colombia. *American Ethnologist* 40 (1): 71–87.

Advancing Rule of Law in a Global Context – Susetyo, Rinwigati Waagstein & Budi Cahyono (eds)
© 2020 Taylor & Francis Group, London, ISBN 978-1-138-32782-5

People's participation, democratic governance, and the ASEAN Economic Community (AEC)

Khoo Ying Hooi
Department of International and Strategic Studies, Faculty of Arts and Social Sciences, University of Malaya, Malaysia

ABSTRACT: Since the introduction of the Association of Southeast Asian Nations (ASEAN) Community, the pillar of economics has been the most vibrant compared to the other two pillars, political-security and sociocultural. As of 2016, ASEAN is the sixth largest economy in the world and the third in Asia, with gross domestic product standing at US$2.6 trillion. ASEAN is expected to become the fourth largest global economy by 2050. Despite these accomplishments, ASEAN economic integration is not immune to key challenges. ASEAN is home to plenty of natural resources, but a 2015 study by the UN Economic and Social Commission for Asia and the Pacific found that Southeast Asia has lost 13% of its forest in the past two decades. The region also faces natural hazards, climate change, income inequalities, lack of social security for masses, lack of institutional capacities, and rapid demographic shifts such as urbanization and migration. To overcome this, a sustainable development model comes into the picture as an ideal structure for ASEAN to move forward. By focusing on two approaches, people's participation and democratic governance, this article intends to explore the challenges of ASEAN's sustainable development model and its way forward.

Keywords: ASEAN Economic Community (AEC), democratic governance, people's participation, sustainable development

1 INTRODUCTION

The Association of Southeast Asian Nations (ASEAN) is considered the premier regional association in Southeast Asia and the most prominent regional grouping in the Third World (Frost, 2012). ASEAN's founding declaration in Bangkok in 1967 called upon its member states to '... ensure their stability and security from external interference in any form or manifestation to preserve their national identities by the ideals and aspirations of their peoples'. ASEAN is regarded as an important factor for stability in Southeast Asia for various reasons. Among these are its cooperative activities and its policies of constant active dialogues with not only the Asia Pacific countries but also with other major key players in the world. Moreover, ASEAN's promotion of wider cooperation forums in East Asia and the Asia Pacific is also significant. ASEAN engages actively with external stakeholders, for example, by upgrading and implementing negotiations on a Regional Comprehensive Economic Partnership Agreement (RCEP) with China, Japan, the Republic of Korea, India, Australia, and New Zealand, as well as strengthening economic cooperation with other important stakeholders, such as the United States, the EU, Canada, and Russia. Significantly, ASEAN is often regarded as constituting a diplomatic, security, and economic and cultural community (Ganesan, 1994).

The year 2017 was a significant milestone for ASEAN as it celebrated its fiftiethth founding anniversary. After five decades, ASEAN is said to be a government affair, as it remains a challenge for ordinary citizens to relate to the concept of an ASEAN Community that in 2007 established three pillars: ASEAN Economic Community (AEC), ASEAN Political-Security Community (APSC), and ASEAN Socio-Cultural Community (ASCC). Since the introduction of the ASEAN Community, the economic pillar has been the most vibrant among the three. As of 2016, ASEAN is the sixth largest economy in the world and the third in Asia, with gross domestic product (GDP) standing at US$2.6 trillion. ASEAN is expected to become the fourth largest global economy by 2050. Despite these accomplishments, ASEAN countries are not immune to key challenges. ASEAN is home to plenty of natural resources, but a 2015 study by the United Nations Economic and Social Commission for Asia and the Pacific (ESCAP, 2015) found that Southeast Asia had lost 13% of its forest in the past two decades. The region also faces natural hazards, the impact of climate change, and rapid demographic shifts such as urbanization and migration.

1.1 *Regional integration in Southeast Asia and the formation of ASEAN*

Geographically, Southeast Asia consists of eleven countries that reach from the south of China and to the east and southeast of India. This region has long been influenced by external sources because of its rich natural resources and strategic location. ASEAN, formed in 1967 by Indonesia, Malaysia, the Philippines, Singapore, and Thailand, brought a new level of regional cooperation to protecting the region so as not to be controlled by external global forces. Traditionally, ASEAN's cooperative approach emphasized mutual respect for national sovereignty, avoiding confrontation, agreement through consensus, and most importantly, that all decisions be made at a pace that all the member states are comfortable with. Brunei Darussalam joined the regional body in 1984. After 1995, four more new members were admitted to the regional body: Vietnam, Laos, Myanmar, and Cambodia, bringing the number of member states to ten at present. These ten countries represent the Southeast Asia region, although the increased diversity within the regional body has at the same time posed challenges to ASEAN in various areas of cooperation.

ASEAN was formerly known as the Association of Southeast Asia (ASA). Together with Thailand and the Philippines, Malaysia, which was known as Malaya at that time, formed ASA in 1961. Established in Bangkok, the idea to form ASA was unsettled due to the vulnerable geopolitics after World War II and an attempt to improve regional ties in Southeast Asia. Symbolically, the initiative was considered a significant step towards greater regional cooperation among the countries, as it allowed the foreign ministers of each country to gather and participate in initiatives to enhance confidence building and form closer regional ties among the members. Back then, one key challenge that the ASA members faced was domestic communist insurgencies. The original aspiration of ASA was to establish an organization similar to the European Economic Community (EEC). However, as there were only three countries among the Southeast Asia countries that participated in the ASA, the organization was eventually created with a much looser structure with a more informal approach. ASA nevertheless did not last long because of the rising diplomatic tensions between the Philippines and Malaya in 1963 over the Sabah issue (Camilleri, 2003).

Later on, another initiative to strengthen regional stability and relationships among countries in Southeast Asia was initiated in 1963. From the same members that initiated the ASA, the greater Malayan Confederation or Maphilindo, a combined name consisting of Malaya, the Philippines, and Indonesia, was formed. However, Maphilindo did not succeed largely because of the failure of these three countries to achieve consensus. One of the key issues at that time was the territorial disagreement between the Philippines and Indonesia with Malaya. That subsequently led to the withdrawal of Malaysia from Maphilindo as a result of Indonesia's confrontation with Malaysia, which at that time was a newly formed federation. With that, the Maphilindo initiative officially ended.

The two failed multilateral initiatives in fostering regional cohesiveness and stability highlighted the diverse ethnic, ideological, and geopolitical environment that still loomed large on the Southeast Asian landscape. Nevertheless, the countries in the region were aware of the political risks that they had to face despite their differences. Working under many constraints, the key regional players continued to find ways to overcome or manage these divisions for two main reasons, to safeguard their internal stability and in some cases, for regime survival purposes (Camilleri, 2003). Realizing the need to enhance regional stability, the resumption of bilateral relations between Malaysia and the Philippines and the end of confrontation between Indonesia and Malaysia later on paved the way for the formation of ASEAN on 8 August 1967 under the Bangkok Declaration with the founding members of Indonesia, Malaysia, the Philippines, Singapore, and Thailand.

When ASEAN was first established, it was the height of the Cold War. Although backed by the major powers at that time, that is, the United States and the then Soviet Union together with their allies, the region proved to be fragile to some extent, as it was the place where the domino theory took hold. With the formation of ASEAN, a new set of confidence in nation-building and economic growth was expected to bring along an extended period of peace and stability in the region. What followed was a tremendous pace of institution building that generated the peace process and fostered regional cohesiveness in terms of social and economic gaps between the member countries.

ASEAN was founded on a loose structure. One important principle associated with ASEAN is the 'ASEAN Way'. Generally, the 'ASEAN Way' refers to mutual respect among its members. When ASEAN was formed in 1967, the Cold War and Vietnam War were at its height. Another concern in the initial years during the formation of ASEAN was the condition of economic development in the region. There were vast differences among the five founding members, with most of the members relying heavily on agriculture-based economies as their primary trading product. Despite the various shortcomings or weaknesses that the founding members of ASEAN needed to face, the members aimed for a model of regional cooperation that allowed diversity, which could maximize the members' diplomatic and political strengths.

ASEAN's model of cooperation could be divided into several phases. The first phase is from its establishment year, 1967, until 1975. During that period, the key focus was more on multilateral dialogues among the five founding members and confidence-building processes. Second is the Kuala Lumpur Declaration in November 1971, which among other things called for the establishment of the Zone of Peace, Freedom, and Neutrality (ZOPFAN) in the region (Camilleri, 2003). The end of the wars in Indochina in 1975 was then accompanied by a sense of uncertainty in the Southeast Asia region (Frost, 2008). That subsequently triggered another phase of ASEAN's model of cooperation in development. The first ASEAN summit was held in Bali in February 1976. Members of ASEAN signed the historical declaration known as Bali Accord II. It was significant as it witnessed the upgrading of both regional dialogues and efforts at economic cooperation among the members. In the 1970s and 1980s, ASEAN gradually gained a substantial regional and international profile when it started to initiate economic cooperation in the region itself. Several incidents took place in which ASEAN played an important political role, for instance, the Indochina refugee crisis from 1978 to 1979 and the conflict over Cambodia after 1978 (Frost, 2008).

Since the late 1990s, ASEAN has made substantial efforts to maintain its profile and prominence. After the end of the Cambodia conflict and the end of the Cold War, ASEAN's membership was expanded to also include Vietnam in 1995, Laos and Myanmar in 1997, and Cambodia in 1999. With that, ASEAN in the late 1990s was able to represent Southeast Asia more holistically. But on the other side of the coin, the expansion of membership poses some challenges to ASEAN. ASEAN's diversity is now becoming wider, posing further challenges in term of economic integration.

1.2 *Pursuing the ASEAN community*

From the late 1990s onwards, the changes in the international landscape have triggered ASEAN members to advance their efforts in enhancing their commitment to cooperation because of several factors. One of the factors is the adverse impact of the Asian financial crisis on many of its members in mid-1997. At the same time, the need to cooperate and compete effectively with the rapidly rising economic power of China and India, whose large markets and low-cost labour have been highly attractive to foreign investors, was recognized (Frost, 2008).

ASEAN has pursued cooperation in several approaches. One of them was the development of the ASEAN Community. The idea of the ASEAN Community was unsettled at the 9th ASEAN summit meeting of heads of government, in Bali in 2003, which has become known as the Bali Concord II. The adoption of the Bali Concord II is significant because it can be seen as ushering in the fifth and latest phase in ASEAN's development. The ASEAN Community involves three key pillars: the AEC, the APSC, and the ASCC (Frost, 2008). The ASEAN Charter was finally adopted at the 13th ASEAN Summit in November 2007. Generally, the ASEAN Charter served as a foundation to achieve the ASEAN Community. The Charter eventually entered into force on 15 December 2008. Such a development signified that ASEAN is tasked to operate under a common legal framework to form its community-building process. The ASEAN Charter supplied provisions to establish ASEAN integration by 2015. To achieve that, three pillars of the ASEAN Community were established:

1. APSC is designed to create a dynamic political-security environment among the ASEAN member states. In the long run, its purpose is to create a rule-based community with shared values and norms to achieve a peaceful and stable region.
2. AEC is aimed to implement economic integration initiatives to achieve a single market in Southeast Asia.
3. ASCC envisions a community that is people-centred to achieve solidarity and unity among the ASEAN peoples.

2 DISCUSSION

2.1 *People's participation and democratic governance: challenges and opportunities*

Southeast Asia's varied historical and geopolitical circumstances created diversified local political structures. Amidst globalization and social transformation, some Southeast Asian countries have adopted democratic systems. However, many more remain as authoritarian or communist regimes. Despite the differences in their political structures, the Southeast Asian countries share one common trait, that is, the existence of state repression, which threatens the civic space in the region. While the region has made some remarkable political transformations with its old political establishments being challenged by the emergence of opposition forces and civil society from time to time, there remain some doubts on the prospect of ASEAN as the regional grouping in providing democratic spaces for dissent voices.

Since the 2007 ASEAN Charter, ASEAN has been pursuing political and democratic reforms under the umbrella of the three pillars within the ASEAN Community, but at a slow pace. Some principles of the Charter have not been adequately implemented and to some extent are almost neglected by some ASEAN member states. This is particularly true when it comes to issues concerning human rights, democracy, fundamental freedoms, good governance, and the rule of law. AEC is the pillar that has advanced the most compared to the other two pillars, political-security and sociocultural. As the regional organization celebrates its fiftieth anniversary and its promise to bring about a rules-based, people-oriented, and people-centred ASEAN, there is increasing concern over the stagnant and, at times, the regressive process of democratization in the region. The disparity in terms of the level of economic development and distrust on several issues among the ASEAN member states are challenges that remained unresolved.

There is a global trend towards involving social activists in deliberative decision-making, but in the region of Southeast Asia, limited space has been provided to social activists to contribute to policy formation or implementation. The ASEAN Civil Society Conference (ACSC)/ASEAN People's Forum (APF), for example, provides a channel for dialogue between civil society organizations (CSOs) and state leaders, but this exchange does not genuinely pluralize decision-making processes because it is highly informal and depends on the wishes of the host government or the mood of ASEAN leaders. The closed nature of Southeast Asian regional regimes limits the ability to address complex transnational concerns such as sustainable growth. According to Acharya (2003), participatory regionalism helps intergovernmental organizations because it provides them with a perspective drawn from the grassroots of society. Also, including CSOs in policy formulation means that overstretched officials can turn to community-based specialists when designing and implementing policy. Widening participation in regional regimes also benefits the populace at large, because individuals affected by the policy can petition CSOs to lobby on their behalf.

Now into its fiftieth year, ASEAN is the most durable regional economic and political grouping in the developing world. However, while there are talks about sustainable development at various levels, people's perception about it is still very low. It is, as Acharya (2003) analysed, the main constraint to creating a meaningful participatory regionalism in ASEAN. McCormick, in his 2013 book, *Why Europe Matters* criticizing the European Union (EU), supposed that the lack of democracy is generated by the inability of people to articulate their interests in the decision-making process. A deliberative approach to democracy is determined by people's ability to influence the decision-making process. According to this approach, ASEAN is considered as democratic only if it supports political participation from internal interest groups within it. But the development of those institutions has turned into a bureaucratic-technocratic form within ASEAN where there is still limited room allocated for civil society to engage in the decision-making process, particularly at the summits, where the political decision still belongs to a representative of the states.

Recognizing that the Millennium Development Goals (MDGs) might have failed certain people and countries, the 2030 Agenda of the Sustainable Development Goals (SDGs) sets out to 'reach the furthest behind first' and concludes with a pledge that 'no one will be left behind'. SDGs are an opportunity for ASEAN. ASEAN should utilize the adoption of SDGs to both strengthen and partly refocus its framework for regional integration, as doing so would better serve sustainable development across the region. In practice, this would mean aligning the overall objectives of the ASEAN Community with those of the SDGs and strengthening this regional framework. In the 2014 Naypyidaw Declaration on ASEAN's post-2015 vision, ASEAN member states decided to 'promote the development of clear and measurable ASEAN Development Goals (ADGs) to serve as ASEAN benchmark for key socio-economic issues'. It is a positive sign to see a commitment in an open regional declaration, but such an effort has not been followed up since then.

There have been suggestions such as incorporating SDGs using pre-existing mechanisms. There are, however, several key challenges. One is financial in terms of big investments. Second is the governance issue. Innovations and improvements in governance will be needed at every level. Legislative and regulatory changes are going to be needed for sustainable development. The whole of government approaches is also needed across economic, social, and environmental decision-making. Hence, one crucial point is multi-stakeholder and people's participation, that is, increasing the engagement and meaningful involvement of non-state actors in ASEAN meetings, which directly reflect the spirit of ASEAN Community. That means development actors of all kinds will need to work collaboratively across the range of interlinked SDGs. Previously, the MDGs had been handled mainly within the ASCC, and institutional links of ASCC with other pillars have been limited. However, in contrast, SDGs should be embedded throughout the three Community blueprints. Without effective institutional coordination mechanisms, ASEAN's institutional response for the SDGs will likely be fragmented and as a result able to provide only very limited support for national and subnational implementation. Therefore, far more attention is needed in addressing governance challenges in terms of implementation and monitoring.

During the Malaysia 2015 chairmanship, the notion of a people-centred ASEAN was framed in such a way that it will be a powerful vehicle for the realization of people's aspirations, good governance, transparency, social development, women and youth empowerment, and providing opportunities for all. It was put forward at a time that ASEAN Community building had been more of a top-down approach, and it is timely that Asian now approaches it from the bottom up, where ASEAN governments should listen to their people. In the Kuala Lumpur Declaration on ASEAN Community Vision 2025, signed by leaders during the 27th Asian Summit in April 2015, the formulation is as follows: 'We resolve to consolidate our

Community, building upon and deepening the integration process to realise a rules-based, people-oriented, people-centred Asian Community, where our peoples enjoy human rights and fundamental freedoms, higher quality of life and the benefits of community building, reinforcing our sense of togetherness and common identity, guided by the purposes and principles of the ASEAN Charter'.

With various ASEAN Plans of Actions and Protocols, ASEAN hopes to eventually move closer towards its goal of building the ASEAN Community and the focus has since shifted to post-2015. Very often, ASEAN is criticized for acting as an elite-driven and state-centric project using a top-down approach. Realizing such shortcomings, various efforts were taken to overcome this, for example, with the slogan-like, people-centred ASEAN. This is then further illustrated by the fact that activities and projects of ASEAN heavily involve only the experts, political leaders, and government officials. Little information is available for the citizens and concerned stakeholders. This low awareness level is one of the key factors identified by ASEAN itself when it comes to assessing its overall achievement in community building. Aside from building a community that values inclusivity, in its relation with ASEAN, Malaysia also prioritises and promotes the practice of effective and responsive governance. At the same time, Malaysia also seeks to provide solutions to soft issues such as strengthening ASEAN institutions and mechanisms, environmental protection, empowering women in societies, and providing an opportunity for all. In this way, people's engagement will contribute toward the development and greater prosperity of the region (Credo, 2015).

The AEC is the most 'productive' pillar. To some extent, the AEC has managed to achieve more goals. Serious efforts at economic integration among ASEAN member states started mainly after the end of the Cold War. Initial obstacles such as differences in ASEAN members' development level, imported substitution policies of industrialization, and the limitation of intra-ASEAN trade posed challenges to the calculation of benefits and costs of integration in the region (Bordi, 2014). But issues related to non-tariff barriers, free flow of skilled workers, and varying levels of development of member states, among others, remain challenges to the regional economic integration in achieving the overall aims of the ASEAN Community.

It is a reality that ASEAN member governments do not have homogeneous political systems, with varying levels of political development and different levels of openness to the concept of people's participation. The relationship between ASEAN and its member governments with civil society has not been harmonious for various reasons, one being suspicion and mistrust. Approaches of different ASEAN governments to the acceptance of civil society are varied, with some consistently resisting civil society participation and engagement. Civil society groups through the ACSC/APF have limited access to key Asian policymakers. Last year, the ASCS/APF did not manage to take place in Laos because of concerns over possible restrictions and limited freedom of expression on key issues of concern to ASEAN, which are inconsistent with the agreed ACSC/APF's modality of engagement. The event finally took place only in Timor Leste, although Timor Leste has not yet been accepted as a member of ASEAN.

The concept of participation in democratic governance focuses attention on the interaction and linkages between state and society and how the people can take part in public life. Both direct and indirect participation is important and there is a growing understanding that development must be linked to the rights of people and the responsibility of the state has repercussions on how to understand and interpret participation. With the growing scope for participation through political institutions in the wake of the recent wave of democratic upsurge, the focus has shifted to include governance structures and political institutions and the degree to which they are open; by way of inclusiveness, transparency, and accountability (SIDA, 2002). A change of focus has been strengthened further towards political decentralization. Political institutions are gaining in importance both at the local level and higher, at intermediate levels, where there previously had been a very limited state presence. It is at this local level of governance that social or community forms of participation are now merging with political participation through decentralization (SIDA, 2002).

Based on the approach of consensus and non-interference, a role for civil society has been limited, and so is the attention to human rights owing to the diversity of national positions. At the same time, the bureaucratic processes of ASEAN are lengthy, and very often, it takes time to shift positions and move a specific agenda forward due to the many push and pull factors to reach consensus. What ASEAN needs is to allow the engagement to happen at multiple levels so that it can further strengthen the quality of the relationship between governments and between governments and their citizens as well as the quality of the process mechanisms. Non-governmental organizations, academic institutions, and think tanks should not dismiss the importance of more proactive participation and continue to work at the national as well as at the regional level to ensure civil society becomes an integral part of the future of ASEAN. The main institutional reference is Article 16 of the ASEAN Charter stating that 'ASEAN may engage with entities, which support the ASEAN Charter, in particular, its purposes and principles' wherein the term 'entities' implies all organizations that are not government bodies including parliamentarians think tanks, academic organizations, business organizations, and CSOs. In doing so, a long-term approach and diversified strategies are required.

Participation is both a result and a process (SIDA, 2002). Effective participation requires an enabling environment. Participation on its own is not enough. It can be constructive only if it occurs within structures that enable it to make at least some minimal impact on events. That is why so much attention is given not just to civil society, but also to democratic processes and decentralization. These three things can provide such structures. They can enable participation to have an impact to influence outcomes. Unless this happens, increased participation will eventually cause those participating to become frustrated because their efforts produce next to nothing (SIDA, 2002). There are also socio-economic and cultural obstacles to participation. For example, poverty and inequality disempower people. In most countries, social, economic, and cultural factors are of great importance for people's participation. Mass poverty, a low level of education, and other evidence of little respect for and realization of economic, social, and cultural human rights, will normally limit the expansion of participation, even in a system with democratic structures for participation.

3 CONCLUSION

For long, sceptics believed that the deadline for the ASEAN Community would not be met. Traditionally, ASEAN has a multitude of old and new declarations, agreements, treaties, conventions, protocols, plans of action, blueprints, concords, and so on to address a growing number of old and new challenges. But the key is to ensure members' compliance (Acharya, 2013) to the ASEAN Charter as committed by the ten member states. ASEAN has been successful in convening major powers in multilateral platforms such as the East Asia Summit and the ASEAN Regional Forum (ARF). Through these platforms, ASEAN can lead other states and is simultaneously given diplomatic leverage with its relations with big powers (Credo, 2015).

ASEAN's mandate has since expanded rapidly. Its functions now cover a range of new transnational or non-traditional security issues, such as climate change, disaster management, counter-terrorism, pandemics, food security, drug trafficking, and many other issues (Acharya, 2013). Like all previous regional efforts at community building, ASEAN with its current structure is expected to be able to overcome the challenges of consolidating its three ASEAN pillars with the aim to maintain the prosperity in the region and also to play a role in balancing the power struggle among great powers such as the United States and China at the international level. Although in reality doubts and mistrusts still exist among the ASEAN member states, there are, however, grounds for optimism that these intra-ASEAN conflicts would not jeopardize the organization. The idea of a people's ASEAN as lauded by Malaysia is promising, but there are areas to be tackled, especially on the question of the rule of law and good governance which thus far has been kept rather silent not only in Malaysia but also in other ASEAN member states. For instance, ASEAN member states have made a tentative commitment to human rights, but this remains constrained by the resilience of the non-interference norm (Acharya, 2013). The importance of participation in the democratic state is based on the idea of political equality. People need to feel that their participation is regarded as valuable and leads to a real influence on the decisions taken by their political representatives. It is, therefore, important to emphasize that not only does the state play an important role in allowing, stimulating, and creating a democratic sphere and an enabling environment for participation, but that it is also the responsibility of the state to involve its people in decision-making to the best of its ability.

REFERENCES

Acharya, A. (2003). "Democratisation and the Prospects for Participatory Regionalism in Southeast Asia." *Third World Quarterly*, *24*(2), 375–390.

Acharya, A. (2013). "ASEAN 2030: Challenges of Building a Mature Political and Security Community. Asian Development Bank (ADB)." Working Paper No. 441.

Camilleri, J. A. (2003). *Regionalism in the New Asia-Pacific Order*, Vol. 2. Cheltenham: Edward Elgar Publishing.

Frost, Frank. (2008). "ASEAN's Regional Cooperation and Multilateral Relations: Recent Developments and Australia's Interests." Research Paper No. 12 (2008-09). Retrieved from http://www.aph.gov.au/About_Parliament/Parliamentary_Departments/Parliamentary_Library/pubs/rp/rp0809/09rp12

Ganesan, N. (1994). "Rethinking ASEAN as a Security Community in Southeast Asia." *Asian Affairs: An American Review*, *21*(4), 210–226.

Khoo, Ying Hooi. (2016). "ASEAN Must Grab the opportunity to avoid becoming a 'pawn'" in *The Malaysian Insider*. Retrieved from http://www.msn.com/en-my/news/other/ASEAN-must-grab-opportunity-to-avoid-becoming-a-'pawn'/ar-BBpuYQt

Leifer, M. (2013). *ASEAN and the Security of South-East Asia* (Routledge Revivals). London: Routledge.

McCormick, J. (2013). *Why Europe Matters: The Case for the European Union*. London: Palgrave Macmillan.

Nischalke, T. (2002). "Does ASEAN Measure Up? Post-Cold War Diplomacy and the Idea of a Regional Community." *The Pacific Review*, *15*(1), 89–117.

Severino, R. C., Thomson, E., & Hong, M., eds. (2010). *Southeast Asia in a New Era: Ten Countries, One Region in ASEAN*. Institute of Southeast Asian Studies.

Simon, S. (2008). "ASEAN and Multilateralism: The Long, Bumpy Road to the Community." *Contemporary Southeast Asia: A Journal of International and Strategic Affairs*, *30*(2), 264–292.

Swedish International Development Cooperation Agency (SIDA). (2002). *Participation in Democratic Governance*. Stockholm: SIDA.

United Nations Economic and Social Commission for Asia and the Pacific (2015). *Economic and Social Survey of Asia and the Pacific* 2015. Part 1: *Making Growth More Inclusive for Sustainable Development*. ESCAP.

Advancing Rule of Law in a Global Context – Susetyo, Rinwigati Waagstein & Budi Cahyono (eds)
© 2020 Taylor & Francis Group, London, ISBN 978-1-138-32782-5

A study of the implementation of working conditions of women workers in transnational corporations (TNCs) in Indonesia

Shanti Haduri Halim
Institute for Human Rights and Peace Studies, Mahidol University, Thailand

ABSTRACT: Indonesia, as a host country of Foreign Direct Investment (FDI), has providedg opportunities for transnational corporations (TNCs) to open their business operations through their subsidiary companies. The goal of TNC business operations is the creation of job opportunities in Indonesia, where women can participate in the labour market. As a host country, Indonesia must compete with others in attracting foreign investors to Indonesia. The loosening of regulations in the host country is a way to attract foreign investors. However, the loosening regulations will have an impact on the workers, whose rights as workers will be jeopardized. The protection mechanisms are important tools to protect workers' rights from abuse by employers. This article focuses on the protection mechanisms and the implementation of working conditions in TNCs in Indonesia by collecting data from women office workers in three different types of TNCs through interviews and questionnaires. The authors of this article found that implementation of working conditions in Indonesia still does not meet the standards of Indonesia Manpower Act No. 13/2003.

Keywords: Foreign Direct Investment, Indonesia, transnational corporations, women workers, working conditions

1 INTRODUCTION

As a state, Indonesia must improve the living standards of its citizens by creating job opportunities without gender discrimination. As one of the countries in the Asia Pacific, Indonesia has a population of 255,46 million, 11.2% of whom still live below the poverty line (ADB, 2016). Eradication of poverty can be achieved by opening job opportunities for all men and women by attracting foreign investors to open their business operations in Indonesia. Globalization has led to opportunities for countries to collaborate with foreign investors through business cooperations that improve the economy. Economic globalization has given developing countries an opportunity to raise their economic growth through the expansion oftransnational corporations (TNCs) in other countries. TNCs are part of Foreign Direct Investment (FDI), in which foreign-owned activities are undertaken in more than one state (Dunning and Lundan, 2008, p. 3). TNCs have played a major role in the economy of a host country, as they can create job opportunities in those countries.. Host countries have sought any effort to attract foreign investors through FDI, which engages TNCs as the business entities in their operation. To attract TNCs and increase economic growth, the state has loosened investment regulations (Aguirre, 2011, p. 132) by providing conveniences for foreign investors to invest in the host country.

After the economic recession in Asia in 1997, the Indonesian government tried to improve the economy by collaborating with foreign investors through FDI. FDI can bring employment opportunities to the host country (Gopalan et al., 2016, pp. 32–33). FDI has brought significant changes to the economic situation in Indonesia by opening opportunities for women to participate in the labour market. According to the Universal Declaration of Human Rights (UDHR) Article 23 part 1, everyone has the right to work, to free choice of employment, to just and favourable conditions of work, and to protection against unemployment. The International Covenant on Economic, Social and Cultural Rights (ICESCR) Article 6 part 1 also recognized the right to work for everyone and the opportunity to gain a living by work.

The protection mechanism for workerss rights in Indonesia is regulated in Indonesia Manpower Act No. 13/2003 that regulates the relationship between employers and employees. The Manpower Act prevents driving working conditions below the appropriate standard (Blackburn, 2006, p. 1). Many TNCs operate in Indonesia and hire local workers, including women, as their employees to work in offices, factories, plantations, and mining. The state as duty bearer must ensure that TNCs do not violate the workers' rights by providing adequate protection mechanisms that apply in Indonesia. The Indonesian government must reinforce the Manpower Act without compromise to reduce the violations of workers' rights including those for office workers, as the Manpower Act applies to all workers. It was commonly thought that the labour standards apply only to factory workers, while office workers might not have been aware that their rights as workers are protected in the Indonesia Manpower Act No. 13/2003.

The collaboration between duty bearers is needed to protect workers' rights. Women, as one of the vulnerable groups in the labour market, should be given more attention . Therefore, to protect workers' rights especially for women office workers, the Indonesia government must ensure labour protection regarding working conditions complies with Indonesia Manpower Act No. 13/2003 and the International Labour Organization (ILO) standards. Indonesia became an ILO member in 1950 and had ratified ILO conventions such as concerning equal remuneration (C100) and discrimination (employment and occupation) (C111).

2 DISCUSSION

2.1 *The existence of TNCs in Indonesia*

Globalization has brought the global economy to several countries with TNC expansion seeking new market opportunities with low costs of operation and a good profit. The economic reformation in Indonesia has made Indonesia an attractive host country for FDI, including TNCs (Prihandono, 2013, p. 1). FDI offers possibilities for the Indonesian government as a host country to open investment opportunities with the third parties doing business in Indonesia. The protection of human rights in doing business must be the main criterion in the international investment agreement. In attracting foreign investors seeking cheap labour for their businesses, the race to the bottom theory has been an issue for host countries.

In operating their businesses, TNCs need office workers, and are required to protect the workers from exploitation by the business enterprises. The Indonesian government should take appropriate measures in ensuring business enterprises as third parties respect workers' rights in Indonesia. The total workforce of women in Indonesia has reached 51% of the country's total workforce (World Bank 2017). The large number of women workers has created opportunities for women in the labour market. Protection of workers' rights must be a major concern for TNCs in doing their business operations because the workers are the main stakeholders. The protection of workers' rights must comply with the domestic and international labour standard: Indonesia Manpower Act Number 13/2003 as a legal provision for TNCs in Indonesia. Internationally, the Universal Declaration of Human Rights (UDHR), International Covenant on Civil and Political Rights (ICCPR), International Covenant on Economic, Social and Cultural Rights (ICESCR), Convention on the Elimination of all Forms of Discrimination Against Women (CEDAW), and the ILO Conventions serve as guidance for TNCs in implementing the labour standards related to working conditions.

The expansion of TNC businesses in host countries opens new opportunities for seeking new markets and profits. The targetted customers can be both domestic and overseas. Overseas customers are already aware of the human rights issues and encourage the exporter companies to comply with the labour standards. There are some labour standards that the exporter company must comply with such as in SA8000, the Business Social Compliance Initiative (BSCI), and ISO26000. While in Indonesia, domestic customers are still not aware of the obligation for the companies as producers to comply with the human rights standards. The low level of awareness in domestic customers has resulted in some companies, including TNCs, trying to avoid compliance with the labour standards because the legal sanction for violating workers' rights based on the Indonesia Manpower Act Article 183–190 still does not apply in Indonesia. The state cannot implement the legal sanction because TNCs have provided benefits for the host country in opening job opportunities and increasing economic growth.

The state must respect and protect people's rights by ensuring businesses respect and fulfil those rights by taking proactive measures (Fukuda-Parr et al., 2015, p. 21). State and business enterprises should know what role they have in protecting and respecting the workers' rights. According to UN Guiding Principles on business and human rights (2011), to identify, prevent, mitigate, and account for how they address their adverse human rights impacts, duty bearers should carry out human rights due diligence. Due diligence can be implemented by assessing the human rights impacts, action plan, and protection mechanism.

The author's research attempts to analyse the mechanisms for protecting women office workers' rights and the implementation of standards for working conditions in Indonesia and to assess whether the implementation of working conditions based on Indonesia Manpower Act No. 13/2003 applies in TNCs through FDI. According to theILO's 2012 working conditions laws report (ILO, 2013a), working conditions consist of a minimum wage, working hours, overtime hours, overtime remuneration, annual leave, and maternity leave. The FDI regulations for foreign investments in Indonesia related to labour provisions will be evaluated, and the implementation of labour standards in TNCs analysed based on the national rules in Indonesia Manpower Act No. 13/2003, and international labour standards based on the ILO conventions. The scope of the research is the women who work in the office in the TNC headquarters office in DKI Jakarta, which consists of fifteen participants from three types of TNC industries. Among these three TNCs, only one TNC has a trade union, and the others do not.

2.2 *Protection mechanisms and the implementation of working conditions in TNCs as the result of FDI*

FDI is a new beginning for countries (Dunning, 1994, pp. 26–27) to collaborate in global economic activities. As non-state actors, business enterprises have played a significant role in providing jobs and recruiting workers in Indonesia. TNCs are hiring men and women to fill positions in offices, factories, and some other work areas. As a state, the Indonesia government must have mechanisms for protecting workers' rights based on the Manpower Act and international labour standards. UN Guiding Principles on business and human rights have outlined state duties to protect workers from human rights abuses by the third parties, including employers, through appropriate policies and regulations. Violations of workers' rights can be avoided by legal enforcement from the state in implementing the policies and regulations.

The Indonesian Capital Investment Coordinating Board (Badan Koordinasi Penanaman Modal or BKPM) statistical data in 2016 show the amount of FDI in Indonesia had reached US$28,964.1 million. Many foreign investors are attracted to investing in Indonesia because of the country's the resources, one of which is labour. As foreign business activities, TNCs must control and manage production and distribution activity in the host countries as the source of FDI (Guru, 2016). TNCs are a tangible entity that opens their business branches in the host countries as FDI recipients (Cohen, 2007, p. 36). In attracting foreign investments, the FDI in Indonesia still lacks labour perspectives. The protection of workers' rights must be started from the beginning stage when an investment trade agreement is made. According to the information from the BKPM staff, there is no requirement for foreign investors to comply with the labour standard. The BKPM do not check the ability of foreign investors to comply with the labour standard that applies to Indonesia. Once the new TNC business operation has been approved by the BKPM, the company must register the employees in the Ministry of Manpower. After TNCs run their businesses in Indonesia, it is the duty of the Ministry of Manpower to monitor the TNCs related to the implementation of labour standards. The Ministry of Manpower, as the state institution in charge of the labour standards, has monitoring mechanisms through employee report forms and labour inspections.

The employee report form, which is provided every year by each company in Indonesia, is based on the format from the Ministry of Manpower by the human resources department. According to Act No. 7/1981 mandating the company to provide employment reports, Article 8 stipulates that the company is obliged to report the information of their employees. The employee form consists of information related to the workers' information. The required information related to working conditions includes the number of workers based on gender and age; minimum wage; the number of permanent workers and contracted workers; the length of working hours; the social insurance; and the industrial relations information. The employee report form has important information related to the worker's condition in a company; however, there is no re-checking from the Ministry of Manpower to validate the data.

The labour inspection is another monitoring mechanism from the Ministry of Manpower. The labour inspectors visit to ensure that the workplace has followed the minimum employment standard that applies to Indonesia. Labour inspectors are assigned based on their expertise and give advice and information to the employers to implement the legal provision relating to the workers' protection. In the implementation, there is lack of coordination in labour inspection among regions, high turnover (ILO, 2013b, p. 49), lack of motivation, and also a lack of technical training for the labour inspectors (Tjandra, 2010, p. 12). Conflict of interest at the district level is also one of the reasons that labour inspections cannot work properly (Dupper et al., 2016, p. 22). The insufficient number of labour inspectors has made it impossible for the state to conduct monitoringoptimally.

In Indonesia, the role of the trade union is that of an independent organization that can promote and protect the rights of the workers because it has an equal position with the company in doing collective bargaining. The trade union also has a function in rules-making and collective bargaining (Ewing, 2005, p. 4) together with the workers and company. The provision of a labour standard is stipulated in the company Collective Labour Agreement (CLA), and the trade union carries a mission to assist workers as trade union members in defending their rights at the bipartite and tripartite levels, and up to the Industrial Relations Dispute Court (Pengadilan Perselisihan Hubungan Industrial or PPHI). When the rights of workers are being violated by the company, the trade union can assist the workers in PPHI against the company. Becoming members of the trade union has enabled workers to have support whenever their rights as workers are being violated by employers. The capability of the trade union in doing collective bargaining has made it possible for its membersto receive favourable working conditions and other benefits.

Meanwhile, not all TNCs have a trade union in their company, especially those that engage in services and non-manufacturing companies. Workers in this situation face difficulties in defending their rights as workers because they are only representing themselves when facing the employers. When no trade union exists in the company, the TNCs as the employers must have protection mechanisms to avoid violations of workers' rights.

Another available protection mechanism is self-regulation on the part of the TNCs reflected in the TNC code of conduct and voluntary commitment to the UN Global Compact for respecting human rights. According to Organisation for Economic Co-operation and Development (OECD) guidelines for TNCs, TNCs are responsible for respecting human rights by providing policies in the host countries in which they operate (2008, p. 14). UN Guiding Principles on business and human rights affirm that business enterprises are expected to be responsible for implementing due diligence to identify, prevent, and mitigate the impact linked to their business operation.

The code of conduct as Corporate Social Responsibility (CSR) is a self-regulated commitment that reflects human rights protection and applies to all TNCs subsidiaries companies (Marrella, 2007, p. 289). The code of conduct is a non-legally binding commitment from the company to respect human rights, including the workers' rights, and applies in all TNC business operation. Although the TNCs have committed to the code of conduct, violations of women office workers' rights are still occurring. Labour violations happen because of shortcomings in the monitoring system and audit process and incomplete information about working conditions in the subsidiary companies although the code of conduct exists (Miles, 2015, p. 14). The subsidiary companies lack responsibility in implementing the code of conduct because the code of conduct is not legally binding (Kinley and Tadaki, 2003–2004, p. 956). Also, the lack of audit from the TNC headquarter office has made the subsidiary companies neglect responsibility for the protection of workers' rights.

UN Global Compact, as the world's largest global corporate sustainability initiative, has reached more than than 8,000 companies in more than 160 countries to meet their commitments (UN Global Compact, 2014, p. 7). The Communication on Progress (COP) is reported annually by each member on their activities related to the ten core principles of the Global Compact. Amnesty International (2003) criticized the UN Global Compact for its inability to enforce the compliance of each member and recommended strengthening the accountability related to the ten principles.The UN Global Compact cannot guarantee that all members respect workers' rights in their business operation because the COP has not been validated.

The protection mechanisms that are available from the state, trade union, and TNC self-regulation should be able to avoid the violations of the workers' rights. In protecting workers' rights, the protection mechanisms must perform well in making each party knows its duty. The implementation of labour standards cannot be done properly without the monitoring and legal enforcement from the state as the duty bearer.

2.3 *Results on working conditions of women office workers*

Based on the analysis of the protection mechanisms that are available in Indonesia, the author's research found that among the participating women office workers, violations of working conditions are occurring in working hours, overtime hours, and overtime remuneration. All the participants receive a wage based on the minimum wage of DKI Jakarta and receive annual and maternity leave based on the Indonesia Manpower Act No. 13/2003. According to the Indonesia Manpower Act, No. 13/2003, the normal working hour is eight hours per day and forty hours per week. More than 53% of the participants work more than eight hours that do not count as overtime hours. In a TNC that has a trade union, the working hours are regulated in the CLA based on the Manpower Act No. 13/2003. According to the Indonesia Manpower Act No. 13/2003, Article 77, employees who work five days in a week are entitled to work eight hours a day. Based on the ILO Forty-Hour Week Convention (C047), the length of working hours is forty hours per week. Long working hours can lead to an impact on women's health and the care of their families (Doheny, 2016). Employers should limit the maximum number of working hours to avoid the double burden of women as workers and caregivers (George and Wofford, 2017). Even though the participation of women in the labour force has increased, the role of women in the household should be considered (UN, 2015, p. 110). The majority of the participants have children, and the length of working hours impacts their role as mothers and caregivers and has led to the participants delaying their time to return home. Based on interviews with the trade union representative, if employees are working more than eight hours a day, the extra hours will be counted as overtime hours and the maximum of the overtime is three hours per day. Meanwhile, the participants who work in TNCs without a trade union are working more than eight hours per day, and the extra hours are not calculated as overtime hours.

The Indonesia Manpower Act, No. 13/2003, Article 78, allows employees to work no more than three hours of overtime per day or fourteen hours per week, and employers are obligated to pay the overtime remuneration. It was found that 62% of the participants are working more than three hours of overtime per day. Participants have to work overtime because of unfinished work, and sometimes the company gives them work that is not part of their job description. Although the participants know that the extra work is not part of their job description, they cannot refuse because they are a concerned about the effect of their performance on their continuation of employment in the company. Of the total participants, only 33% who receive overtime remuneration work in a TNC that has a trade union. It was also found that a transportation facility is not provided by employers for the women office workers who have worked overtime. Forty-seven per cent of the women office workers had to take care of their transportation by themselves to return home at the risk of their safety. General Recommendation No. 25 of CEDAW stipulates the substantive equality that men and women can enjoy equal rights, only women have different biological, social, and cultural needs.

Meanwhile, women have an equal opportunity to work overtime as do men, based on substantive equality that women can be treated differently. The obligation for employers to provide return or roundtrip transportation for women workers is stipulated in Indonesia Manpower Act No. 13/2003 in Article 74, part 4. The result shows that the TNCs that do not have a trade union do not provide transportation facilities for women office workers.

The author's research found that the most violations in working conditions for women office workers occur in TNCs that do not have a trade union. Violations are still occurring in working hours, overtime hours, and overtime remuneration, as well as in providing transportation facilities for women office workers. The existence of trade unions can promote and monitor the implementation of workers' rights in the company. Without the existence of the trade union in the company, the women office workers do not dare to object to the company for fear of being dismissed. In this situation, the women office workers are representing themselves in facing the company. The condition would be different if trade union existed in the company. The role of trade unions in promoting and protecting the workers' rights is seen in the low number of violations of working conditions among women office workers. The effectiveness of protection mechanisms that exist in Indonesia at the moment derives only from the trade unions. The author's research found that without the existence of the trade union in TNCs, violations of working conditions for women office workers are still occurring because of a lack of a protection mechanism in Indonesia.

3 CONCLUSION

FDI as a source of economic development and engagement of foreign investors has brought changes in the economic situation of the host country. FDI in Indonesia has brought TNCs that open their business operations by recruiting local workers, including women. The issue of compliance of TNCs' business operations in Indonesia with labour standards has raised concerns for the implementation of measures for the protection of their workers, including women office workers. It has been shown that in implementation of labour standards related to working conditions of women office workers there is a different level between the TNCs that have a trade union and those that do not. Women office workers in TNCs that do not have a trade union are still vulnerable to violations of working conditions. A strengthening of the implementation of the protection mechanisms in applying the labour standards in Indonesia is needed to protect workers' rights, including those of office workers, from abuse by business enterprises. In Indonesia, the protection mechanisms from the state and self-regulation are still not optimal; only the the trade unions can promote and protect the workers' rights. Law enforcement pertaining to the TNCs that operate in Indonesia must be implemented effectively to preventthe TNCs fromviolating workers' rights. Legal sanctions against TNCs that have violated workers' rights must be implemented in Indonesia as the host country to prevent violations of the workers' rights, including those of office workers. The state as the duty bearer must protect workers' rights in Indonesia by applying legal enforcement to TNCs that operate in Indonesia. Attracting foreign investment must be done in responsible ways by not ignoring the labour standards in the host country.

REFERENCES

Act. (Number 13/2003). Concerning Manpower.
Act. (Number 7/1981). Concerning Compulsory Company Manpower Report.
ADB, & ILO. (2006). *Core Labour Standards Handbook*. Manila: Asian Development Bank. Retrieved from https://www.adb.org/sites/default/files/institutional-document/33480/files/cls-handbook.pdf

Aguirre, D. (2011). Corporate Liability for Economic, Social and Cultural Rights Revisited: The Failure of International Cooperation. *California Western International Law Journal, 42*(1), 123-148.

Amnesty International. (2003, April 7). *Letter to Louise Frechette raising concerns on UN Global Compact*. Retrieved August 21, 2017, from https://www.globalpolicy.org/component/content/article/177/31749.html

BKPM. (2016). *Statistic of foreign direct investment realization based on capital investment activity report by sector, Q4 2016*. Retrieved July 20, 2017, from http://www.bkpm.go.id/en/investing-in-indonesia/statistic

Blackburn, D. (2006). *The role, impact and future of labour law*. Retrieved June 3, 2017, from http://www.ilo.org/wcmsp5/groups/public/—ed_dialogue/—actrav/documents/publication/wcms_111442.pdf

Cohen, S. D. (2007). *Multinational Corporations and Foreign Direct Investment: Avoiding Simplicity, Embracing Complexity* (1st ed.). New York: Oxford University Press.

Dunning, J. H. (1994). Re-evaluating the Benefits of Foreign Direct Investment. *Transnational Corporations, 3*(1), 23-52.

Dunning, J. H., & Lundan, S. M. (2008). *Multinational Enterprises and the Global Economy* (2nd ed.). Cheltenham and Massachusetts: Edward Elgar Publishing.

Dupper, O., Fenwick, C., & Hardy, T. (2016). *The Interaction of Labour Inspection and Private Compliance Initiatives: A Case Study of Better Work Indonesia*. Jakarta: Better Work Indonesia.

Ewing, K. (2005). The Function of Trade Unions. *Oxford Industrial Law Journal, 34*, 1-22. doi:10.1093%2Filj%2F34.1.1

Fukuda-Parr, S., Lawson-Remer, T., & Randolph, S. (2015). *Fulfilling Social and Economic Rights* (1st ed.). Oxford: Oxford University Press.

George, E., & Wofford, D. (2017, July 11). *Women's health in global supply chains - Re-envisioning the business role*. Retrieved July 29, 2017, from https://www.ihrb.org/other/children/womens-health-in-global-supply-chains-re-envisioning-business-role

Gopalan, S., Hattari, R., & Rajan, R. S. (2016). Understanding foreign direct investment in Indonesia. *Journal of International Trade Law and Policies, 15*(1), 28-50.

Guru, S. (2016). *Role of Multinational Corporations (MNCs) in Foreign Investments*. Retrieved June 30, 2017, from http://www.yourarticlelibrary.com/microeconomics/foreign-investment/role-of-multinational-corporations-mncs-in-foreign-investments/38224/

ILO. (1951). *C100 - Equal Remuneration Convention (No. 100)*. Retrieved November 13, 2016, from http://www.ilo.org/dyn/normlex/en/f?p=NORMLEXPUB:12100:0::NO::P12100_ILO_CODE:C100

ILO. (1958). *C111 - Discrimination (Employment and Occupation) Convention (No. 111)*. Retrieved November 13, 2016, from http://www.ilo.org/dyn/normlex/en/f?p=NORMLEXPUB:12100:::NO:12100:P12100_ILO_CODE:C111:NO

ILO. (2013a). *Labour and Social Trends in Indonesia 2013: Reinforcing the role of decent work in equitable growth*. Jakarta: International Labour Organization.

ILO. (2013b). *Working Conditions Laws Report 2012, a global review*. Geneva: International Labour Organization.

Kinley, D., & Tadaki, J. (2003-2004). From Talk to Walk: The Emergence of Human Rights Responsibilities for Corporations at International Law. *Virginia Journal of International Law, 44*(4), 931-1024.

Marrella, F. (2007). Huma rights, arbitration, and corporate social responsibility in the law of international trade. In W. Benedek, K. D. Feyter, & F. Marrella (Eds.), *Economic Globalisation and Human Rights* (pp. 266-310). Cambridge: Cambridge University Press.

Miles, L. (2015). The 'integrative approach' and labour regulation and Indonesia: Prospects and challenges. *Economic and Industrial Democracy, 36*(1), 5-22.

Morrison, J. (30 May 2017). *Why the G20 Labour and Employment Statement Matters*. Retrieved June 17, 2017, from https://www.ihrb.org/other/governments-role/why-the-g20-labour-and-employment-statement-matters

OECD. (2008). *OECD Guidelines for Multinational Enterprises*. OECD.

Prihandono, I. (2013). Transnational Corporations and Human Rights Violations in Indonesia. *Australian Journal of Asian Law, 14*(1), 1-23.

Tjandra, S. (2010). Disputing Labour Dispute Settlement: Indonesian Workers' Access to Justice. *Law, Social Justice & Global Development*, 1-26.

UN. (1979). *Convention on the Elimination of All Forms of Discrimination against Women*. Retrieved from http://www.ohchr.org/Documents/ProfessionalInterest/cedaw.pdf

United Nations. (1948, December 10). *Universal Declaration of Human Rights*. Retrieved October 20, 2016, from United Nations: http://www.un.org/en/universal-declaration-human-rights/

United Nations. (1966, December 16). *International Covenant on Civil and Political Rights*. Retrieved October 20, 2016, from United Nations High Commissioner for Human Rights: http://www.ohchr.org/en/professionalinterest/pages/ccpr.aspx

United Nations. (1966, December 16). *International Covenant on Economic, Social and Cultural Rights*. Retrieved October 20, 2016, from United Nations High Commissioner for Human Rights: http://www.ohchr.org/EN/ProfessionalInterest/Pages/CESCR.aspx

United Nations. (2004). *CEDAW General Recommendation Number 25 on temporary special measures*. Geneva.

United Nations. (2011). *Guiding Principles on Business and Human Rights: Implementing the United Nations "Protect, Respect and Remedy" Framework*. Retrieved October 17, 2016, from United Nations Human Rights Council: http://www.ohchr.org/Documents/Publications/GuidingPrinciplesBusinessHR_EN.pdf

United Nations. (2015). *The World's Women 2015: trends and statistics*. New York: United Nations Department of Economic and Social Affairs.

United Nations Global Compact. (2014). *Guide to corporate sustainability. Shapping a sustainable future*. New York: United Nationa Global Compact.

World Bank. (2017). *Labor force participation rate, female (% of female population ages 15+)*. Retrieved July 1, 2017, from http://data.worldbank.org/indicator/SL.TLF.CACT.FE.ZS

Shanti Haduri Halim is a Master's Student in Mahidol University's in Institute for Human Rights and Peace Studies in Master of Human Rights and Democratisation (MHRD) program. Her areas of interest are in Business and Human Rights and Corporate Social Responsibility (CSR).

Advancing Rule of Law in a Global Context – Susetyo, Rinwigati Waagstein & Budi Cahyono (eds)
© 2020 Taylor & Francis Group, London, ISBN 978-1-138-32782-5

Optimization of Hajj fund development through infrastructure-based sukuk

Iffah Karimah & Shafira Iskandar
Durham University, UK

ABSTRACT: The Hajj Fund in Indonesia has great potential to be developed because of the high demand for Hajj in Indonesia. Along with the establishment of the Hajj Fund Management Agency (BPKH), the government is considering investing Hajj funds to infrastructure development in Indonesia. This article examines the possibilities of infrastructure-based sukuk as an alternative investment to optimize Hajj fund management by comparison with Lembaga Tabung Haji management in Malaysia and looks at the issues of Hajj management in Indonesia. The authors conclude that the Hajj fund can be maximized through infrastructure-based sukuk as long as it considers the needs of Hajj pilgrims themselves as the investors.

Keywords: Hajj fund, infrastructure, Lembaga Tabung Haji, long-term investment, sukuk

1 INTRODUCTION

As one of the pillars of worship in Islam, Hajj is obligatory for Muslims who are able to participate. Every year large numbers of of Muslims from all over the world come to Mecca and Medina in Saudi Arabia as Hajj pilgrims. As the country with the highest Muslim population, Indonesia sends the largest number of Hajj pilgrims every year. In 2017, Indonesia sent more than 200,000 people to Mecca and Medina as Hajj pilgrims.

Furthermore, *The Jakarta Post* (2017) reported that because of a high demand for the Hajj, the waiting list for the Hajj from Indonesia is fifteen to twenty years. Consequently, the deposit from the numerous waiting Hajj pilgrims is stored in the Hajj fund for several years. This will create a huge potential to develop the Hajj fund.

In July 2017, the president of Indonesia, Ir. Joko Widodo, stated that the Hajj fund should be invested to build infrastructure in Indonesia (Ihsanuddin, 2017), raising debate in society. Some members of the Muslim population rejected the idea because of political sentiment toward President Joko Widodo and the perception that the policy is not favorable to some Muslims. Ulemas and Islamic economists in Indonesia as well are split on the issue: those who support and those who areagainst this plan for the use of the Hajj funds. Some Islamic economists support that idea because previously most of the Hajj funds had been managed by conventional banks, and this is a good time to bring the Hajj fund back to halal investment. Others think investing the Hajj fund in infrastructure is risky because it is a long-term investment. There is also the probability of loss in the investment that will be borne by the prospective pilgrims (Yudhistira, as cited by Primadonna, 2017).

Of course, there are socioeconomic aspects that should be taken into account before implementing a policy. There are at least three aspects that should be considered: validity of investment according to Islamic law and positive law, a suitable mechanism to manage Hajj funds, and allocation of the results of the management of Hajj funds.

This article tries to examine the issue and proposes an ideal mechanism to optimize infrastructure-based sukuk as an alternative to Hajj fund management, based on society's needs.

2 DISCUSSION

2.1 *Management of Hajj funds in Indonesia*

Every year, the number of Indonesian citizen registering to be Hajj pilgrims constantly rises though the Hajj quota is limited. Consequently, Hajj pilgrim's waiting list is increasing. This will lead to an accumulation of Hajj funds. Furthermore, as stated in considerations of Law No. 34 of 2014 about Hajj fund management, accumulation of Hajj funds can potentially be enhanced by management to upgrade the quality of Hajj management itself.

According to Law No. 34 of 2014 about Hajj fund management,

Hajj funds are funds includes the payment of hajj pilgrimage fees, hajj efficiency funds, endowment of the ummah, and the value of benefits which controlled by the state to organize the pilgrimage and the implementation of the program activities for the benefit of Muslims. (Article 1 paragraph 2)

The Hajj fund has huge investment potential because of its massive amount and its funds will be long-lasting while waiting for Hajj pilgrims to depart. To be included on the waiting list, a Hajj applicant should deposit around 25 billion rupiahs (Hariyanto, n.d.). In June 2017 alone, as stated by a representative from the Ministry of Religious Affairs, the Hajj fund collected 100 trillion rupiahs (Fachrudin, 2017).

Based on the Ijtima by Ulema Indonesian Council (MUI) in 2012 about the Status of Ownership Hajj Fund in Waiting List, "Fund deposits for candidates in Hajj waiting list in the account of the Minister of Religious Affairs may be invested (tasharuf) in productive matters (as cited by Mubarok and Hasanuddin, 2013). Consequently, Hajj funds can be invested in various forms, including placement of deposits in Sharia banking or invested in sukuk. By the basic principles of Islamic law, all muamalah activities are allowed as long as they do not contain those explicitly forbidden in Al-Qur'an and Sunnah (maysir, gharar, riba).

Previously, the Hajj fund in Indonesia was managed by the Ministry of Religious Affairs (MORA) and invested in several instruments, yet only 40% of the Hajj fund has invested in Islamic finance institutions or the Islamic finance instruments namely Surat Utang Negara (Government bonds), Shariah deposits, and Surat Berharga Syariah Negara (known as sukuk or Islamic bonds). The rest of the Hajj fund has placed deposits in conventional banks (Fauzi, 2017).

Where does the benefit from the investment go?

Prospective pilgrims have to pay two fees; the first payment is deposited at the time of registration, and the second payment is deposited at the time the pilgrims go on the Hajj. During the interval between registration and departure, the money paid by prospective pilgrims is invested. Profits from the investments are incorporated into the Hajj Funds Optimization Account (separated from the first deposit account). Upon departure, the prospective pilgrim only pays (1) the flight cost, (2) accommodation in Mecca and Medina, and (3) living cost. Other costs required for service of pilgrims such as the cost of some rental lodgings in Mecca and Medina; consumption during the Armina, Medina, and Mecca; issuance and completion of passports and visas; Hajj pilgrimage insurance; accommodation and meals in a Hajj dormitory, etc. are charged to benefits from the Hajj Funds Optimization Account. From the beginning, prospective pilgrims are not charged with the full cost, but only half of the real cost of Hajj. The rest of the cost is covered by the profit from Hajj fund investment (Anon., 2017).

According to the new regulations of Law No. 34 of 2014 regardingHajj fund management, the Hajj fund managed by the Hajj Fund Management Agency (BPKH) and the Hajj fund must be placed and invested in Islamic banks (Article 46). This aligns with the aim of the Masterplan of Islamic Finance Architecture in Indonesia (2015) to place Hajj funds in Islamic banking accounts. Also, placement and investment of the Hajj fund may be made in banking products, securities, gold, direct investment, and other investments as long as they are in accord with Shariah principles (Article 48). Therefore, the investment of the Hajj fund in infrastructure development using sukuk as an instrument is possible.

However, there are several parties who rejected the idea of investing the Hajj fund in infrastructure-based sukuk for several reasons.

First, because infrastructure-based sukuk is a long-term investment, it will take several years to achieve a return. As stated by Bhima Yudhistira, a researcher in the Institute for Development of Economics and Finance (INDEF), commercial infrastructure projects usually yield returns in the long term. Moreover, in the case of noncommercial projects such as bridges or highways, it is certainly rather difficult to gain a profit. This type of investment also has operational risk when the infrastructure project is constrained by technical matters, causing the completion of the project to be delayed or even completely terminated. In Indonesia, many infrastructure projects are delayed because several phases, e.g., land acquisition, have not been completed (Primadhyta, 2017).

Consequently, the result of the investment will be a loss. Therefore, it is possible prospective pilgrims would bear the loss; even worse, the Hajj fund may not be sufficient to send them to Hajj. This is contrary to the main purpose of Hajj fund investment, which is supposed to be directed to productive sectors to give high returns.

Second, there are still several problems with Hajj management and Hajj facilities in Indonesia. In reality, Hajj pilgrims from Indonesia face problems that recur every year because the solution has not yet been found. The problems that always arise begin with the registration for Hajj, the cost of Hajj, the accommodation and transportation of pilgrims, the management of Haj funds (Dana Abadi Ummat), and the failure of pilgrims to participate in Hajj (Zubaedi, 2016).

Consequently, the results of Hajj fund management must be prioritized for the sake of Hajj pilgrims themselves. This is in agreement with the statement ofIjtima of Indonesian Ulema Council in 2012 about the Status of Ownership Hajj Fund in Waiting List that the result of investment belongs to prospective pilgrims (who own the Hajj fund in the form of the waiting list deposit), for instance, to reduce the real cost of Hajj (as cited in Mubarok and Hasanuddin, 2013). INDEF Executive Director Enny Sri Hartati suggests that the Hajj fund would be better used to improve the quality of Hajj management itself, such as facility development for prospective pilgrims (Riyandi, 2017). This is in accord with the aim of Hajj fund management in the law about Hajj fund management to enhance the excellence of Hajj management, and its aim to give larger benefit (*maslahah*) for ummah Muslims as stated in the Hajj Fund Management Act (Article 3).

However, in practice, the results from Hajj fund management will not be received directly by the prospective pilgrims, because the benefit from the management of the Hajj funds received from prospective pilgrims on the waiting list will be used as an operational fund of Hajj implementation in the current year. In other words, prospective pilgrims on the waiting list help the pilgrims who perform Hajj in the current year (Mubarok and Hasanuddin, 2013).

2.2 *Hajj fund management in Malaysia: Lesson learned from Dana Tabung Haji*

Managing Hajj funds is important not only in Indonesia but also in other Muslim countries such as Malaysia. This is because the Hajj quota in Malaysia in 2017 will be higher by 10,000 pilgrims compared to the previous three years, which means the demand of citizens to fulfill their obligation for pilgrimage has increased sharply year by year (Bernama, 2017).

Increased numbers of pilgrims have motivated Royal Professor UngkuAziz to establish a Sharia-based fund management that can facilitate prospective Malaysian pilgrims to deposit their money through Lembaga Tabung Haji (LTH), also known as The Pilgrims' Management and Fund Board of Malaysia (Mannan, 1996). Lembaga Tabung Haji is designed to provide investment services and opportunities for the pilgrims, with a vision as "The Pillar of Ummah's Economic Success" and supported by one of the missions which is "to strengthen the ummah's economy" (Yahaya et al., 2016). This means Lembaga Tabung Haji has tried to manage the Hajj funds through the implementation of Islamic management principles to ensure the pilgrims obtain their Mabrur Hajj with a considerable amount of money.

Lembaga Tabung Haji was created to increase the efficiency of the management of the Hajj fund in Malaysia to support the needs of the pilgrims, especially in terms of flights, accommodations, foods, medical supplies, and performance of Hajj rituals (Bernama, 2017). This means the objective (*maqasid*) of this institution is to maximize the value to customerds as one of its stakeholder groups, hence stimulating the achievement of spiritual, moral, and socioeconomic life and success in the hereafter (*Falah*).

The main activity of LTH is investing (Anon., 2017). Therefore, the funds that have been deposited by the prospective pilgrims would be invested in halal and competitive investments. Based on investment activities of LTH, the funds are invested in equity, fixed income, and property (Anon., 2017). These investments are beneficial for both the pilgrims and the country because they ensure the prosperity of the pilgrims as well as promoting the Islamic money market to society that might increase the number of Muslims who invest in this market.

Lembaga Tabung Haji will invest the Hajj funds into equity that has long-term maturity with a competitive return, which means the dividend generated from the investment would be in line with the risks of the investment (Anon., 2017). The selection of the equity is cru-ci0al, as the institution has to ensure that the investment has net present value (NPV) > 0, which means the investment is profitable and hence could reduce the Hajj cost for the pilgrims. Also, the institution has to deliberately choose the maturity of the investment, so the return from the investment could be received before the pilgrims are scheduled to depart.

It is recommended that Indonesia consider entirely adopting Sharia-based management that has been applied by Lembaga Tabung Haji Malaysia in managing its Hajj funds. This is because it would assist the pilgrims (depositors) in knowing how the deposits are invested and would benefit them because the return would be translated to the reduction of Hajj cost, proper meals during Hajj activities, convenient accommodation in terms of location, better medical service, and greater performance of Hajj rituals. The practice of this management could support poor people to fulfill their obligation to perform a pilgrimage to Mecca at a lesser cost, at the same time obtaining more Hajj benefits from the return on the investment.

Also, this could be the implementation of the previously stated theory thatthat "the Hajj funds should be used for the sake of Hajj pilgrims itself," as the purpose of investing the Hajj fund is to gain a high return that is used to improve and maximize the Hajj services. This management is more reliable because the allocation of the funds is more transparent, as the pilgrims would be informed regarding the amount of dividend of the year. Sharia-based management might attract more people to register for a pilgrimage, as they know the system has been set according to Sharia law.

2.3 *Infrastructure-based sukuk as an alternative to manage the Hajj fund*

A sukuk is an Islamic bond, also known as Islamic investment certificate (Wilson n.d.). This Islamic-based security is different from the normal interest-based securities as the term "interest" does not exist for sukuk. From the perspective of Islamic law, every Islamic financial transaction must be based on the following conditions (Afshar, 2013):

1. The payment or reception of interest (riba) from a loan is not permitted.
2. Gharar, where the trading of the asset does not state or hide the consequences arising from a financial transaction, is prohibited.
3. Based on Islamic law, money does not classify as an asset, so it works as a measurement of value. Institutions or solo traders cannot generate income from money, as it would cause "riba."

4. The yield of the investment is based on profit or loss sharing or the agreed price before the transactions.
5. No transaction is allowed that involves beverages, alcoholic drinks, pork, gambling, pornography, etc.

The purpose of the issuance of sukuk in Indonesia, as mentioned in explanation of Law No. 19 of 2008 on State Sharia Bonds (SBSN), is (1) to expand the funding base of the state budget or the company, (2) to encourage the development of the Sharia financial market, (3) to create benchmarks in the Sharia financial market, to diversify the investor base, (4) to develop alternative investment instruments, and (5) to optimize the utilization of state property or company, public funds that have not been netted by conventional bond and banking systems.

There are three types of underlying assets in sukuk: an asset of the country, service, and government projects. Sukuk that has been used since 2000 is based on underlying service, while sukuk infrastructure is based on underlying government projects such as the development of public infrastructure (Haura, 2010).

The use of sukuk for infrastructure is allowed as long as it does not violate the aforementioned conditions. The Hajj funds are suitable to be invested in sukuk as it is based on Sharia law, and the return that might be given to the pilgrims is free from riba and gharar. This investment also offers ans advantage for sukuk holders (investors) because they hold ownership in the "sukuk asset" (Thomas et al., 2005, p.154).

There are several advantages of putting Hajj funds in sukuk (Haura, 2010):

1) To avoid risks from the banking system. It will prevent risks from banking collapse because the fund is managed by the government.
2) Sharia-compliant investment.
3) Investment free from default risk (the failure of payment). The payment to the investor is guaranteed by the government through Law No. 19 of 2008 on State Sharia Bonds (SBSN).

Sukuk has been used not only in Muslim majority countries but also in Western countries such as the United Kingdom (UK). Britain was the first non-Muslim country to issue sovereign sukuk (Anon., 2014). The establishment of sukuk in the UK has led it to be the hub for Sharia-compliant finance. This decision might bring a great deal of profit for the UK's Islamic market, as more than 3 million Muslims live in England (Beckford, 2016). The need for halal investment might increase in line with the growth of the Muslim population.

The mechanism of sukuk in the UK is that the investors invest their money to buy sovereign sukuk (Al-Ijara), as this Sukuk is an asset-based system, equal to the purchase of an asset by the investor (Isaac, 2017).

It is suggested that Indonesia start to manage the Hajj funds into infrastructure-based sukuk. The main reasons are that this type of investment fulfills the criteria of halal investment (does not contain maysir, gharar, and riba), offers a higher benefit than investments in bank deposits, and has lower default risk as the government secures the investment with an underlying asset. Moreover, it tends to be suitable for prospective pilgrims who are risk-averse, as the risk would be shared because the funds to be invested are collected from all the pilgrims, which means anyinvestment loss would be shared among the pilgrims. Furthermore, the use of sukuk in Indonesia could promote the Indonesian Islamic market to attract more investors and lead the country to becoming the largest sukuk market in the world. Last but not least, as Hajj funds contain a huge amount of money; hence if we could maximize the use of these funds through profitable investments, it could benefit both the government and pilgrims.

Object: Hajj Fund from Hajj Pilgrims on theWaiting List	Invested in infrastructure-based sukuk.	Return for pilgrims' benefit, e.g., reduce cost

Figure 1. Recommended scheme for infrastructure-based sukuk.

The first step of this Hajj funds management is that BPKH collects the funds from the prospective pilgrims on the waiting list and then invests it into a sukuk infrastructure. It will be safer if the funds are invested only in the government's projects, as the government would be able to secure the investment through an underlying asset (in this case, an asset of infrastructure project). Also, based on Law No. 19 of 2008 on State Sharia Bonds (SBSN), the government guarantees that investors will be obtaining a return, resulting in a lower default risk is lower because the government could liquidate the underlying asset if it does not have the capacity to pay back the initial investment as well as the return.

Next, the other important issue is that BPKH has to conduct project analysis to assess whether the project shows a positive return and whether the investment period (maturity) ends before the departure of the pilgrims. This is to ensure the pilgrims will be able to cover all of the Hajj cost using the return on the investment and will depart as scheduled.

The profits that come from the sukuk return would be given back to the pilgrims as the investors, to reduce the cost of Hajj and improve the facilities in Indonesia (pre-departure service), Mecca, and Medina (during the Hajj rituals). It should be used to fulfill the needs of the pilgrims. Hence the wisest decision is to use the returns to solve the problems in Hajj management that arise every year, such as the development of accommodations and transportation near the Holy Mosques in Mecca and Medina. This will improve the quality of Hajj management and thus will benefit the Hajj pilgrims. It is consistent with the aim of Hajj fund management: to fulfill *maslahah* of the ummah.

3 CONCLUSION

The management of Hajj funds is important to maximize the use of money through sukuk investment to fulfill the public's needs. Development of infrastructure using the returns of the investment is one of the approaches. The mechanism of Hajj fund management should be efficient and focused on the benefits for the pilgrims so that the fund is available to use when it is needed. The government has to prioritize the interest of the ummah, such as the development of facilities needed before departure and during Hajj such as transportation and accommodation. Lastly, evaluation regarding the properness of the facility that supports the Hajj activities has to be conducted so that the allocation of the fund is effective and efficient.

REFERENCES

Afshar, T. (2013). "Compare and Contrast Sukuk (Islamic Bonds) with Conventional Bonds: Are They Compatible?" *The Journal of Global Business Management*, 9(1), 44–45.

Anon. (2014). "Government Issues First Islamic Bond." - GOV.UK. Gov.uk. Retrieved October 21, 2017, from https://www.gov.uk/government/news/government-issues-first-islamic-bond

Anon. (2017a). Portal Tabung Haji. Tabunghaji.gov.my. Retrieved October 20, 2017 from http://www.tabunghaji.gov.my/servis-haji

Anon. (2017b). Portal Tabung Haji. Tabunghaji.gov.my. Retrieved October 20, 2017 from http://www.tabunghaji.gov.my/maklumat–2

Anon. (2017c). *Portal Tabung Haji. Tabunghaji.gov.my*. Retrieved 20 October 2017, from http://www.tabunghaji.gov.my/pelaburan-th

Anon. (2017d, August 1). "Five Controversial Things About Hajj Fund." *BBC Indonesia* Retrieved from http://www.bbc.com/indonesia/indonesia-40778194

Anon. (2017e, August 7). "Indonesia Aims to Have World Largest Haj Fund in 10 Years." *The Jakarta Post*. Retrieved from http://www.thejakartapost.com/news/2017/08/07/indonesia-aims-to-have-world-largest-haj-fund-in-10-years.html

Beckford, M. (2016). "Muslims in the UK Top 3 Million for the First Time with over 50% Born Abroad." Mail Online. Retrieved October 21, 2017 from http://www.dailymail.co.uk/news/article-3424584/Muslims-UK-3-million-time-50-born-outside-Britain-Number-country-doubles-decade-immigration-birth-rates-soar.html

Bernama. (2017a). "Malaysia's Haj Quota Is 30,200 This Year." Themalaymailonline.com. Retrieved October 19, 2017 from http://www.themalaymailonline.com/malaysia/article/malaysias-haj-quota-is-30200-this-year#6AEkuqmHxT3k5ivp.97

Bernama. (2017b). "Tabung Haji Staff Urged to Give Their Best in Serving 'Guests of Allah'." These daily.my. Retrieved October 19, 2017 from http://www.thesundaily.my/news/2017/08/22/tabung-haji-staff-urged-give-their-best-serving-guests-allah

Fachrudin, F. (2017, September 26). "Where Does the Government Invest Hajj Fund Nearly 100 Trillion Rupiahs?" Kompas.com. Retrieved from http://nasional.kompas.com/read/2017/09/26/17152421/ke-mana-pemerintah-investasikan-dana-haji-hampir-rp-100-triliun

Fauzi, I. (2017, August 1). "MUI Request to Move Hajj Fund to Islamic Banks." Metrotvnews.com. Retrieved from http://news.metrotvnews.com/read/2017/08/01/737678/mui-minta-dana-haji-dipindah-ke-bank-syariah

Hariyanto, E. (n.d.). "Investment of Hajj Funds in Sukuk Infrastructure." Retrieved from http://www.djppr.kemenkeu.go.id/page/loadViewer?idViewer=6910&action=download

Haura, A., (2010). "The Management of Hajj Funds on the Sukuk Dana Haji Indonesia (SDHI)." Retrieved from http://repository.uinjkt.ac.id/dspace/bitstream/123456789/2399/1/ARIE%20HAURA-FSH.pdf

Ihsanuddin. (2017, July 26). "Jokowi Wants Hajj Fund Invested in Infrastructure." Kompas.com retrieved from http://nasional.kompas.com/read/2017/07/26/12145401/jokowi-ingin-dana-haji-diinvestasikan-untuk-infrastruktur

Ijtima' Ulemas 2012. "Resolution Compilation of Ijtima' Ulemas of Fatwa Commission in Indonesia IV Year 2012," as cited in J.

Mubarok and M. Hasanuddin, M. (2013). "Fatwa on Financing the Management of Hajj Funds and Status of Prospective Hajj Pilgrims in Waiting List." *Al-Ittihad: Journal of Islamic Economics*, 5(1), 36.

Isaac, A., (2017). "Can the UK Make the Most of Sukuk, the Islam-Compliant Bond?" *The Telegraph*. Retrieved October 21 2017 from http://www.telegraph.co.uk/business/2017/10/15/can-uk-make-sukuk-islam-compliant-bond/

Law No. 34 of Year 2014 about Hajj Fund Management.

Mannan, M. (1996). "Islamic Socioeconomic Institutions and Mobilization of Resources with Special Reference to Hajj Management of Malaysia." Islamic Development Bank (IDB), *40*, 21. Retrieved from http://ierc.sbu.ac.ir/File/Book/ISLAMIC%20SOCIOECONOMIC%20INSTITUTIONS%20AND%20MOBILIZATION%20OF%20RESOURCES%20WITH%20SPECIAL%20REFERENCE%20TO%20HAJJ%20MANAGEMENT%20OF%20MALAYSIA_47279.pdf

Masterplan of Islamic Finance Architecture in Indonesia. (2015). Jakarta: Badan Perencanaan Pembangunan Nasional. Retrieved from https://www.bappenas.go.id/files/publikasi_utama/Masterplan%20Arsitektur%20Keuangan%20Syariah%20Indonesia.pdf

Mubrak, J., and M. Hasanuddin. (2013). "Fatwa tentang Pembiayaan Pengurusan Dana Haji dan Status Dana Calon Haji Daftar Tunggu." *Al-Ittihad: Journal of Islamic Economics*, 5(1), hlm. 36.

Primadhyta, S. (2017, August 2). "Advantages and Disadvantages of Investment of Hajj Fund in Infrastructure." *CNN Indonesia*. Retrieved from https://www.cnnindonesia.com/ekonomi/20170802121306-78-231888/untung-dan-buntung-investasi-dana-haji-ke-infrastruktur/3/

Riyandi, S. (2017, August 1). "Five Attacks on Jokowi after Asking for Hajj Funds to Build Infrastructure." Merdeka.com. Retrieved from https://www.merdeka.com/uang/lima-serangan-ke-jokowi-usai-minta-dana-haji-bangun-infrastruktur/harus-sesuai-uu-dan-aturan-agama.html

Wilson, R. (n.d.). "Islamic Bonds: Your Guide to Issuing, Structuring, and Investing in Sukuk Overview of the Sukuk Market." http://www.euromoneylearningsolutions.com/. Retrieved from http://www.alhudacibe.com/images/Presentations%20on%20Islamic%20Banking%20and%20Finance/Sukuk/Structuring%20and%20investing%20in%20Sukuk%20by%20Prof.%20Rodney%20Wilson.pdf

Yahaya, H., M. Majid, A. Talaat, M. Zulkifli, and N. Talaat (2016). "Tabung Haji Malaysia as a World Role Model of Islamic Management Institutions." *International Journal of Business and Management Invention*, 5(II), 45–46. http://dx.doi.org/2319 – 8028.

Zubaedi, Z. (2016). "Problematical Analysis of Hajj Management in Indonesia (Restructuring Hajj Management Model Towards Modern Hajj Management)." *MANHAJ*, 4(3), 194.

Advancing Rule of Law in a Global Context – Susetyo, Rinwigati Waagstein & Budi Cahyono (eds)
© 2020 Taylor & Francis Group, London, ISBN 978-1-138-32782-5

The penalty policy of criminal blasphemy in the plural society in the framework of renewal of the criminal law

Ajie Ramdan
Faculty of Law, University of Padjadjaran, Indonesia

ABSTRACT: The verdict of the Constitutional Court No. 140/PUU-VII/2009 rejects the petition of the petitioner for all, in its consideration Paragraph 3.61. Existence of religion that has been recognized by the state becomes an obligation for the state to protect it from misuse. According to the Court, there is no authority for the state to not recognize the existence of a religion because the state must guarantee and protect the religions embraced by the people of Indonesia. Law No. 1 of PNPS 1965, which has been examined by the material in the Constitutional Court in its consideration of Paragraph 3.33, is still urgently needed and substantially relevant, although it needs to be formally improved. The regulation of religious crime is important because it is the embodiment of the first precept in Pancasila – belief in the One Supreme – which means that in Indonesia, religion is the main aspect of society. Ahok's speech, which caused turmoil in the community, was categorized as a crime of religious blasphemy by the Central Jakarta District Court with a verdict of two years in prison. The Constitutional Court has a stake in guarding against the blasphemy of religion in a pluralistic society by defending Law No. 1 of PNPS 1965. Due to the importance of this law, the government and the parliament must follow up by updating the penal law.

Keywords: Blasphemy, Constitutional Court, Protection, Revision

1 INTRODUCTION

Criminal is a legal classification in which sanctions are granted based on existing legal norms. A human action is said to be an offense because the rule of law imposes a sanction as a consequence of the act. Such an action is a criminal offense if it has criminal sanctions, and it is a civil offense if it has a civil sanction as a consequence. Based on the view of positive law, no other criteria can define an act as an offense other than sanctions according to the rule of law. There is no offense without any sanction, and therefore there is no offense for the act itself. In traditional criminal law theory, there is a distinction between *mala in se* and *mala in prohibited*. *Mala in se* is an act that is evil in itself, while *mala in prohibited* is an act that is called evil because it is prohibited by the positive legal order. This distinction is a typical element in the doctrine of natural law, which cannot be applied in positive legal theory (Asshiddiqie and Safa'at, 2012, pp. 46–49). Legislation plays a very strategic role as the state's basis on which to achieve its goals. In the case of determining a prohibited act or a criminal offense, criminal law policy is applied. The process of determining whether an act is criminal ends with the formation of a law in which the act is threatened with a criminal sanction. The policy of preventing crime through the creation of criminal law, according to Barda Nawawi Arief, is an integral part of social welfare politics. Related to criminalization and penal policies, in Barda Nawawi Arief's view, are two central issues that should be noted, especially in the formulation stage (Arief, 2011, p. 30):

1. The issue of determining what actions should be criminalized;
2. The issue of determining which sanctions should be used or imposed on the offender.

The question that arises in Case No. 140/PUU-VII/2009 is whether the criminal penalty contained in Articles 1–3 of the Prevention of Blasphemy Law and Article 156a a–b of the Criminal Code added by Article 4 of the Religious Defamation Prevention Law is a form of criminalization of freedom of thought. This question arises because abuse or defamation of religion is very difficult to prove so that the ruling regime can use it to criminalize other religious minorities, against the principle of the rule of law.

Law No. 1/PNPS/ is designed, first, to secure the state and society against the ideals of national development in which the misuse of or defection from religion is seen as a revolutionary threat. This law is meant, second, to prevent the emergence of schools of thought that are considered contrary to religious law. Such streams of thought are considered to have violated the law, broken national unity, and tarnished religion, so national awareness is raised by issuing this law. This rule is intended, third, to prevent the misappropriation of religious teachings that are regarded as principal teachings by the religious scholars concerned, and this rule protects against blasphemy and the doctrine of not embracing belief in the One Supreme God (Mudzakir, 2011, p. 62).

The decision of Constitutional Court No. 140/PUU-VII/2009, which rejected the petition of the petitioner for all, must be followed up by revising Law No. 1/PNPS/1965. With the preservation of the law, the current criminal act of religious blasphemy can be processed through the criminal justice system. Examples of cases that have received widespread attention include a speech by Basuki Tjahaja Purnama, alias Ahok, as governor of DKI Jakarta, which was uploaded by Buni Yani and circulated in social media sourced from the PR Jakarta provincial government. During a working visit to the Thousand Islands, Ahok gave the following speech (www.youtube.com/watch?v=bTAKjnCBUMw):

> So do not believe with the people, can it be a little in the heart of the father of my mother can not choose me, right? Lied to wear letter Al-Maidah 51, all expression. That's the right of the father's mother, so the father's mother can not choose ya because I am afraid to go to hell, fooled so yes.

This statement offended Muslims, especially scholars who teach the Qur'an. On October 7, 2016, the Muhammadiyah Youth Force, the Advocates of Love of the Homeland, the Anti-defamation Forum (FAPA), and Irena Handono reported Ahok to the criminal investigation unit of the national police headquarters (BARESKRIM POLRI) with allegations of defamation as stipulated in Article 156a of the Criminal Code (Penal Code). When we look back to the Indonesian Constitution of 1945, it clearly guarantees the independence of Indonesian citizens to embrace their respective values and to worship according to their religions and beliefs. With his words, Ahok, as the governor of Jakarta Jakartatelah, hurt Indonesian citizens who embrace Islam. In the future, the number of forms of religious blasphemy will grow rapidly, so the revision of Law No. 1/PNPS/1965 is indispensable for pluralistic Indonesian society.

2 DISCUSSION

2.1 *Criminal policy*

Policies or efforts to combat crime are an integral part of social defense and welfare. Therefore, it can be said that the ultimate goal of criminal politics is the protection of society. Criminal politics is an integral part of social politics, that is, policies or efforts to achieve social welfare. Crime prevention efforts need to be pursued with a policy approach (Arief, 2011, pp. 3–26):

1. There is integrity between criminal politics and social politics.
2. There is integrity between penal and non-penal crime prevention efforts.

Sudarto argues that if criminal law is to be involved in attempts to overcome the negative aspects of the development/modernization of society, it should be seen in the overall relationship of criminal politics and social defense planning, and this should also be an integral part of the national development plan. Marc Ancel argues that penal policy is both a science and an art that ultimately has a practical purpose of enabling positive law to be formulated better and to provide guidance not only to lawmakers but also to courts that apply the law and to the organizers or the executors of court decisions. In essence, criminal law policy is not merely legal engineering work that can be done in normative and systematic jurisdictions. In addition to the normative juridical approach, criminal law policy also requires a factual juridical approach that may be sociological, historical, and comparative, and that requires a comprehensive approach from other social disciplines. Understanding of the policy or politics of criminal law can be gleaned from legal politics as well as from criminal politics. According to Sudarto, political law is:

a. Attempts to realize good rules according to current circumstances and situations;
b. The policy of the state through competent bodies to establish the desired rules for society.

Sudarto further states that implementing criminal law policies means holding elections to achieve the best results in the sense of meeting the requirements of justice and efficiency. The implementation of political law means realizing the current rules of criminal legislation. Therefore, the revision of Law No. 1/PNPS/1965 on Prevention of Abuse and/or Blasphemy should aim to protect the community. Criminal politics is essentially an integral part of social politics, that is, policies or efforts to achieve social welfare.

2.2 *Criminal law renewal policy*

Efforts to reform the penal law (penal reform) are in essence subject to the penal policy field that is part of and closely linked to "law enforcement policy," "criminal policy," and "social policy." This means the renewal of the penal law in its essence (Arief, 2005, p. 62):

1) It is part of the policy (rational effort) to update legislation to make law enforcement more effective.
2) It is part of the policy (rational effort) to combat crime in the framework of community protection.
3) Is part of the policy (rational effort) to overcome social and humanitarian problems to achieve/support national goals (i.e., "social defenses" and "social welfare").
4) It is an effort to review and reassess (reorientation and reevaluation) basic ideas or sociophilosophical, sociopolitical, and sociocultural values that currently underlie criminal policy and law enforcement. It is not reform of the criminal law if the intended orientation is tantamount to the value orientation of the old penal code of the colonial inheritance (Old Criminal Code or WvS).

Thus, the renewal of criminal law for religious defamation should be pursued by a policy-oriented and a value-oriented approach. Criminal law enforcement should be conducted by exploring and analyzing the sources of written law and legal values that apply in pluralistic societies, such as in religious and customary law.

2.3 *Action and formulation of delik blasphemy according to decision of the constitutional court no. 140/PUU-VII/2009*

The question that arises in Case No. 140/PUU-VII/2009 is whether the criminal penalty contained in Articles 1–3 of the Prevention of Blasphemy Law and Article 156a a–b of the Criminal Code added by Article 4 of the Religious Defamation Prevention Law is a form of criminalization of freedom of thought. This question arises because abuse or defamation of religion is very difficult to prove so that it can be used by the ruling regime to criminalize other religious minorities, against the principle of the rule of law.

According to Mudzakkir in the submission of this matter, expert information states that: "The provision of Article 4 of the Law on the Prevention of Blasphemy is an amendment of the Criminal Code, which is to add Article 156a. The criminal law norm in Article 156a in letter is a legal norm which determines the sanction for evil, whose evil nature is attached to a prohibited act, while the criminal nature arises because it is evil. Its evil nature is enmity, abuse and desecration of religion."

To punish a person while meeting the demands of justice and humanity, there must be an act contrary to the law that can be blamed on the perpetrator. In addition to these conditions, the relevant offender must be a responsible person (*toerekeningsvatbaar*) or *schuldfhig*. General terms of the offense are the nature of unlawful (*wederrechtelijkheid*) errors (*Schuld*) and accountability (according to criminal law) (*toerekeningsvatbaarheid*) (Remmelink, 2003, pp. 85–86).

Also, according to Andi Hamzah in this expert statement: The formulation of the offence of religious defamation must be by the principle of legality, *nullum delictum nulla poena sine praevia lege penal*, there must be a provision of legislation before a criminal act. A new person may be sentenced if there is a law that prohibits such criminal conduct first, common offences in all countries. That is, the offence is neutral, such as theft, but there is a non-neutral delict, namely: religious offence and moral offences, such as religious blasphemies and pornography. In China, people are free to blaspheme religion. The new Dutch Criminal Code also regulates the desecration of religion, and the offence of ideology, because it is in the realm of the mind.

The Constitutional Court, in its decision, page 287, section [3.51], states that the Law on the Prevention of Blasphemy does not forbid anyone to individually interpret religious teachings or conduct religious activities that resemble a religion embraced in Indonesia. The forbidden is intentionally publicly recounting, advocating, or seeking general support, to make interpretations of a religion held in Indonesia or to carry out activities that resemble activities of that religion, which interpretations and activities deviate from the points of religious teachings (Article 1 of the Law on the Prevention of Blasphemy). If it is not regulated, then it is feared to cause conflict, and horizontal conflict can cause unrest, splits, and hostility in society.

To prove the existence of *actus reus* and *mens rea* in criminal acts referred to in Article 1 of Law Number 1/PNPS/1965, in particular relating to elements of interpretation and activities deviating from the points of religious teachings, it is certain that the judge, with reference to the explanation of the quo law, will request information from the clergy and/or the Ministry of Religious Affairs, who have a tendency to certain denominations. Criminalization based on religious abuse or blasphemy should be informed by clerics and the Ministry of Religious Affairs. In this case, it requires information from religious scholars (Rukmini, 2007, p. 132). Article 186 of the Criminal Procedure Code (KUHAP) states that the testimony of an expert witness is what an expert states in court. The information of an expert may also be given at the time of examination by the investigator or the public prosecutor who is outlined in a report form and sworn to oath at the time of his acceptance of office or occupation (Elucidation of Article 186 of the Criminal Procedure Code). With the rejection of the petition of the petitioners by the Constitutional Court, the Constitutional Court has maintained Law No. 1/ PNPS/1965 on Prevention of Abuse and or Blasphemy.

In order to revise national criminal law, the excavation of law that comes from religious law (especially the Islamic religion) should be directed to the principles and main points of thought related to the main issues in the field of criminal law: criminal matters and criminal liability issues. Therefore, the author describes the principles in the field of criminal law that occurred in the case of religious defamation committed by Ahok.

2.4 *Crime and criminal accountability*

Criminal liability is not punishable if there is no mistake (*Geen strap Zonder Schuld; actus non-fact rum nisi men's sit res*). Persons who cannot be blamed for a criminal act may not be criminally charged. The exception is the existence of certain psychic circumstances in the person committing the criminal act and the relationship between the situation and the actions done in such a way that the person may be reproached for doing the deed (Moeljatno, 2002, pp. 153–158). Determination of whether someone is held accountable, according to Jan

Remmelink, is very dependent on the situation and social conditions that exist, including the nature and context of a criminal act that is concretely done. The existence of capability is the basis of a blameworthy plea. The absence of this ability as a variant of the absence of error (*afwezigheid van all schuldlavas*) is psychic oversight caused by mental illness. In our society, no single human group is incapable of being held accountable, which can be contrasted with other groups capable of being held accountable. Even those who are mentally ill can be held accountable (Remmelink, 2003, p. 191). The element of responsible ability can be equated with the element of unlawful nature. Both are absolute conditions, one for the prohibition of action (the existence of the unlawful nature) and the other for the errors (Moeljatno, 1980, p. 168). The requirement of criminal liability is essentially identical to the criminal punishment/penalty requirement (Arief, 2005, p. 137).

Ahok's speech caused a commotion and anxiety for Muslims in Indonesia. Was it a criminal offense? The indictment made by the public prosecutor was an alternative indictment. In this case, each of the indictments excludes each other. The judge can choose to state that the second indictment has been proven without first deciding on the first indictment. Was Ahok's speech an offense under Article 156 or 156a of the Criminal Code for which he ought to be held accountable? Article 156 of the Criminal Code is contained in Chapter V on Crimes against Public Order, which reads as follows:

> Anyone who publicly expresses feelings of hostility, hatred or contempt for any or some of the Indonesian people, is punishable by a maximum imprisonment of four years or a fine of up to three hundred rupiahs. The words of the group in this article and the next article mean, every part of the Indonesian people, which is different from one or more other parts because of his race, his native country, his religion, his place of origin, his descendants, nationality or status, nationality or status according to constitutional law.

Article 156a reads as follows:

Sentenced to a maximum imprisonment of 5 years whoever intentionally publicly expels a feeling or commits an act:

a. Which is essentially hostile, abusive or defamatory of a religion held in Indonesia;
b. With the intention that people do not embrace any religion that is also conceived of the One Godhead.

The controversy generated from Ahok's speech was related to the mention of Al-Maidah verse 51 of Al-Qur'an that Ahok made during his visit to the Thousand Islands in support of fisheries programs developed by residents. Ahok delivered his comments intentionally in public. The feeling he expressed is considered an insult to or blasphemy of Islam that is against Al-Qur'an and Ulama, according to the religious opinion the Indonesian Ulama Council presented on October 11, 2016. Therefore, his speech amounted to blasphemy. The expression of feelings is punishable by four years' imprisonment under Article 156 and five years' imprisonment under Article 156a of the Criminal Code. Let us describe the formulation of the offense of religious blasphemy by Ahok.

The error Ahok committed was to convey his opinion about Al-Maidah verse 51 in public, which is considered defamation of Islam. Such statements are considered unlawful under Articles 156 and 156a of the Criminal Code. This refers back to the accountability for capability disclosed by Jan Remmelink. The situation and social conditions that occurred after Ahok spoke included the fact that Muslims in Indonesia took offense, and they staged massive demonstrations on October 14, 2016, November 4, 2016, and December 2, 2016.

2.5 *Purposely*

The Memorie van Toelichting (MvT), when applying Crimineel Wetboek of 1881 which became the Indonesian Criminal Code of 1915, stated, among other things, that a criminal must consciously willing to commit a particular crime (*de bewuste richting van den will op even bepaald midriff*). Regarding the MvT, Satochid Kartanegara points out that what is meant by *opzet Willens en weten* (desired and known) is (Marpaung, 2012, p. 13): "A person who commits an act deliberately must want (will) the act and must *menginsafi* or understand the consequences of the action." Some experts define *de will* as desire or will. Thus, a criminal act is the exercise of the will. The will can refer to:

a) Prohibited acts
b) Prohibited effect

Criminal law contains two theories of intent: the theory of will and the theory of knowledge. The theory of will is directed toward the realization of an act (*de op verwerkelijking der wettelijke omschrijving Berichte will*), whereas according to the theory of knowledge, intent is the will to do by knowing the necessary elements (*de will tot handle big voorstelling van de tot de wettlijke omschrijving behoorende bestandelen*). Furthermore, Pompe writes about the two theories that the difference does not lie in the intent to engage in the behavior (positive or negative) itself, both of which are called wills, but in the deliberate opposition to other elements (insofar as they have to be overwhelmed) and the circumstances that accompany it. Concerning these elements, one speaks of knowledge (has a description of reality, so knows and understands) while others address the will (Moeljatno, 1980, p. 116).

According to Moeljatno, the theory of knowledge is more satisfactory, because it only relates to the elements of the deed done. There is no causal relationship between motive and deed. The only question is whether the defendant knows (*menginsyafi*) or understands his actions, as well as the consequences and circumstances that accompany it (Moeljatno, 1980, p. 117). Ahok's speech was a form of feeling or opinion issued intentionally in public under the theory of deliberate acceptance as knowledge. By embracing this theory, we can take two paths – namely, to prove a causal relationship within the defendant's mind between motives and purpose, or to prove the existence of *penginsyafan* or understanding of what was done and the consequences and circumstances that accompany it. Therefore, Ahok should have understood that his words would tarnish the religion of Islam and that they would have consequences. The consequences are evidenced by the large demonstrations of Muslims who demanded justice.

2.6 *Responsibility*

According to Van Hamel, the ability to be responsible (legally) is a condition of psychic maturity and normality that includes three other abilities: understanding the direction of the factual objectives of one's own act, the awareness that the act is socially forbidden, and free will with regard to the act (Remmelink, 2003, p. 213). According to Moeljatno, there is a capacity for accountability (Moeljatno, 2002, p. 165), which means the ability to discriminate between good and bad, lawful and unlawful deeds, and the ability to determine their will according to conviction about the good and bad of the deed.

The first is the intellectual factor that can distinguish between permissible and unlawful acts. The second is the factor of feeling or the will (volitional factor) that can adjust behavior with conviction on behalf of what is allowed and what is not. As a consequence, of course, the person who cannot determine his will according to the conviction about the good and bad of the deed has made no mistake when committing a criminal act. Such a person cannot be held accountable. According to Article 44 of the Criminal Code, such incapacity must be caused by mental illness or physical deformity. Along with the existence of rules relating to legal capacity as formulated in Article 44 of the Criminal Code has arisen an unspoken principle: an individual cannot be punished if he or she made no mistake. Responsibility is an elemental error and it must be proven. Therefore, since in general people are normal in their minds, and capable of being responsible, then this element is thought to be always present unless there are signs indicating that the defendant may be an abnormal soul. In this case, the judge shall order a special examination of the defendant's soul, even if not requested by the defendant. If the result is, indeed, that his or her soul is not normal, then, according to Article 44 of the Criminal Code, criminal sanctions cannot be imposed. If the results of the examination are doubtful for the judge, legal capacity cannot be proven and criminal sanctions cannot be imposed, according to Van Hattum (Moeljatno, 2002, p. 168).

According to Moeljatno, this element of legal capacity can be equated with the element of unlawful nature. Therefore, both are absolute requirements, one for the prohibition of action (the existence of the nature of the law), and the other for the existence of mistakes. About both of them, in the Criminal Code there is a reason for the abolition of criminal sanctions in Articles 49, 50, and 51 (justification) and in Article 44 (Moeljatno, 2002, p. 168). Ahok as a non-Muslim does not have the competence to speak on Al-Maidah verse 51. Because Ahok is not a Muslim, let alone a scholar, Ahok never realized what he was saying would offend Muslims, especially scholars who teach the Qur'an. Public officials who often upload their activities to social media need to be aware and mindful of such public concerns.

Although not a Muslim, Ahok certainly can discriminate between good and bad deeds, and he also can determine his will. Of course his mind is also normal. With Ahok's capabilities, the element of error and unlawful elements exists. Therefore Ahok can be held legally responsible for blasphemy against Islam.

2.7 Revision of religion and religious delegation in the rkuhp for national criminal law reform

Excavation of religious and traditional law is normal and can even be said to be the demand of the times, especially in Indonesia. For the Indonesian people, it is clearly a national burden and even a national obligation and challenge because it has been mandated and recommended in various national legislations and seminars so far. The problem is how to explore, transform, and actualize the values of traditional (customary) and religious law so that it can be accepted into an integrated norm within the national legal system. To examine the values of religious law, several things should be considered (Arief, 2012, pp. 322–333):

a. Since the objective of legal extraction is to fulfill and realize a national legal system, legal excavation should aim to strengthen the national legal system. This means, first of all, there must be a common understanding of what is meant by the national legal system.
b. Since national law must protect all countries in all aspects of life, legal excavation must be based on common national insights in the field of national law development.
   1. National Insight
   2. Insight Archipelago
   3. Insights of Unity in Diversity

In order to accomplish the renewal of national criminal law, the extraction of law that comes from the values of religious law (especially the Islamic religion) should be directed to the principles and ideas related to the main issues in the field of criminal law: criminal matters and criminal liability issues.

The Criminal Code (WvS) has no special chapter on religious offenses, although some offenses the WvS lists can be categorized as religious offenses. *Delik* religion is widely spread in the Criminal Code, such as murder, theft, fraud/cheating, insult, slander, and moral deception (adultery, rape, etc.). The subparagraphs (a) in the Criminal Code are not necessarily the same and do not include all sins according to the teachings or norms of religious law. *Delik* religion in the meaning of subparagraph (b) is seen primarily in Article 156a (defamation of religion and committing acts so that people do not adhere to religion). Professor Omar Senoadji also includes the offense in Articles 156–157 (insults against religious groups, known as group libel) in the subgroup of religious offenses (b) as well, while religious offenses in the meaning of subparagraph (c) in the Criminal Code are spread among others in Articles 175–181 and 503 to 2, which include:

- hindering religious meetings and burial ceremonies (Article 175 of the Criminal Code);
- interfering with religious meetings and funerals (Article 176 of the Criminal Code);
- laughing at religious officers carrying out their duties (Article 177 1 of the Criminal Code);
- insulting objects of worship (Article 177 of the Second Criminal Code);

- blocking a corpse's transport to the grave (Article 178 of the Criminal Code);
- staining/damaging graves (Article 179 of the Criminal Code);
- digging, taking, or removing bodies (Article 180 of the Criminal Code);
- hiding/removing corpses to conceal death/birth (Art. 181); and
- making noise near a building of worship or at worship.

In the draft of 2006/2012, religious offenses are organized in Chapter VII on Criminal Acts against Religion and Religious Life (Arts. 341–348). The holding of this special chapter started with the first concept of Book II of 1977, known as the BAS (concept compiled by Team Basaroedin). It included Article 181 s.d. 196 Chapter VI up to the development of the concept in 1993–1998, but was later included in Chapter VI (as Articles 257–264). The subsequent concepts included in Chapter VII of Book II are, namely, Articles 290–297 of the 2000–2002 concept, Articles 336–343 of the 2004 concept, in Articles 342–345 of the 2005 concept, and Articles 341–348 of the 2006/2012 concept. The scope of this chapter of Criminal Acts against Religion and Religious Life, in Concept 2006/2012, which is not different from previous years, is as follows:

- This chapter consists of two parts. Part One, on Criminal Acts against Religion, and Part Two, on Criminal Acts against Religious Life and Religious Means. Part One (Criminal Acts against Religion) (Articles 341–344) consists of:
  - Paragraph 1: Disgrace against Religion
    - Stating feelings or performing acts of contempt for religion held in Indonesia (Article 341);
    - Insulting the majesty of God, His word, or His nature (Article 342);
  - Paragraph 2: Damage to Places of Worship (Article 348) – namely tainting or unlawfully damaging or burning buildings where worship takes place or holy objects are worn (Article 348).

Although most of these are similar to those outlined in the Criminal Code, there are also some interesting differences:

a. The offenses in Articles 175 s.d. 176, 178 s.d., 181, and 503 of the Criminal Code relating to burial rites and other deeds related to corpses/graves are not included as religious offenses within the concept but still appear in the Criminal Code, in Chapter V on Crime against Public Order.
b. In addition to the public defamation offenses formulated in Article 257 of the 1993 concept or Article 341 of the 2006–2012 concept (such as Article 156a of the Criminal Code), there is also a more detailed formulation of the offense concerning blasphemy or *Godslatering*, in the form of contempt against God and deeds mocking, desecrating, or degrading religion, apostles, prophets, scriptures, teachings, or religious worship. This more explicit and more specific formulation is not found in the Criminal Code.
c. Criminal religious life as set in the concept is still very limited on the issue of religious freedom, especially in running religious services and ceremonies. Therefore, the acts that are threatened with the crime in the concept are:
  - interfering with, obstructing, or dissolving religious ceremonies/meetings;
  - making noise near a building of worship;
  - mocking a worshipper or a religious officer performing a duty (among other *mubaligh*); and
  - negating people's belief in religion.

Although this act aims to protect freedom of religion (worship and belief), it is also indirectly intended to prevent unrest and clashes among religious communities. Thus, in this case, it aims to protect the harmony of religious life. This goal is also seen with the prohibition of desecration of buildings of worship and of blasphemy. However, it must be admitted that not all deeds related to the issue of religious harmony are governed in the concept.

One problem that is quite vulnerable concerning the issue of religious harmony is broadcasting religion to other religious people. An explicit formulation of this is not found in the concept. In the concept there is only Article 261/1993 (Article 345/2006–2012) concerning sedition to exclude beliefs. In this article, it is affirmed that the forbidden is instigated in any form, so it can be questioned whether religious broadcasting can be included here. The phrase "instigating" in any form is quite extensive, so if the broadcasting is intended to prevent the other person from believing in his/her religion, then such a thing may be included in Article 260. Only Article 260 has a provision that is not easy to apply in the aforementioned case. Therefore, it is only natural that anyone can propose a separate article on this matter. As mentioned in Article 261, the concept of incitement to abolish belief in religion can be said to mean to incite other apostates. So the punishable (as a delicacy) act here is the act of the agitator, not the works of the apostate.

3 CONCLUSION

1. Ahok has committed blasphemy against Islam that must be criminally accounted for. Although Ahok is a non-Muslim and does not have the competence to speak about Al-Qur'an Al-Maidah verse 51, he should realize what he said will offend Muslims and especially scholars who teach the Qur'an. Ahok can be held legally responsible, and thus the element of error and the unlawful element are considered to exist. Therefore Ahok is responsible for blasphemy against Islam. The reasons for the abolition of criminal sanctions in Articles 49, 50, and 51 of the Criminal Code (justification) and in Article 44 of the Criminal Code (incapable of being liable) cannot be used to remove criminal charges against Ahok.
2. Revision of Law No. 1/PNPS/1965 on Prevention of Abuse and or Blasphemy united into the RKUHP has been very urgent. Because the act of blasphemy at any time can arise, if such action has not been regulated in Indonesia's positive law, such an act will cause tension in pluralistic Indonesian society.

4 SUGGESTIONS

1. The government and the People's Legislative Assembly should complement the shortcomings of criminal defamation that have not been included in Law No. 1/PNPS/1965.
2. Revision of the crime of blasphemy contained in the RKUHP should be encouraged to obtain ratification into the Criminal Code (Penal Code). The ultimate goal of criminal politics is the protection of society. Criminal politics is an integral part of social politics, that is, policies or efforts to achieve social welfare.

REFERENCES

Arief, Barda Nawawi. (2008). *Kebijakan Hukum Pidana*. Jakarta: Kencana.
Arief, Barda Nawawi. (2011). *Pembaharuan Hukum Pidana Dalam Perspektif Kajian Perbandingan*. Bandung: PT Citra Aditya Bakti.
Asshiddiqie, Jimly and Ali Safa'at. (2012). *Teori Hans Kelsen Tentang Hukum*. Jakarta: Konpress.
Hamzah, Andi. (2004). *Hukum Acara Pidana*. Jakarta: Sinar Grafika.
Marpaung, Leden. (2012). *Asas Teori Praktik Hukum Pidana*. Jakarta: Sinar Grafika.
Moeljatno. (1980). *Asas-Asas Hukum Pidana*. Yogyakarta: Universitas Gadjah Mada.
Moeljatno. (2002). *Asas-Asas Hukum Pidana*. Jakarta: Rineka Cipta.
Mudzakir, Dkk. (2011). *Analisis dan Evaluasi Undang-Undang Nomor 1/PNPS Tahun 1965 Tentang Pencegahan Penyalahgunaan Dan/Atau Penodaan Agama*. Jakarta: Kementerian Hukum dan HAM.
Remmelink, Jan. (2003). *Hukum Pidana Komentar atas Pasal Terpenting dari KUHP Belanda dan Padanannya Dalam KUHP Indonesia*. Jakarta: Gramedia.
Rukmini, Mien. (2007). *Perlindungan HAM Melalui Asas Praduga Tidak Bersalah dan Asas Persamaan Kedudukan Dalam Hukum Pada Sistem Peradilan Pidana Indonesia*. Bandung: Alumni.

Advancing Rule of Law in a Global Context – Susetyo, Rinwigati Waagstein & Budi Cahyono (eds)
© 2020 Taylor & Francis Group, London, ISBN 978-1-138-32782-5

Plan the development from the constitution: Expanding constitutional directives in the 1945 constitution of Indonesia

Indra Perwira, Ali Abdurahman & Mei Susanto
Constitutional Law Department, Faculty of Law, Universitas Padjadjaran, Indonesia

Adnan Yazar Zulfikar
Center for State Policy Studies, Faculty of Law, Universitas Padjadjaran, Indonesia

1 INTRODUCTION

At the end of its 2009–2014 term, the People's Consultative Assembly of the Republic of Indonesia (Majelis Permusyawaratan Republik Indonesia) (MPR RI) issued MPR Decision No. 4/MPR/2014 containing recommendations for the upcoming term. One of the recommendations was "to reformulate the national development planning system based on [the] Broad Guidelines on State Policy [Garis-Garis Besar Haluan Negara] [GBHN] model as a guideline for state administration." The GBHN was a directive guideline for state policy in the era before the amendment of the 1945 Constitution of the Republic of Indonesia (Undang-Undang Dasar Republik Indonesia 1945) (UUD 1945). At that time, the MPR was the supreme state body authorized to appoint and dismiss the president of the Republic of Indonesia. In other words, the president was appointed to implement the GBHN as a guideline of state policy established by the MPR itself. If the president failed to do so, then the MPR could dismiss him from office.

The recommendation to bring back the GBHN development model was motivated by the unsustainability in development under the Law of the Republic of Indonesia No. 25/2004 Concerning the National Development Planning System (Sistem Perencanaan Pembangunan Nasional) (SPPN). Under the SPPN model, development was unsustainable because the five-year Medium Term Development Plan (Rencana Pembangunan Jangka Menengah) (RPJMN), as a more strategic plan than the twenty-year Long Term Development Plan (Rencana Pembangunan Jangka Panjang) (RPJP), had been created only based on the vision and mission of the president, who is elected through a general election as a part of the presidential form of government adopted after the 1945 Constitution was amended. Under these circumstances, the development direction could easily change following presidential elections. This did not happen under the GBHN model, wherein the president was appointed by the MPR to perform its mandates. Furthermore, the direction of development had become very president-centric as the president's vision had become the program of government in the form of presidential regulations, without any participation even from the House of Representatives.

Unfortunately, many parties, including constitutional law scholars, opposed the recommendation to reformulate the national development planning system based on the GBHN model. Constitution amendments to reinstate the GBHN model, they argue, can be viewed as reestablishing the MPR as the supreme state body. That kind of change seems like an attempt to restore the Indonesian constitutional system to what it was during the New Order Era. Moreover, such changes also would revive a parliamentary form of government and therefore would violate the agreement of the amenders of the constitution to establish the presidential form of government as permanent legal policy under the 1999–2002 constitutional amendment process.

The alternative policy, to revise the SPPN development planning model, formalizes the principal contents of the development plan as provisions under the 1945 Constitution. By formalizing a more permanent legal policy regarding development plans, whether they involve social or economic policy, it is expected that the negative side effects of presidential elections will not lead to instability in development planning. Thus, elected presidents will not make any new development plans that differ extremely from those of the previous president, as both presidents must base their plans on the same constitutional principles.

This paper examines the alternative to implementing such a constitution-based development plan under the 1945 Constitution of Indonesia. This paper consists of three parts. First, this paper examines the theoretical perspective regarding what a constitution should contain. This approach looks at the ideological background of a constitution, whether liberal countries have a different constitutional order than do socialist ones. Second, this paper identifies alternative forms of development planning under a constitution and their possibility of enforcement. Then, this paper studies the structure of the 1945 Constitution in term of clauses addressing development. Does the 1945 Constitution have a development plan clause? If it does, is it sufficient as a basis for stabilizing the development plan system?

2 DISCUSSION: WHAT COULD THE CONSTITUTION DO FOR NATIONAL DEVELOPMENT?

2.1 The content of the constitution

To answer the question "what could the constitution do for national development?" we begin by answering first "what is the content of a constitution?" J. G. Steenbeek argues that a constitution has three basic features (Soemantri, 1986, p. 51):

1. Protection of human rights and citizen rights;
2. Fundamental constitutional structure; and
3. Distribution and limitation of fundamental constitutional authority.

No country in the world has the same constitution as any other. As K. C. Wheare contends, "there is no one form of Constitution which is practicable or suitable or eligible for all communities" (Wheare, 1966, p. 34). Bagir Manan identifies several factors that cause the content a constitution to differ between one country and another (Manan, 2011, pp. 10–11):

1. Differentiation on philosophical and ideological bases;
2. Differentiation on theoretical and conceptual bases;
3. Differentiation of cultural background;
4. Differentiation of historical background;
5. Differentiation in the form of state (federal or unitary), form of government (republic or monarchy), and system of government (presidential or parliamentary).

We analyzed if whether a constitution contains development planning depends heavily on the first factor, the philosophical or ideological basis adopted by its state. Socialist countries tend to formalize development policy, either on economic or on social matters. On the contrary, liberal countries rarely do so as their economic activity often depends on the market mechanism.

These relations can be explained historically. James Buchanan (1994) explains that the classical liberals of the eighteenth century, such as David Hume and Adam Smith from the Scottish Enlightenment or Alexander Hamilton and James Madison as founding fathers of the United States of America, were highly skeptical about the capability and willingness of politicians to further the interests of ordinary citizens. In Federalist Paper No. 51, James Madison cautions that governments are a necessary evil, institutions to be protected from, but made necessary by the elementary fact that all persons are not angels. Madison further states that "ambition must be made to counteract ambition." Governments, along with those persons who are empowered as their agents of authority, are not to be trusted. For classical liberals, constitutions were necessary as a means to constrain authority. Classical liberals feared state power, and the problem of constitutional design was ensuring that such power would be effectively limited. Adam Smith presented an unflattering image of politicians as "insidious and crafty" and prone to influence by special interests (Tamanaha, 2004, p. 532).

Buchanan claims that in the classical liberal conception, constitutions instruct governments as to what they might and might not do rather than how governments do whatever it is that they do. In the classical liberal constitutional order, governments have a limited role in framework maintenance; the activities of government are functionally restricted to the parameters of social interaction. Governments, ideally, are constitutionally prohibited from direct action aimed at "carrying out" any of several basic economic functions: (1) setting the scale of values, (2) organizing production, and (3) distributing product. These functions are to be carried out beyond the conscious intent of any person or agency. They are performed through the operation of the decentralized actions of the many participants in the economic nexus, as coordinated by markets, and within a framework of "laws and institutions" that are appropriately maintained and enforced by government. This constitutional order is highly influenced by the principle of laissez-faire, a French term that means "leave alone," and asserts that government should be less involved in economic matters. Laissez-faire is a key principle of free market capitalism.

In the context of the US Constitution, Erwin Chemerinsky contends that the function of a constitution is to create a national government, to separate power between the federal and state governments, and to protect individual liberties (Chemerinsky, 2006, pp. 1–6). These functions can be seen in the US constitutional framework in the US Constitution's Preamble, Article 1 about the legislative branch, Article 2 about the executive branch, Article 3 about the judicial branch, Article 4 about the relations among the states and between each state and the federal government, and Articles 5–7 about the Constitution itself – the amendment mechanism, the supremacy clause, etc. All of these clauses are constitutional devices for liberals to accomplish the purpose of limiting government and protecting individual freedom. These clauses are political in substance and do not discuss the economic nexus in the least.

What should a constitution contain? Wheare answers this question with the statement "the very minimum, and that minimum to be rules of law." The liberal constitutional order, as discussed earlier, is already the rule of law, but we could say that that constitutional order is only the formal conception of the rule of law.

The socialist constitutional order could be seen as a reaction against the classical liberal constitutional order. James Buchanan states that this reaction stems from the generalized unwillingness of participants in the body politic to accept the spontaneous allocative and distributive results generated in the operation of a market economy (Buchanan, 1994, pp. 4–5). Socialists do not take these results to be "natural"; they do not understand them to be the working out of the whole complex of separate choices made by persons in their many capacities. From this point of view, the results of the market process were taken to be artifactual – produced rather than emergent – and hence subject to direct manipulation, change, and redirection by politicized collective action.

Buchanan further identifies two separate sources that reacted against classical liberalism (Buchanan, 1994, p. 5). The first of these is Karl Marx, who successfully isolated, identified and publicized elements in the operation of market capitalism that seemed vulnerable. Marx concentrated on the vulnerability of capitalism to financial crises, on its tendency toward concentration in industry, and on the alleged distributive exploitation of the proletariat. The second of these sources is political idealists who saw that any failures of markets could be fully corrected by directed political action. Those views helped socialists to build a vision that rejects a self-regulating or non-politicized economy; on the contrary, a controlled or regulated economy becomes necessary. This shift from a self-regulating economy to a controlled or regulated economy may be directly related to issues of constitutional dictation concerning how governmental controlling and regulating functions are to be performed.

The socialist constitutional order necessarily extends the scope for politicization well beyond the narrowly defined limits of collective authority under the classical liberal order. In terms of the content of a constitution, the socialist constitutional order indicates a more substantive rule of law than does the classical liberal order. In the substantive conception of the rule of law, the government has a larger role in organizing public interest to achieve the aims of the state. These aims usually include general welfare; in other words, the substantive rule of law is highly related to the conception of the welfare state (*verzorgingstaat*).

The most classical socialist constitution was the constitution of the Russian Socialist Federative Soviet Republic (RSFSR). Under the RSFSR, economic activity was not market-based, but was regulated, directed, and controlled forcefully by the constitution as supreme law. The third clause of Article 1 of the RSFSR Constitution enumerates the economic policies of the RSFSR: private property was declared national property to be apportioned on the principles of egalitarian land tenure; all forests, mineral wealth, and waters of national importance were the property of the nation.

At present, there is a convergence tendency between liberal and socialist countries in terms of their economic constitutional orders. In the first instance, the liberal constitutional order represented by the US Constitution says virtually nothing about economic policy. Conversely, the socialist constitutional order represented by the Russian Constitution regulates many economic aspects. According to Jimly Asshiddiqie, after World War II, the idea of a welfare state that, to some extent, justifies state intervention in the economic realm began to affect liberal countries in Western Europe (Asshiddiqie, 2016, p. 114). Such influence continued to rise until the liberal countries were regulating economic policy in legislation, even in their constitutions. Ireland was the first country to formalize economic policy under its constitution of 1937. In Ireland's constitution, the provisions on economic policy are formulated in its chapter entitled "Directive Principles of State Policy," which contains the general directive principles of the economic policy of Ireland.

Indonesia, under its original, unamended 1945 Constitution, from the very beginning has been enumerating provisions on economic policy. First of all, the preamble of the 1945 Constitution confirms that a purpose of the state is to improve public welfare. Furthermore, social and economic policies were formulated in Chapter XIV, entitled "Social Welfare," which state that the economy shall be organized based on the family system, important production sectors and natural resources shall be controlled by the state, and poor and destitute children shall be cared for by the state.

This provision could be seen as an expression of revolutionary expectations about social and economic justice (King, 2011, p. 233). This constitutional framework is highly influenced by socialist values. Mohammad Hatta calls the Indonesian economic system based on Article 33 of the 1945 Constitution religious socialism (Ruslina, 2012, p. 70). Other socialist values take the form of Soepomo's "Integralistic State" concept, which resembles traditional village societies. Soepomo's Integralistic State concept argues that the state does not guarantee individual or class importance, but it does guarantee the importance of society in its entirety as a unit (Nasution, 2011, p. 13). Without any direct relation, Article 33 of the 1945 Constitution contains a clause similar to the third clause of Article 1 of the RSFSR Constitution, cited earlier.

After several constitutional amendments between 1999 and 2002, the 1945 Constitution has at least three forms of economic provisions. The first is the preamble, as this part of the constitution remains unchanged. The second is new, more detailed clauses about human rights guaranteed under Chapter XA, entitled "Human Rights." Several clauses contain economic rights as the right of the fulfilment of basic needs (Article 28C (1)), the right to work and receive fair and proper remuneration and treatment in employment (Article 28D (2)), the right to housing, the right to a healthy environment and to obtain medical care (Article 28H (1)), and the right to own property (Article 28H (4)).

The last is an amended clause about social welfare in Chapter XIV of the 1945 Constitution. The title of the chapter was changed from "Social Welfare" to "The National Economy and Social Welfare" to reaffirm this article as the basis of the national economic system. Article 33 itself has a new clause about the principles of the organization of the national economy under the basis of democracy; the principles are togetherness, efficiency in justice, continuity, an environmental perspective, self-sufficiency, and keeping a balance in the progress and unity of the national economy. The amendment also changed Article 34 with several new clauses about the obligation of the state to develop a system of social security and the obligation of the state to provide sufficient medical and public service facilities.

2.2 *Form of development plan in the constitution*

According to the explanation given earlier, we could categorize formalization of a development plan in a constitution as follows:

2.2.1 *On the preamble*

The preamble of a constitution usually expresses the purpose of either the establishment of a government or the enactment of a constitution. A preamble could set that constitution's goals for the future, and the goals could address foreign policy, human rights, peace, or national development. Preambles as an expressive text are sometimes enforceable, but in some countries they are not viewed as formally operative (Ginsburg, Foti, and Rockmore, 2013, p. 102). The preamble of the 1945 Constitution itself is considered enforceable through the authority of the Constitutional Court of Indonesia. Often, in constitutional review cases, the preamble becomes a basis either for an application or for consideration of judges in a court decision. Constitutional Court Decision No. 28/PUU-XI/2013 declared that Law of the Republic of Indonesia No. 17 of 2012 Concerning Cooperation unconstitutional as the law is contradictory to improving the public welfare an established in the preamble of the 1945 Constitution.

2.2.2 *On human rights provision*

The formalization in constitutions of social and economic rights, including rights of development, has been a polemic among scholars. The polemic covers two main issues. First, such formalization of economic and social rights appears to require significant political decision-making directly related to resources and economic priorities. Second, in terms of justiciability, economic and social rights are regarded as nonjusticiable human rights. Treating these rights as justiciable would burden the government with obligations it actually cannot fulfill. The situation is different with civil and political rights that only require the government to fulfill the obligation of noninterference; economic and social rights burden the government to take some positive action to fulfill basic human needs at the very least, which requires the government to have sufficient resources.

Prioritizing civil and political rights as the first generation of rights and economic and social rights as the second generation of rights is misleading, however, including for reasons that both generations of rights have different methods to be respected, protected, and fulfilled. Both generations of rights must be treated as indivisible and interdependent; both are necessary for a "good life" and human flourishing. Crucially, the effective enjoyment of first-generation rights depends on the realization of second-generation rights: one needs certain resources to effectively exercise freedom in the civil and political sense. Sandra Fredman notes that freedoms of speech or assembly are of little use to a starving or homeless person (Fredman, 2008, p. 67). What use is freedom of association if someone cannot get to a meeting because they work fourteen hours a day in a call center or garment factory? (IDEA, 2014). Therefore, including civil and political rights in a constitution without including economic and social rights would only leave half of the obligation fulfilled.

Katherine Aldrich (2010) has identified several modes of recognition of economic and social rights: the first mode of recognition is informal recognition, which Aldrich describes as recognition by informal actors such as private donors or nongovernmental organizations. The second mode of recognition is formal recognition, which could be either on an international level by international treaties or on a domestic level by statutes or constitutions. In terms of implementation, formal recognition on a domestic level is more effective for individuals, because only the government has the national structure, resources, and power to help realize such rights. Afterward, on a domestic level, constitutional protection for such rights is considered more powerful than statutory or ordinary law protection as it is highly dependent on political will, and it could be easily changed by political power. This is contrary to the notion of human rights, as rights should not be dependent on political will for their realization. Tadeusz Zielinski argues that unless socioeconomic rights are elevated to the constitutional level, the authorities will have full discretion to disregard, or even to further reduce citizens' social entitlements (Sadurski, 2002, p. 7).

Aldrich (2010) further classifies constitutional protection for economic and social rights into two types: aspirational recognition or justiciable recognition. It is aspirational if the protections are purely directing political action, but it is justiciable if the protections can be redressed through judicial action. Even in an aspirational capacity, such recognition is important as it may help shape government priorities.

The amendment of the 1945 Constitution provides economic and social rights protection under the human rights chapter. Four clauses contain economic rights: the right to fulfill basic needs (Article 28C (1)); the right to work and receive fair and proper remuneration and treatment in employment (Article 28D (2)); the right to housing, the right to a healthy environment, and the right to obtain medical care (Article 28H (1)); and the right to own property (Article. 28H (4)). These clauses are followed by a clause upholding governmental responsibility for these rights in Article 28I (4), which states: "The protection, advancement, upholding and fulfilment of human rights are the responsibility of the state, especially the government."

2.2.3 The obligatory directive principles
The specific development directive principles meant here are principles that guide government policies on present and future development. The rationale for such provisions is planning for a state to fulfill its obligation to improve public welfare. The provisions could be called directive principles of state policy, directive principles of social policy, or simply directive principles. These specific development directive provisions have two key features, according to Lael K. Weis (2017, p. 6). First, directive principles are obligatory as they place binding constitutional obligations on the state to promote particular social values. They are different from constitutional statements regarding particular values, which merely declare fundamental values. Directive principles should create state obligations concerning particular fundamental values. They contain imperatives such as "the state shall" or "the state must." Directive principles are intended to set standards of achievement for the legislature and the executive (Medawatte, 2012, p. 1).

Directive principles are ideals that the state must consider in the formulation of policies and laws to secure social, economic, and political justice to all. Directive principles specify the programs and mechanics of the state to attain the constitutional goals set out in the preamble. They are standards of achievement that guide all government organs in running their business. They are both means and ends to attain socioeconomic and political justice in the constitution (Gebeye, 2015, pp. 5–6).

The second feature of directive principles is that they are contraindicative, according to Weis (2017, p. 7). This means that courts are not permitted to define the scope and content of the values in question, to define the legal norms necessary to promote those values, and then to enforce those norms against the state (for example, by requiring the state to take an action to fulfill the norm, or by striking down legislation that is incompatible with the norm).

This contraindicative character has two aspects. First, the content of the relevant obligations and their subject matter is typically nonjusticiable or else not judicially enforceable. Second, these obligations require implementation through nonjudicial means. This is predominantly, although not exclusively, directed legislation (Weis, 2017, p. 7).

Directive principles and socioeconomic rights are closely related. Directive principles had been originally conceived instead of socioeconomic rights. Directive principles as an innovation in constitutional principles were necessitated by the pragmatic challenges of constitutional rights and the quest for progress, on one hand, and the need to protect the vulnerable sections of society from politics, on the other (Gebeye, 2015, p. 13).

It is not controversial that directive principles, though not judicially enforceable, are judicially cognizable and therefore available for courts to look to for interpretive assistance in resolving legal ambiguities. Similarly, if the conduct that counts as a violation of a nonjusticiable directive principle also affects a justiciable fundamental right, it may give rise to indirect justiciability. At least some of the rights and principles embraced by directive principles should be viewed as having the normative force of freestanding justiciable constitutional rights. When decisions regarding directive principles made by other branches fail to meet the relevant constitutional standards that are implicit within directive principles, then the judiciary, after a proper and impartial examination of the decision in light of the fundamental law of reason, should feel empowered to intervene and order a remedy (Atupare, 2014, pp. 86–87).

Indonesia's 1945 Constitution has several clauses on development that could be categorized as directive principles, such as Chapter XIII Concerning Education and Chapter XIV Concerning the National Economy and Social Welfare. Chapter XIII contains obligations for the government to undertake basic education, including the obligation to fund it (Article 31 (2)). Even more, it contains the obligation to prioritize the budget for education to a minimum of 20% of the state budget and of regional budgets to fulfill the needs of national education (Article 32 (4)).

To enforce these provisions, the Constitutional Court of Indonesia has declared laws on state budgets unconstitutional several times, because they did not fulfill the minimum 20% budget allocation. These decisions are Constitutional Court Decision No. 13/PUU-VI/2008, No. 24/PUU-V/2007, No. 026/PUU-IV/2006, No. 026/PUU-II/2005, and No. 012/PUU-III/2005.

2.3 *The expanding development directive under the 1945 Constitution*

Based on the aforementioned categorizations of formalized development plans, the 1945 Constitution contains all three forms, in the preamble, in the clause on human rights provision, and in the obligatory directive principles provisions. The question "is it sufficient to be a basis for a national development system?" can be answered immediately with "no."

The 1945 Constitution is a short constitution. Hence its texts are openly interpretive, including the directive principles for development mainly represented by Chapter XIV Concerning the National Economy and Social Welfare. The preamble only mentions the purpose of improving public welfare. The human rights provisions only artificially recognize some kinds of economic and social rights, plus imposing the obligations on the government to respect, protect, and fulfill those rights. Yes, those human rights provisions could be claimed, but they are absent in terms of how the government could primarily fulfill economic and social rights positively.

The obligatory directive principles provisions under Chapter XIII Education and Chapter XIV The National Economy and Social Welfare cannot be considered sufficiently detailed to serve as the basis of a national development plan. They can be compared to directive principles in the constitutions of other countries such as the Directive Principles of Social Policy in Ireland's 1937 Constitution, which consist of ten clauses (under one article); the Directive Principles of State Policy in India's Constitution, which consist of nineteen clauses; the Declaration of Principles and State Policies in the 1987 Constitution of the Philippines, which consist of twenty-eight clauses; or even the Social Order in the 1988 constitution of Brazil, which consists of eight chapters with thirty-eight articles and more clauses. Our constitution provides relatively limited content on the development plan; there are only two chapters in the constitution on directive principles, the chapter on education and the chapter on the national economy and social welfare with only three articles.

Even worse is that our constitution could be interpreted in multiple ways. For example, Article 33 points (2) and (3) says, "Sectors of production which are important for the country and affect the life of the people shall be under the powers of the State. (3) The land, the waters and the natural resources within shall be under the powers of the State and shall be used to the greatest benefit of the people." These provisions authorize the government to take a powerful role in determining the use of natural resources and other sectors of production that are important for the country and affect the lives of the people. Private parties are prohibited from taking such powers. These provisions have been interpreted variously by government and legislators, and they have been used to establish various laws regarding natural resources, ranging from extremely liberal laws to socialistic ones. Moreover, these provisions have also been interpreted inconsistently by the courts. In its constitutional review of Law No. 20 of 2003 Concerning Electricity, the Constitutional Court ruled that the implementation of the unbundling system in the electricity sector is a liberalization of public sectors; hence it is a violation of Article 33 of the 1945 Constitution. In its constitutional review of Law No. 22 of 2001 Concerning Oil and Gas, the Constitutional Court ruled that the implementation of the unbundling system in the oil and gas sectors is constitutional. Both sectors are important for people's livelihood. The Constitutional Court acknowledges that the interpretation of "controlled by the State" in Article 33 of the 1945 Constitution depends on the dynamics of development of those sectors. The Constitutional Court has therefore interpreted these provisions case by case (Magnar, Junaenah, and Taufik, 2010, p. 168).

The development and economic directives in the 1945 Constitution should be expanded. The 1945 Constitution should provide more comprehensive direction regarding the development plan, not only in economic and natural resources but also in other fields such as public health, labor, and marine and fisheries industries. It should provide more clear directives for guiding state policy.

3 CONCLUSION

The 1945 Constitution can be classified as an economic and a social constitution; it originated from a socialistic view. The 1945 Constitution provides obligations and foundations for the government to plan development regarding social and economic policies. Three kinds of development provisions can be identified under the 1945 Constitution: the preamble, human rights provisions and obligatory directive principles regarding education (Articles 31 and 32), and the clauses on national economic and social welfare (Articles 33 and 34).

However, these three provisions are deemed insufficient and often are too unclear to be the basis for a national development plan. These provisions ultimately are constitutional directive principles that contain guidance for the government on how it should plan development regarding social and economic policies; they need to be expanded to provide more detailed, clearer directives. That expansion could be facilitated through the amendment power of the People's Consultative Assembly (MPR).

LEGISLATION

1945 Constitution of Republic of Indonesia
1945 Constitution of Republic of Indonesia Amended
Constitution of the Russian Socialist Federative Soviet Republic
Constitution of the United States of America
Constitutional Court of Republic of Indonesia Decision No. 13/PUU-VI/2008
Constitutional Court of Republic of Indonesia Decision No. 28/PUU-XI/2013
Constitutional Court of the Republic of Indonesia Decision No. No. 026/PUU-IV/2006
Constitutional Court of the Republic of Indonesia Decision No. No. 026/PUU-II/2005
Constitutional Court of the Republic of Indonesia Decision No. No. 012/PUU-III/2005
Law of Republic of Indonesia No. 17 of 2012 Concerning Cooperation
Law of Republic of Indonesia No. 25 of 2004 Concerning National Development Planning System
MPR Decision No. 4/MPR/2014 Concerning Recommendation for the MPR Period of 2014–2019

REFERENCES

Aldrich, K. (2010). Constitutionalizing Economic, Social, and Cultural Rights in the New Millennium. Thesis, Dissertation, Professional Papers. Paper 241. ScholarWorks at the University of Montana.
Asshiddiqie, J. (2016). *Konstitusi Ekonomi.* Jakarta: Kompas Media Nusantara.
Chemerinsky, E. (2006). *Constitutional Law Principles and Policies.* New York: Aspen Publishers.

Fredman, S. (2008). *Human Rights Transformed*. Oxford: Oxford University Press.

Gebeye, B. (2015). The Potential of Directive Principles of State Policy for Judicial Enforcement of Socio-economic Rights: A Comparative Study of Ethiopia and India. Thesis. Budapest, Hungary: Central European University.

King, B. (2011). "Government and Politics." In W. Frederick and R. Worden (eds.), *Indonesia: A Country Study* (pp. 229–306). Washington, DC: Library of Congress.

Sadurski, W. (2002). *Constitutional Courts in the Process of Articulating Constitutional Rights in the Post-Communist States of Central and Eastern Europe, Part I: Social and Economic Rights*. Italy: European University Institute.

Soemantri, S. (1986). *Prosedur dan Sistem Perubahan Undang-Undang Dasar*. Bandung, Indonesia: Alumni.

Wheare, K. (1966). *Modern Constitutions*. Oxford: Oxford University Press.

Atupare, A. (2014). Reconciling Socioeconomic Rights and Directive Principles with a Fundamental Law of Reason in Ghana and Nigeria. *Harvard Human Rights Journal 27*, 71–106.

Buchanan, J. (1994). "Notes on Liberal Constitution." *Cato Journal 14*(1), 1–9.

Magnar, K., Junaenah, I., and Taufik, G. (2010). "Tafsir MK Atas Pasal 33 UUD 1945: Studi Atas Putusan MK Mengenai Judicial Review UU No. 7/2004, UU No. 22/2001, dan UU No. 20/2002." *Jurnal Konstitusi 7*(1), 111–180.

Manan, B. (2011). "Kewajiban Hakim Memahami, Memelihara dan Menerapkan UUD 1945". *Varia Peradilan*, no. 309.

Ruslina, R. (2012). "Makna Pasal 33 Undang-Undang Dasar 1945 dalam Pembangunan Hukum Ekonomi Indonesia." *Jurnal Konstitusi 9*(1), 49–82.

Tamanaha, B. (2008). "The Dark Side of the Relationship between the Rule of Law and Liberalism." *NYU Journal of Law and Liberty 3*, 516–547.

Ginsburg, T., Foti, N., and Rockmore, D. (2013). "We the Peoples: The Global Origins of Constitutional Preambles." Coase-Sandor Institute for Law and Economics Working Paper, No. 554.

International Institute for Democracy and Electoral Assistance (IDEA). (2014). *Social and Economic Rights*, 1–21.

Medawatte, D. (2012). Non-enforceability of Directive Principles of State Policy: Real Barrier or Fake. Retrieved from: www.researchgate.net/publication/233980362.

Nasution, A. (2011). "Towards Constitutional Democracy in Indonesia." *Southeast Asian Constitutionalism 1*, 1–44.

Weis, L. (2017). "Constitutional Directive Principles." Preprint Version of *Oxford Journal of Legal Studies*. Retrieved from: https://ssrn.com/abstract=3000179.

The bail-in policy: solution or problem?

Lily Evelina Sitorus, Anna Erliyana & Yunus Husein
Faculty of Law, Universitas Indonesia, Depok, Indonesia

ABSTRACT: The bail-in policy is a policy of handling the financial system crisis based on Law No. 9 of 2016 on the Prevention and Handling of the Financial System Crisis. The policy was born as a result of the failure of the bailout policy in handling the financial system crisis in 2008. This paper will try to analyze the bail-in policy from the perspective of constitutional law, whether it is by the legal system applicable in Indonesia. If the bailout policy in America is regarded as a policy of government intervention on the free economic system, whether the bail-in policy in Indonesia can be interpreted as a result of government negligence in the national economy. The analogy method will be used in this paper to find the best solution that fits the legal system applicable in Indonesia.

1 INTRODUCTION

Century Bank bailout policy issued by Stability Committee of the Financial Sector (KSSK) which has attribution authority from Government instead of Law (PERPU) No. 4 the Year 2008 about Financial System Safety Net (JPSK). The discretion occurred when the bailout decision was made, the chairman of KSSK chose to bail Century Bank, and the choice is within its authority as chairman of KSSK. The choice taken by the KSSK chairman is discretion as understood by Dworkin because the discretion is within the scope of the legal principles so that the accountability of choice is the authority of the judiciary under the Administrative Court Law. In the decision, it is also explained that the subsequent handling of the Bank Century bailout policy lies within the authority of Deposit Insurance Corporation (LPS).

Therefore, it can be understood that in the doctrine of State Administration Law, KSSK has granted delegation authority to LPS to follow up the handling of Century Bank. The delegation's authority then makes accountability of Bank Century passed to LPS. In this context, Century Bank's bailout policy is very important to study because it has a point of interest with some legal regimes in Indonesia. The Law of State Administration had the greatest role when Century Bank's bailout policy was first issued. Any party who feels aggrieved over the exit of the policy should immediately demand the cancellation of the policy to the Administrative Court. If it is done, then the problem of abuse of authority has been proven by the Administrative Court.

However, unfortunately, the bailout policy is considered a controversial policy because it raises the risk of state losses. Therefore, the government with Law No 9 of 2016 intends to correct the policy with the new policy of bail-in. This paper will try to see whether the bail-in policy is a better policy than a bailout or just simply adds to the problem.

2 DISCUSSION

2.1 *The bailout policy*

In answering the position or status of Century Bank's bailout policy, the position of the Indonesian state as state law or rechsstaat must be understood. The concept of a legal state or 'rechtsstaat', is used in Indonesia as a counterpart to the concept of 'the rule of law' in the West. The true meaning of 'rechtsstaat' can be found in the Explanation of the 1945 Constitution. However, the Explanation does not explain the meaning of 'rechtsstaat' but merely states that Indonesia is a state based on the law (*rechtsstaat*). The Indonesian state based on law (*rechtsstaat*) is not based on mere power (*machtsstaat*). In the amendment of the 1945 Constitution 1999-2002, the explanation was deleted, but it was not explained further about the concept of the legal state in question (Hosen, 2010).

The establishment of policy rules in the practice of governance is a common thing. According to Philips M. Hadjon ", the daily administration of government shows how the body or administrative officials often take certain policy steps, among others, creating what is now often called the rule of *belief* (policy belief). Policy regulation is a formal form of policy established by state administrators based on the discretionary principle (Hadjon, 1999).

If previously the task/purpose of the government is only to make and keep the law or in other words: just keep order and tranquillity alone then now the goal/task of government not only implement the law or realize the will of the state but also organize the public interest. With the increase of government activities that almost enters the entire field of community life, according to Hadjon, the stance that the actions of the authorities can not be judged by the court including the state administrative court is no longer relevant (Hadjon, 1987).

Century Bank's bailout policy is a multidimensional policy. From a policy perspective, the legal dimension of state administration can assess whether this policy has legitimacy as a legal product or not. In the context of PERPU No 4 the Year 2008 on JPSK, then the dimension of constitutional law is more appropriate to assess. Further development of alleged corruption in this policy also gives the dimension of criminal law related to the element of state loss. Also, the economic law dimension is also strongly related to this policy due to the onset of the global economic crisis.

The clash of legal dimensions embedded in Century Bank's bailout policy is seen in the controversy that occurred after the policy was issued. The finance ministry represents the executive, the House of Representatives representing the legislature and judges representing the judiciary. Related to that context, then the legal interpretation of the policy becomes an important thing to do. That is because, each executive, legislative and judicial institution also has its method of interpretation. The President interpreted the world economic crisis as a compelling crunch, thus deciding PERPU No. 4 of 2008 on JPSK. The House of Representatives rejected the PERPU so that it could be interpreted the absence of crunch force in the event of a global economic crisis. Judicial interpretation is another matter.

Century Bank bailout policy if tested based on legality, then the first to be seen is the legal basis. The policy is derived from the decision of KSSK No. 04/KSSK.03/2008. To analyze the decision of KSSK, it must also see the legal basis for the issuance of the decision, namely Article 18 paragraph (1) PERPU No 4/2008 concerning Financial System Safety Net (JPSK) which reads, "in the case of banks declared as Failed Banks identified by Systemic Impact by Bank Indonesia, KSSK decided Bank Failed to Impact Systemic or Not Systemic Impact ".

The decision of KSSK is the authority of the Minister of Finance as the chairman of KSSK, as stipulated in Article 6 PERPU No 4 the Year 2008 on JPSK which reads as follows: "KSSK serves to establish policies in the context of crisis prevention and handling". Therefore, if viewed from the principle of legality, then the decisions issued by KSSK can be judged by applicable regulations. In measuring whether abuse has occurred, it should be evident that officials have used their authority for other purposes. The occurrence of abuse of authority is not due to an omission. The abuse of authority is made consciously by diverting the purpose that has been given to his authority. Transfer of destination is based on personal interests, either for the benefit of himself or others.

2.2 The bail-in policy

The controversy of Century Bank's bailout policy was also adopted in the Law No nine the Year 2016 so that the handling of the financial system crisis is no longer using the bailout policy, but with the concept of bail-in. According to the Minister of Finance, the concept of bail-in is the handling of problems using its bank resources and banking contributions. This is because strong coordination among agencies in the prevention and handling of financial system crises is necessary for the financial system to function effectively and efficiently. Here are some important points regarding the financial system crisis settlement protocol based on the Law:

a. Strengthening the roles and functions and coordination between the four institutions that joined the KSSK. This Committee is regulated in Chapter II, namely Articles 4 to 15. KSSK consists of the Ministry of Finance, Bank Indonesia, the Financial Services Authority, and the Deposit Insurance Agency. The existence of KSSK to prevent and handle the financial system crisis. The four institutions are each by their duties to implement the maximum effort to prevent the financial system crisis;
b. Encourage crisis prevention through strengthening the banking supervision function, especially banks that are designated as systemic banks. The application of a list of systemic banks should be done from the beginning when financial stability is in normal condition;
c. Handling bank problems by prioritizing the concept of bail-in, namely handling liquidity and solvency problems of banks using the resources of the relevant banks coming from shareholders and creditors of banks;
d. The method of handling liquidity and solvency issues of banks is regulated in a comprehensive and comprehensive manner, which is done earlier, among others, through the implementation of the bank's sanitation action plan (recovery plan) which has been prepared by the bank;
e. The President, as head of state and head of government, holds full control in handling the financial system crisis. The President acts as the final determinant to decide the condition of financial system stability under normal conditions or financial system crises after receiving a recommendation from KSSK. The President may also decide or end the bank restructuring program in case of banking problems that endanger the national economy.

The concept of bail-in was introduced by Credit Suisse, which says that the best way of handling systemic troubled banks is to force creditors, not taxpayers, to bear bank losses. At the banker's dinner in late January 2011, the Governor of BI issued a bail-in concept to replace the bail-out in addressing national banking problems. According to the Governor of BI, bail-out raises new turmoil, both from the economic side, political complications and legal issues. Therefore, the bail-out paradigm needs to be replaced by bail-in, i.e. the banks themselves must have a buffer to absorb risks and shocks in the event of a crisis (Sitompul, 2011).

If bail-in is to be applied in Indonesia, under applicable law, one of the approaches that can be used is the bankruptcy approach. Under the Bankruptcy Law, the debtor who is a bank may only be petitioned for bankruptcy by BI. Large bank creditors are depositors or depositors and banks that provide facilities to the bank through the interbank money market. During this time, BI has never used the bankruptcy approach in solving problem banks. The reason is, firstly, the bankruptcy procedure through the court extends the time when the settlement of problem banks needs a short time, especially for the payment of depositors. Second, revocation of the bank's business license quickly can help keep the bank's asset value for the benefit of the creditor and at the same time maintain the credibility of the regulator so that in turn it reduces the risk of systemic risk. Third, the need for immediate payments to customers is difficult through bankruptcy proceedings in courts. It can not be denied; customer payments are quickly important to prevent rush (Sitompul, 2011).

According to Sitompul (2011), the bail-in concept needs to be accompanied by three pillars: supervision, internal governance and market discipline. Supervision by the central bank must be complemented by internal discipline of the banking and external disciplines (markets).

Without such discipline, supervision will not be able to keep pace with the speed of liberalization, globalization and technological advances in financial instruments. By involving internal governance, it means that banking itself should be the best place to manage and maintain sound management practices. The presence of market discipline is necessary because, without a competitive and punitive market for failure to compete in the market, there is not enough incentive for bank owners, managers and customers to make sound financial decisions. Market discipline requires a conducive climate of openness. Therefore, it is necessary to review the provisions concerning openness applicable to banks.

2.3 *Comparative cases*

The decision to conduct a systemic rescue often sparked a debate between pro and contra parties. Policymakers are always faced with difficult choices, often even facing a dilemma to take action or not. In the case of the crisis in Europe in 2010 in Spain, the Spanish Central Bank took over a small local bank called Bank CajaSur. This bank does not include banks that are categorized as having systemic impacts when viewed from the linkage and scale, so it should not cause too big to fail problems. According to Chatib Basri, the reason is simple in a panic situation, then the closure of one bank can be systemic. Furthermore, Indonesia can learn from the mistake of recommendation made by International Monetary Fund (IMF) in 1997 on closing 16 banks that resulted in bank rush, and finally destroy the country's banking system (Prasetyantoko, 2010).

According to Eko Prasojo, Bank Century's bailout policy is the government's discretion under the Minister of Finance in overcoming the current economic turmoil. Furthermore, the legal basis for such discretion should be reviewed in detail to the Minister of Finance and the BI Governor, particularly regarding the existence of KSSK by PERPU No. 4 the Year 2008 on JPSK. Therefore, Eko suggested that the Bank Century issue is not only reviewed through the perspective of criminal law and state administration but rather focus on state administrative law because, in the perspective of state administration, every government decision must be accountable legally and professionally (Soehino, 2000).

According to Todd Zywicki, a bailout that occurred due to the monetary crisis in America in 2008 also problematics. One of them is excessive executive power that threatens the rule of law principle. Also, there is a shift in the concept of a pure market mechanism as state capitalism because of the government's interference with the market. The most important example is the bailout conducted on General Motors and Chrysler, which became the success story of the Obama administration is nothing but a fairy tale to Zywicki (2011).

Bailout policy does not need to be done by the US government. Ford, in comparison, has made various improvements, including renegotiating problem contracts and getting rid of parts that are considered unprofitable companies. As a result, when the crisis occurs, Ford does not need to be bailed out because it has first solved the problem. Unlike General Motors and Chrysler, both are burdened with pensions and healthcare funds so they can not afford to defend themselves in times of crisis. Under such conditions, according to Zywicki (2011), the government should let the two companies go bankrupt. However, what happens is that both companies are applying for a bailout to the government because the burden will be borne by the government if they let their company be liquidated. Both companies are large automotive companies in America that have millions of employees guaranteed by government pension funds so that if there is a massive layoff will cause greater losses because directly their pension funds will be borne by the taxpayers.

The legal issues arising from this auto bailout policy also widened to the issue of clear legal interpretation of the TARP language to help financial institutions with the intent of lawmakers. According to Zywicki (2011) it can be understood if Congress becomes ambiguous on this issue, because on the one hand if they do not allow TARP funds to be used then they are considered unwilling to help the automotive industry while on the other hand, people who disagree bailout become angry because the government uses their funds.

Zywicki (2011) concluded that the real bailout policy did not need to happen, and as a result, the Bush and Obama administration had made a bad precedent with its involvement. Bush's role, in this case, led to the executive's power being in the spotlight, for the reason that a crisis without legality could create a problem policy. However, Zywicki sees Obama's role is more problematic because the policies made in normal situations without legality are considered to be able to change the long-standing market financial system. The free market mechanism becomes bound due to excessive government intervention. Businesses who do not want to compete prefer to create connections with certain politicians so that the policies made can be tailored to their interests. This is bad for small companies and can not compete healthily.

2.4 *Lesson learned*

In her statement in the House of Representatives, former Chairman of KSSK, Sri Mulyani, reiterated her opinion that Bank Century's bailout policy is focusing on the whole banking system when under severe pressure, so out of context for the problem in Bank Century itself, the situation must be maintained to prevent a crisis. Therefore, the Bank Century case is only one cause, but the fundamental to the pressure of the banking crisis is a very real indication. According to Mulyani, with the right decision, then the crisis can be prevented so that all the predictions of economists about the crisis of the second volume did not happen (Mulyani, 2010).

The then BI governor, Boediono, when asked for information about the Bank Century bailout policy, also stated the same thing. In a crisis, the psychology of the community is so exclusive that even the smallest bank can trigger a wider crisis like what happened in 1997 when demand for 16 banks closed. At that time, small banks that closed only 2% less than the total deposit, but all collapsed because of the exclusive situation. Calculation of capital in a crisis process is so dynamic that it is impossible to determine very accurately on the spot. Therefore, the data available at that time is already the best data (Boediono, 2009).

In other cases, Richard M. Salsman still questioning bank bailouts as the official U.S. policy. For all the criticisms of the U.S. government's handling of bank insolvencies in 2008-2009, whether by the left (which correctly opposed the bailouts) or right (which correctly opposed the TARP's partial nationalizations) – the consistent position of defenders of genuine, laissez-faire banking was opposition to any politicized handling of bank failures in the first place, and any subsidies. Salsman (2013) argued more than two decades ago, for a banking system without too big to fail (TBTF), but that means its need a banking system simultaneously devoid of unfair subsidies and free of overly-harsh regulations. The latter follows inexorably from the former. Bank insolvencies would become rarer in a freer system, but they'd be handled in an objective, judicious, and legal context, with the scrutiny of creditors' rights and the hierarchy of creditors' claims. There'd be no question of TBTF, political favouritism, mistreatment of bondholders, or moral hazard.

The exemption of financial institutions from the U.S. bankruptcy code originated in the passage of the FDIC act in 1934 when the U.S. first became an indirect creditor to the banks; the general exemption was strengthened and extended to bank holding companies by comprehensive laws passed in 1956 and 1970. According to Salsman (2013), there are two basic alternatives to bankruptcy for insolvent banks, each morally and practically inferior to bankruptcy: bail-ins and bailouts. In a "bail-in" a government compels depositors and bondholders to pay a tax and become equity holders in the failed institution to which they've lent their funds. In contrast, a government "bail-out" takes income or wealth from taxpayers in general (or from currency holders, by the hidden tax of inflation due to fiat-money creation) – most of whom have no relationship to the failed bank – to assist the failed bank's depositors, bondholders, shareholders, and executives.

Each method entails compulsion, which is improper, and neither ensures safe or sound banking in the first place. Bail-Ins should be more preferred, from the standpoint of justice and efficiency alike, versus bail-outs, because bail-ins involve those who've chosen to involve themselves with a weak or failing institution. Bail-outs, on the other hand, corral and punish innocent bystanders who are un-involved in insolvent banks. Advocates of bail-outs claim to worry about "contagion effects," yet they push a failure resolution method which expands the collateral damage resulting from financial fallout. In contrast, advocates of bail-ins are justified in arguing that contagion is best contained when resolution costs are restricted to those parties most related to (or responsible for) a particular financial insolvency (Salsman, 2013).

3 CONCLUSION

The latest development of the presence of Law No. 9 of 2016 on PPKSK can also be the basis for consideration for the future crisis handling policy. This is because the Act is a follow up of PERPU No. 4 of 2008 on JPSK. The comparison that has been made to the two regulations shows the basic difference which is the basis of the refusal of PERPU No. 4 of 2008 on JPSK which is then adapted by Law No. 9 of 2016 on PPKSK. Changes to these provisions would have an impact on the receipt of Law No. 9 of 2016 on PPKSK as a legal protocol of financial system crisis. The adaptation also proves that the Law of State Administration is recognized as an integral part of law enforcement in Indonesia. The provision also proves that the resolution of the financial system crisis must first see whether there is an abuse of authority in decision-making before other legal regimes are involved.

Model of financial crisis system settlement based on Law No. 9 of 2016 is divided into 2, prevention and handling. In the crisis prevention section of the financial system is subdivided into 3, handling systemic liquidity issues of the systemic bank, systemic bank solvency and bank problems other than a systemic bank. It is related to the bail-in model, which is the solving of system solvency problem of the systemic bank. It can be found in the Elucidation of Article 21 Paragraph (1) as follows: "Solvency problem handling includes, among other things, the conversion of systemic bank obligations into capital (bail-in)".

The controversy of Century Bank's bailout policy should be completed when Law No. 11 of 2015 on the Revocation of PERPU No. 4 of 2008 on JPSK, is ratified. In Article 2 of Law, No 11 the Year 2015 explained that the decision that has been determined by KSSK to carry out its duties and authority based on PERPU No. 4 of 2008 remains legitimate and binding. The bail-in model is certainly adopted as a result of the controversy of state financial losses related to Century Bank bailout policy. The model is assumed to reduce the financial loss of the state.

REFERENCES

House of Representatives, "Minutes of the House of Representatives Meeting of the Committee on the Rights of Parliament Questioning the Century Bank Case, Meeting of Examination with Prof. Dr Boediono ". Jakarta: House of Representatives, 2009.

House of Representatives, "The Minutes of the Meeting of the Committee on the Rights of Angket of the House of Representatives of the Republic of Indonesia concerning the Investigation of the Century Bank Case, the Examination Meeting with Sri Mulyani. Jakarta: House of Representatives, 2010.

Hosen, Nadirsyah. "Emergency powers and the rule of law in Indonesia" in Emergency Powers In Asia Exploring the Limits of Legality, eds. Victor V. Ramraj and Arun K. Thiruvengadam. Cambridge: Cambridge University Press, 2010.

Hadjon, Philips M. Legal Protection for the People in Indonesia, a Study of Its Principles, Handling by Courts in the General Courts and Establishment of State Administration Courts. Surabaya: PT Bina Ilmu, 1987.

Hadjon, Philips M. Introduction to Indonesian Administrative Law. Yogyakarta: Gajah Mada University Press, 1999.

Indonesia, Government Regulation instead of Law of Financial System Safety Net, PERPU No. 4 the Year 2008.

Indonesia, the Financial Crisis Prevention and Treatment Law, Law No. 9 of 2016.

Prasetyantoko, A. Ponzi Economy: The Prospect of Indonesia in the Middle of Global Instability. Jakarta: Kompas, 2010.

Salsman, Richard M. Bankruptcies, Bail-Outs & Bail-Ins: The Good, Bad & Ugly Of Bank Failure Resolution. https://www.forbes.com/sites/richardsalsman/2013/05/01/bankruptcy-bail-ins-bail-outs-the-good-bad-ugly-of-bank-failure-resolution/#7223115341cc, accessed Oktober, 19, 2017.

Soehino, Principles of State Administration Law. Yogyakarta: Liberty, 2000.

Zulkarnain, Sitompul. Bail-in: Increasing Liability of Bank Owners and Creditor. https://zulsitompul.wordpress.com/2011/02/23/bail-in/, accessed Oktober, 19, 2017.

Zywicki, Todd. "The Auto Bailout and The Rule of law", National Affairs Journal, Issue Number 7 (Spring, 2011).

Advancing Rule of Law in a Global Context – Susetyo, Rinwigati Waagstein & Budi Cahyono (eds)
© 2020 Taylor & Francis Group, London, ISBN 978-1-138-32782-5

Opportunities and challenges for Indonesian insurance companies in the ASEAN Economic Community

Kornelius Simanjuntak

Insurance Law, Faculty of Law, Universitas Indonesia, Indonesia

ABSTRACT: The insurance business has been rapidly growing in ASEAN Member Countries due to the increase of the economic middle class and the development of both manufacturing and infrastructure sectors in all ASEAN countries. The insurance business is one of the financial business sectors which has entered the free market in the ASEAN Economic Community (AEC) Blueprint of which as a consequence all insurance companies have free access to enter and sell their insurance products/policies in all AEC Member Countries. Indonesia's insurance market is the most attractive for foreigners, including AEC insurance companies, as Indonesia has the largest population and economy in the AEC. This paper analyses what Indonesian insurance companies should do to tap the opportunities and challenges in the insurance sector created by the AEC. What regulatory support can the Indonesia Financial Authority (OJK) provide to protect Indonesian insurance companies from their competitors from other AEC Member Countries, and at the same time to compete with and penetrate other AEC Member Countries' insurance markets. This paper concludes that Indonesian insurance companies must strengthen their capitalization and improve their human resources, expertise, and services. The OJK should provide regulatory support without breaching the AEC Blueprints and commitments.

Keywords: ASEAN Economic Community (AEC), Challenges, Insurance, Integration, Liberalization, Opportunities

1 INTRODUCTION

The Association of South East Asian Nations (ASEAN) was established on 8 August 1967 as a regional political and economic cooperation. The current Member/States Countries of ASEAN are Brunei Darussalam, Cambodia, Indonesia, Lao PDR, Malaysia, Myanmar, Philippines, Singapore, Thailand, and Viet Nam (ASEAN, 2008, p. 2).

On 20 November 2007 in Singapore, on the occasion of the fortieth anniversary of ASEAN and the thirteenth ASEAN Summit, ASEAN Member Countries adopted the ASEAN Economic Blueprint to transform ASEAN into a single market and production base by 2020, in what is a highly economically competitive region, a region of equitable economic development, and a region fully integrated into the global economy. As a single market and production base, the AEC has five core elements: (1) free flow of goods; (ii) free flow of services; (iii) free flow of investment; (iv) free flow of capital; and (v) free flow of skilled labour. However, before 2020, AEC Member Countries agreed that the insurance sector, one of the financial services, would be integrated at the end of 2015 (ASEAN, 2008, p. 6).

The people of AEC Member Countries have less insurance awareness than do the people of developed countries such as the nations in the European Union and the United States of America. The level of insurance penetration is not the same among the AEC Member Countries. Singapore, Malaysia, and Thailand have higher insurance penetration than other AEC Member Countries. However, insurance will further develop in the years to come in AEC Member Countries due to the following factors: (1) the large population; (2) the more developed a country, the larger the insurance penetration; (3) the potential is significant as the current insurance penetration is still low; (4) the higher the level of education of the people, the more insurance minded they are; (5) mandatory insurances by states for providing protection to their people are still low; and (6) the physical and infrastructure developments in various sectors are still going on.

Indonesia is the largest AEC Member Country in terms of population and territory; it also has the largest insurance business potential, and the brightest prospects for the future as Indonesia has the largest territory both in land and on the seas because it is one of the largest archipelago countries in the world. Indonesia has more than 13,000 islands based on the latest survey by the government of Indonesia – the largest in population among AEC Member Countries, the richest in natural resources, and the most exposed to natural disasters such as tsunamis, earthquakes, volcanic eruptions, and floods. All of these factors mean that Indonesia has the largest insurance potential and the best prospects among AEC Member Countries.

The current insurance penetration is estimated at 20% of the potential market. Insurance's contribution to the Indonesian GDP is still low, about 2%, however, the growth of insurance in terms of premiums is one of the highest in insurance markets worldwide.

The integration of financial services into the AEC free markets creates opportunities and challenges for Indonesian insurance companies as well as for all insurance companies of the AEC. One of the advantages of the integrated insurance market for Indonesian insurance companies is that they can freely enter, perform, and compete for the insurance business in other AEC Member Countries outside Indonesia. On the other hand, all insurance companies of other AEC Member Countries can also do the same in the Indonesian insurance market. The huge insurance potential and prospects in Indonesia make Indonesian insurance markets very attractive to enter for insurance companies of other AEC Member Countries.

1.1 *Research questions*

Given these concerns, the purpose of this paper is to analyse two related research questions. First, what should Indonesian insurance companies do to tap the opportunities and respond to the challenges in the free market insurance sector created by the AEC? Second, what regulatory support can the Indonesia Financial Authority (OJK) provide to bolster and protect Indonesian insurance companies from their AEC competitors, and at the same time to allow them to compete with and penetrate other AEC Member Countries' insurance markets?

This paper is organized as follows: Section 2 presents research methodology; Section 3 examines insurance in the global ASEAN Economic Community (AEC) and in the Indonesian market; Section 4 studies the integration of the ASEAN insurance industry and the role of ASEAN insurance regulatory agencies; Section 5 presents opportunities and challenges for Indonesian insurance companies in the ASEAN Economic Community; and Section 6 presents the conclusion and recommendations.

2 METHODOLOGY

This methodology employed in this study utilizes literature and qualitative legal research. The research questions are addressed by analysing secondary data of the Indonesian insurance industry, AEC Member Countries, and the advanced markets. The data include the total insurance premiums, insurance acts, and regulations related to the research questions.

3 DISCUSSION

3.1 *Insurance in the global, ASEAN Economic Community (AEC), and in the Indonesia market*

3.1.1 *Global insurance premium life and non-life*

The growth of the insurance premium is always affected by the growth of the economy. In 2016 the growth of the global economy (GDP) was 2.5%. Total global direct insurance premiums grew by 3.1%, equal to USD 4.732 billion, down from 4.3% in 2015. The global non-life or general insurance premiums increased by 3.7%, with total premiums of USD 2.115 billion in 2016; the average growth rate of the past ten years was 2% each year. The non-life premium growth of 2016 was less than last year's growth, which was 4.2%. The growth of the non-life insurance premiums for the past ten years was 2% on average (Sigma No. 3/2017, pp. 1–2).

The emerging market of non-life insurance premiums grew at 9.6% in 2016; this growth was above the ten-year average growth of 8.3%. Among the emerging insurance markets, China's non-life insurance premiums had the biggest growth, which was 20% in 2016. Other Asian emerging insurance markets grew by 7.7% (Sigma No. 3/2017, pp. 7–9).

Global life insurance premiums in 2016 grew by 2.5%, which amounted to USD 2.617 billion, which was smaller than the previous year, which was 4.4%. The ten-year average growth was 1.1%. In the advanced insurance markets, life insurance premiums declined by 0.5% in 2016, with a total amount of USD 2.110 billion (Sigma No. 3/2017, p. 8).

3.1.2 *Life and non-life insurance premiums in emerging markets and AEC member countries*

In 2013 the penetration or contribution of insurance premiums (life and non-life) to the GDP of the six largest AEC Member Countries (Indonesia, Malaysia, Philippines, Singapore, Thailand, and Vietnam) as part of the emerging market was 3.2%; this was above the emerging insurance markets' average of 2.7% in the same year. The insurance penetration rates of AEC Member Countries were: Singapore 5.9%, Thailand 5.5%, Malaysia 4.8%, Indonesia 2.1%, Philippines 1.9%, Vietnam 1.4%, and Myanmar 0.05% (A.M.Best, 2014, p. 8).

In 2015 the total insurance premiums (life and non-life) of the AEC (509 insurance companies) were USD 87.9 billion. The growth of the insurance premiums was 8.1% from the previous year, 2014. The Singapore insurance market contributed USD 24.2 billion to the premiums, followed by Thailand with USD 19.1 billion, and Indonesia with USD 12.9 billion (OJK, 2016, p. 1).

The proportion of the total AEC insurance premiums of USD 87.9 billion by types of insurance was dominated by non-life insurance at 63%, followed by life insurance at 22%, professional reinsurance at 9%, composite insurance at 5%, and state-owned insurance at 1% (OJK, 2016, p. 2).

In 2015 total AEC insurance assets were USD 388.1 billion, of which the life insurance sector contributed 83%, equivalent to USD 322 billion, while the non-life insurance sector contributed 17%, equivalent to USD 66.1 billion. The Singapore insurance industry had the biggest assets among ASEAN Member Countries, which amounted to USD 148.84 billion, followed by Thailand with USD 83.95 billion, Malaysia with USD 55.70 billion, and Indonesia with USD 45.42 billion (OJK, 2016, p. 1).

3.2 *Integration of the ASEAN insurance industry and the role of ASEAN insurance regulatory agencies*

The ASEAN Blueprint and the ASEAN Vision for 2025 emphasize three strategic goals in financial sector integration: (1) financial integration, (2) financial inclusion, and (3) financial stability.

In the original plan, the integration of insurance in the AEC was scheduled to begin in the year 2020 as part of the financial services industry, including the banking industry. However, it was mutually agreed that the integration of the insurance sector would start earlier. Therefore the integration of the insurance sector began at the end of 2015. The insurance integration liberalized the insurance markets in all AEC Member Countries, which allows insurance companies, including reinsurance companies of an AEC Member Country, to enter, do, and penetrate the insurance business in all AEC Member Countries' insurance markets.

To make the integration and liberalization of the insurance sector successful, the role of the insurance regulatory agencies (regulators) of the ASEAN Insurance Council (AIC) is highly important. The insurance regulators of the AIC have a regular yearly meeting to discuss various related subjects, including the integration of the insurance sector. On 20 November 2016, the ASEAN Insurance Regulators' Meeting (AIRM) was held in Yogyakarta, Indonesia. During the AIRM, there was also a meeting of the Working Committee of Financial Service Liberalization (WC-FSL) of the AEC financial services sectors. The AIRM also established the ASEAN Insurance Forum (AIFo) for accelerating the integration of the ASEAN insurance sector. The AIFo also held meetings during the AIRM meeting in Yogyakarta (OJK, 2016, p. 3).

The success of AEC insurance integration and liberalization highly depends on the positive response and support of all AEC Member Countries, and the insurance regulators and/or ministers in charge of insurance industries' supervision play important roles. The success of the insurance integration and liberalization will be rated from the achievement of one of the ultimate objectives of the AEC in the financial services industry and the insurance sector in particular – to create the AEC as a competitive market and to provide better security and services in insurance products to aid in achieving prosperity for the AEC's people.

Several issues urgently need the attention of AEC insurance regulators in supporting the insurance integration and liberalization in the AEC; all of them are the tasks of the insurance regulators, but some issues require cooperation with insurance industries:

(1) Ensuring the implementation of all commitments of the AEC Member Countries for the integration and liberalization in the insurance industry;
(2) Ensuring all AEC Member Countries will benefit from the integration and liberalization irrespective of the level of advancement of their economies;
(3) Increasing substantially insurance awareness, literacy, and utilization among the AEC's people;
(4) Raising the level of capitalization and the solvency of insurance and reinsurance companies;
(5) Upgrading the types and scope of insurance policies provided by the insurance companies in the AEC so as to avoid gaps in quality;
(6) Mandating third-party liability for motor vehicle insurance across the AEC Member Countries' borders, like the blue card system in the European Union;
(7) Synchronizing insurance acts and regulations across AEC Member Countries to achieve best practices, a solvency ratio for insurance companies, and a sufficient level of capitalization to ensure the financial strength of the insurance companies;
(8) Regulating professional skills, expertise, competence, and qualifications in the insurance industry;
(9) Providing alternative dispute resolution (ADR) for insurance claims disputes and policyholder/customer protection (Act No. 21 of 2011, Article 4.c; Act No. 40 of 2014, Articles 53 and 54).

This list shows that the insurance regulators and/or ministers in charge of insurance supervision have many tasks ahead in support of the implementation of the insurance integration and liberalization in the AEC.

3.3 Opportunities and challenges for indonesian insurance companies in the ASEAN economic community

3.3.1 Opportunities
Among AEC Member Countries, Indonesia has the largest population and economy and is one of the largest emerging markets in the world. In 2016 the population of Indonesia was

258.7 million; this represents 40.6% of the total AEC population of 635 million (ASEAN, 2017, p. 8).

Indonesia has the largest economy among AEC Member Countries, with a GDP of USD 931.2 million in 2016; this represents 36% of the total AEC GDP of USD 2.553 billion in 2016. Indonesia is one of the largest economies of the emerging markets in the world. The growth of the Indonesian economy/GDP is estimated at 5% yearly from 2016 to 2020. Therefore the Indonesian GDP by 2020 is estimated to reach above USD 1 trillion. It is projected that by 2030 the GDP of Indonesia will be more than USD 3.7 trillion; this will position Indonesia well in global economic forums such as the United Nations and the G-20 (ASEAN, 2017, p. 8).

In addition to these assets, Indonesia also has, among AEC Member Countries, the largest territory both in land and on the seas, and abundant natural resources such as mines, oil, and gas.

The promising growth of the Indonesian economy is supported by the increasing middle class, the high consumer demands of a big population, infrastructure investments and construction projects, manufacturing investments and projects, and development in other fields such as agriculture, plantations, fishery, trading, and maritime industries.

The Indonesian insurance market is one of the fastest growing in the world, and it is projected to be the second highest in non-life insurance premiums in the world with an estimated average annual growth rate of 9.5% between 2014 and 2020. The life insurance market of Indonesia has even bigger average annual growth in premiums: 14.9% for the same period (A.M. Best, 2014, p. 5). All of these factors make the prospects for insurance huge in Indonesia, and they attract investors and insurance companies from overseas.

The huge insurance potential in Indonesia and the AEC markets creates big opportunities for Indonesian insurance companies. Considering Indonesia's large share of the AEC population (40%) and economy (36%) as described earlier, Indonesia has the best insurance potential among AEC countries. The domestic insurance market is vast for all Indonesian insurance companies. Logically, Indonesian insurance companies are in a better position to tap the Indonesian insurance prospects, as they are the local players and have been servicing the markets for many years for some insurance companies and at least a few years for newly established insurance companies. They should know better their market. The Indonesian insurance market potential is already big for them.

In Indonesia in 2017, there were seventy-six non-life insurance companies and fifty-five life insurance companies, five reinsurance companies, and three insurance companies providing mandatory insurance for workers: (1) BPJS Ketenagakerjaan for workmen's compensation, (2) BPJS Kesehatan for medical insurance, and (3) Jasa Raharja for accident insurance covering public transportation (AAUI, 2017, p. 4). The total Indonesian insurance premiums for both non-life and life insurance increased by 19.5% to IDR 247.29 trillion in 2015. The penetration ratio of insurance in Indonesia by comparison to its (gross) premiums with the GDP increased to 2.36% in 2015 from 2.35% in 2014. The total assets of the Indonesian insurance market (life and non-life) in 2015 were IDR 853.42 trillion, of which the life insurance sector contributed 44.29%, equal to IDR 378.03 trillion; social life insurance IDR 226.92 trillion, and general/non-life insurance IDR 124.01 trillion (OJK, 2016, p. 2).

In the non-life/general insurance market, in 2016, there were 162 insurance brokers, 31 reinsurance brokers, and 26 loss adjusters (AAUI, 2017, p. 4). The brokers handle the intermediary (Baker and Logue, 2013, p. 77) while the loss adjusters do the claims survey, investigation of claim causes, and adjustment/calculation of the amount of claims submitted by the insured or policyholder (Jess, 2001, p. 471).

There are twenty-one non-life/general insurance companies with foreign ownership in the form of joint ventures; however, there is no foreign ownership of the reinsurance companies as all five reinsurance companies are owned by domestic investors. Two insurance companies and two reinsurance companies are owned by the state/government of Indonesia.

Figure 1 shows the structure and the ownership of non-life insurance companies and reinsurance companies as of June 2017 (AAUI, 2017, p. 4).

Description	2014	2015	2016	SMT1-2017
General Insurance Companies	79	76	76	76
Reinsurance Company	5	6	5	5
Insurance Broker	157	162	169	169*
Reinsurance Broker	31	31	40	40*
Loss Adjuster	26	26	28	27*

* Data as of Q1 2017

Ownership of Insurance and Reinsurance Companies as of SMT1-2017

Description	State Owned	Private	Joint Venture
General Insurance Companies	2	53	21
Reinsurance Company	2	3	-
Total	4	56	21

Figure 1. Non-life (General) insurance market structure and ownership.
Source: General Insurance Association of Indonesia

To succeed in tapping the huge insurance potential, Indonesian insurance companies (including reinsurers) have to address and respond to the following challenges.

3.3.2 Challenges and regulatory support from the OJK
Insurance companies in Indonesia face several challenges to tapping the huge potential of the AEC insurance markets as follows:

3.3.2.1 Low insurance awareness and literacy
Despite the huge prospects for insurance in Indonesia, insurance penetration and literacy are still low. The contribution of the insurance sector to the Indonesian GDP is still small – less than 3% (OJK, 2016, p. 2). It was estimated that insurance has been tapped currently at below 20% of its potential. To increase insurance awareness, insurance literacy, and the utility of insurance, insurance companies need to do more in a joint effort with the OJK. The introduction of insurance to students is a good way of making the young generation know insurance and its benefits in their lives today and in the future when they are working. The development of insurance awareness among the Indonesian people depends much on the serious efforts of the insurance industry players and the OJK in dissemination, introduction, and promotion of insurance (Insurance Act No. 40 of 2014).

The OJK can support these efforts by providing and enhancing programs of insurance literacy and their implementation by a collaboration between the OJK and the insurance industry. The OJK can approach the Ministry of Education about making the subject of insurance part of the subjects of economy and/or geography in the lower, higher, and senior schools.

3.3.2.2 Improvement in human resources expertise
Indonesian insurance companies need urgently to invest more in increasing the quality of their human resources. They need human capital with good and excellent expertise, skills, competence, and professional qualifications of high or international standards in various fields of knowledge and science. The insurance companies also need these resources to compete against the insurance companies from other AEC Member Countries entering the Indonesian insurance market. Without quality human capital, the local or domestic insurance companies will not be on the same playing field as other AEC Member Countries. They also need to ensure that employment in insurance companies is attractive for young people, and those just graduated from the universities. At present, working in insurance is not yet a high preference, and it

still ranks below the banking industry. Employment in insurance can be made more appealing by increasing the fringe benefits in insurance companies in Indonesia, and also by providing scholarship or financial support for those employees who want to take professional training and or courses in the insurance field.

The OJK can provide support in addressing these challenges by issuing related regulation and closely monitoring current regulation on this matter, and also by pushing the insurance industry to establish insurance institutes in Indonesia which will provide continuous courses in collaboration with international insurance institutes.

3.3.2.3 Lack of capital and capacity

The insurance business is a highly capital-intensive financial services industry. The current level of capitalization of Indonesian insurance and reinsurance companies is not in a high position in comparison to the level of capital of insurance companies in AEC Member Countries, as shown in Figure 2.

Without having good financial strength, Indonesian insurance companies will not be in a strong position to compete with insurance companies from other AEC Member Countries entering the Indonesian insurance market. To enable the Indonesian insurance companies to provide big coverage and to insure objects with a huge value, they need adequate capitalization. Without big capital and good financial strength, the Indonesian insurance companies will not be able to provide the capacity for covering big assets. To increase the current capital of Indonesian insurance companies is a challenge to their shareholders/owners.

Country	Minimum Capital Requirement
Brunei	Life and Non-life: BND 1 million (USD 789,270)
Cambodia	Life and Non-life: SDR 5 million* (USD 7.7 million) Composite Insurer: SDR 10 million* (USD 15.4 million)
Indonesia	Life and Non-life: IDR 100 billion (USD 8.65 million) by Dec. 31, 2014 Sharia Insurer: IDR 50 billion (USD 4.32 million) Reinsurer: IDR 200 billion (USD 17.3 million) by Dec. 31, 2014 Sharia Reinsurer: IDR 100 billion (USD 8.65 million)
Laos	N.A.
Malaysia	Life and Non-life: RM 100 million (USD 30.9 million)
Myanmar	Life: MMK 6 billion (USD 6.1 million) Non-life: MMK 40 billion (USD 40.9 million) Composite: MMK 46 billion (USD 47 million)
Philippines	Existing Life & Non-Life: PHP 250 million (USD 5.7 million) by June 30, 2013 PHP 550 million (USD 12.6 million) by Dec. 31, 2016 PHP 900 million (USD 20.6 million) by Dec. 31, 2019 PHP 1.3 billion (USD 29.7 million) by Dec. 31, 2022 New Life and Non-life: PHP 1 billion (USD 22.8 million)
Singapore	Insurers with investment-linked policies or short-term accident & health policies: SGD 5 million (USD 4 million) Other direct insurers: SGD 10 million (USD 8 million) Reinsurers: SGD 25 million (USD 20 million) Captive insurer: SGD 400,000 (USD 320,000)
Thailand	Life Insurer/Reinsurer: THB 500 million (USD 15.3 million) General Insurer/Reinsurer: THB 300 million (USD 9.1 million)
Vietnam	General and Health Insurers: VND 300 billion (USD 14.2 million) Life Insurers: VND 600 billion (USD 28.4 million) Reinsurers (General and Health): VND 400 billion (USD 18.9 million) Reinsurers (Life and Health): VND 700 billion (USD 33.1 million) Reinsurers (Multiline): VND 1,100 billion (USD 52.1 million) Branches of Foreign Insurers: VND 200 billion (USD 9.5 million)

*SDR = International Monetary Fund special drawing rights. As of Jan. 1, 2014,
SDR 1 = USD 1.54.
Source: AON Benfield Asia Pacific Solvency Regulation 2013, Norton Rose Fulbright; Insurance Regulators

Figure 2. ASEAN non-life and life (General), and reinsurance companies minimum capital requirement by regulation.

Source: A.M. Best Market Review ASEAN Non-life and Life, 2014

As it is not easy for most of the shareholders of insurance companies to increase their capitalization up to a certain level, and also it is not easy to get new investors for that purpose, on this matter the OJK can issue a regulation for a mandatory increase of capital. In cases of existing shareholders not having enough funds, the OJK can play an important role in pushing the insurance companies to carry out mergers and acquisitions. This is one of the solutions to increase capital (AON, 2017, p. 4). Without strong capital, the Indonesian insurance companies will not be able to successfully tap into the insurance market.

3.3.2.4 Insurance products

Indonesian insurance companies must also be able to design and provide updated, advanced insurance products to satisfy their customers. Such insurance products must cope with consumers' very dynamic demands. The variety of insurance products, the scope of coverage, and pricing must not be less than those of foreign insurance companies. Indonesian insurance products should be of high quality, provide a wide scope of coverage, and offer competitive pricing.

The OJK can encourage Indonesian insurers to keep improving their insurance products by regulation. The OJK can support these efforts by speeding up the approval of applications for insurance products.

3.3.2.5 Improvement of the insurance application and claims-handling process

Indonesian insurance companies need to improve their services to the policyholders or insureds. They urgently need to better the insurance application and claims-handling processes by using information technology and digitalization. Today, customers (the policyholders and insureds) want to receive excellent services beyond their expectations. Excellent services in the insurance business are highly important. The customer's perception of insurance depends very much on claims handling and payments. All advertisements will be of no effect in improving the good image and reputation of insurance if the process, handling, and payment of insurance claims are not good and satisfactory to the customers. Indonesian insurance companies must provide excellent services to their customers; they should be able to deliver such services better than their AEC competitors.

The OJK can help to overcome this challenge by reviewing the current procedures and related regulations and then implementing revised, improved ones.

3.3.2.6 Alternative dispute resolution body or agency

The Indonesian insurance industry has had its alternative dispute resolution (ADR) since 2006, namely Badan Mediasi dan Arbitrase Asuransi Indonesia (BMI), the Indonesian Mediation and Arbitration Centre. The writer was the head of the working committee for the establishment of the BMI (Simanjuntak, Lamury 2016, p. 32). The BMI was the first of the permanent ADRs established in the financial services industry in Indonesia. The BMI is an independent and impartial ADR for insurance claims dispute resolution. The mediation and adjudication services are provided free of charge by the BMI; the policyholders, the insureds, or the customers have no need to pay any costs for these services. However, for arbitration services, the disputing parties have to pay the costs, which is cheaper than the cost of arbitration services by the Indonesian National Arbitration Centre (Badan Arbitrase Nasional Indonesia/BANI). Indonesian insurance companies need to further promote the BMI to the public insureds and policyholders, and also to encourage and support the BMI to establish branch offices outside Jakarta to expand BMI services in the provinces throughout Indonesia.

On this challenge, the OJK can push the insurance industry and the BMI to expand the operation of the BMI by issuing a regulation for that purpose. It is also necessary for the BMI to harmonize with similar ADRs in other AEC Member Countries for assurance that the BMI is not left behind in terms of its services and operations.

4 CONCLUSION

The establishment of the AEC creates opportunities and challenges for Indonesian insurance companies and all other insurance companies operating in AEC Member Countries. The integration of the insurance sector had begun by the end of 2015. Indonesia is the largest AEC Member Country in terms of population and economy, and it offers the largest potential in the insurance business. The AEC creates larger insurance markets for Indonesian insurance companies as they have huge domestic prospects, and in addition to that, the insurance markets of other AEC Member Countries are opened to them. The Indonesian general (non-life) insurance market was projected to become the second largest market in the world by 2030. This is a huge opportunity for Indonesian insurance companies. However, in order to tap the huge insurance opportunities, Indonesian insurance companies must face a number of challenges: (1) low insurance awareness and literacy, (2) lack of human resources expertise, skill, competence, and professional qualification up to international standards, (3) lack of capital and capacity, (4) poor-quality insurance products, (5) the need to improve the insurance application and claims-handling processes, and (6) the need to expand ADRs.

In order for Indonesian insurance companies to tap the huge insurance business potential, this paper suggests, they need to address and respond to all of these challenges. The OJK can provide support by issuing the necessary regulations and effective supervision and by doing regular reviews and evaluations for further improvement. The OJK can issue related regulations to protect Indonesian insurance companies without breaching the framework and commitments of the AEC.

ABOUT THE AUTHOR

Kornelius Simanjuntak

Dr Kornelius Simanjuntak is a senior lecturer (*Lektor Kepala*) for insurance law at the Faculty of Law, Universitas Indonesia.

He graduated from the Faculty of Law, Universitas Indonesia: Bachelor of Law (1986), Master of Law (2003), and Doctor of Law (2016).

His doctoral/PhD dissertation: 'Natural Disaster Insurance: The Study Concerning the Necessity of Natural Disaster Insurance Scheme in Providing Relief and Compensation for the Victims of Natural Disaster in Indonesia'.

Insurance and insurance law is his area of expertise.

He was the initiator and concept head of the establishment of the Indonesia Insurance Mediation and Arbitration Centre (BMI) as well as the Indonesia Earthquake Reinsurance Pool, now known as PT Reasuransi MAIPARK Indonesia (MAIPARK Re) after its transformation to a limited liability company.

He was the chief of the working committee proposing to change the old Indonesian bankruptcy law representing the insurance industry.

He served as the chairman of the General Insurance Association of Indonesia for two terms (2008–2011 and 2008–2014), chairman of the Insurance Council of Indonesia (Dewan Asuransi Indonesia) (2011–2014), and chairman of the ASEAN Insurance Council (AIC) (2011–2013).

REFERENCES

AAUI. (2017). 'Indonesia General Insurance Market Update 2017'. Presentation of the Chairman of AAUI (Asosiasi Asuransi Umum Indonesia/General Insurance Association of Indonesia) in the International Insurance Gathering (Indonesia Rendezvous), Bali, 12 October 2017.

A & Best. Special Report: ASEAN Life & Non-life. (2014). Hong Kong: A & Best.

Act No. 21 of 2011 Concerning Otoritas Jasa Keuangan (OJK).

Act No. 40 of 2014 concerning Insurance.

AON Benfield. (2017). 'Reinsurance Market Outlook September 2016'. Accessed 27 October 2017. http://thoughtleadership.aonbenfield.com/documents/20160911-ab-analytics-rmo.pdf.

ASEAN. (2008). *ASEAN Economic Blueprint*. Jakarta: ASEAN Secretariat.

ASEAN. (2017). ASEAN Economic Integration Brief No. 01/June 2017. Jakarta: ASEAN Secretariat.

Baker, Tom and Kyle Logue. (2013). *Insurance Law and Policy: Cases, Materials, and Problems*. New York: Wolters Kluwer Law & Business Publisher.

Jess, Digby. (2001). *The Insurance of Commercial Risks: Law and Practice*. London: Sweet & Maxwell.

OJK. (2015). *Insurance Statistic*. Jakarta: Otoritas Jasa Keuangan (OJK).

OJK. (2016a). *Insurance Statistic*. Jakarta: Otoritas Jasa Keuangan (OJK).

OJK. (2016b). Press Release. 'OJK Hosts 19th ASEAN Insurance Regulators Meeting in Yogyakarta'. Yogyakarta: Otoritas Jasa Keuangan (OJK).

Simanjuntak, Lamury and Sendra. (2016). *Badan Mediasi dan Arbitrase Asuransi Indonesia*. Jakarta: BMI, STIMRA-LPAI.

Swiss Re. (2017). *Sigma No. 3/2017*. Zurich: Swiss Re Management and Swiss Re Institute.

Advancing Rule of Law in a Global Context – Susetyo, Rinwigati Waagstein & Budi Cahyono (eds)
© 2020 Taylor & Francis Group, London, ISBN 978-1-138-32782-5

Regulating disruptive innovation in Indonesian digital business

Qur'ani Dewi Kusumawardani
Ministry of Communication and Information Technology, Central Jakarta, Indonesia

ABSTRACT: Electronic commerce (e-commerce) in Indonesia is growing rapidly. Scholars have estimated that Indonesia will add up to $150 billion to its annual gross domestic product by 2025 because e-commerce continues to evolve as corporations pursue efficiency. Digitalization is expected to boost productivity by introducing practicality, ease of access, convenience, and lower costs. Digital business has also introduced a new sharing economy business model. However, these economic changes have also led to market instability because they have unleashed new sources of supply. Every business faces risks and benefits, and digital technology has brought major disruptions for the past decade. This phenomenon has happened not only in Indonesia but also across the whole world. The authors of this paper employ a qualitative method using descriptive analysis to determine the challenges faced by Indonesian regulators while confronting new technologies or digital business practices that do not fully comply with existing regulatory frameworks. The results obtained from this study examine how regulators have established a new approach and regulatory strategy, which is needed to prevent conflict and disruption of existing business.

Keywords: Digital Business, Disruptive Innovation, Regulation

1 INTRODUCTION

The rapid movement of the cloud computing platform has generated a wide range of technological innovations and new business models. Technology has diversified the patterns of public consumption and resulted in a new business model called the *sharing economy*. The sharing economy has several different names: *collaborative economy*, *online gig economy*, *peer-to-peer economy*, *on-demand economy*, and *collaborative consumption* (individuals use platforms to connect with each other).

The Oxford Dictionary of English defines the sharing economy as "an economic system in which assets or services are shared between private individuals, either for free or for a fee, typically utilizing the Internet." Wosskow classifies the sharing economy as "online platforms that help people share access to assets, resources, time and skills." This definition underlines an important characteristic of the sharing economy: the significant level of disintermediation it allows in transactions between providers and final customers (Wosskow, 2014).

The sharing economy involves new forms of production, transaction (mostly spot transactions), and consumption. They may be regarded as examples of "disruptive innovations" in that they compete with traditional ways of producing, distributing, and consuming goods and services, through the use of technological innovations such as smartphones, digital content, and online distribution that may be considered disruptive (ECORL, 2015).

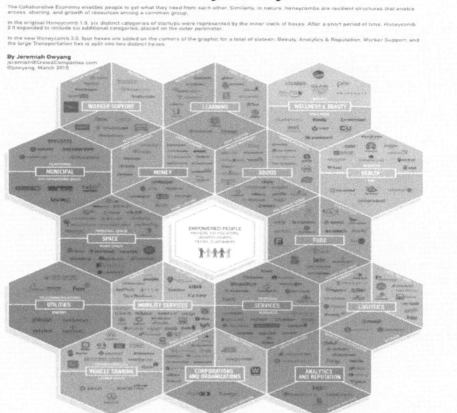

De Streel and Larouche (2015) define disruptive innovation as "a technological innovation that takes place outside the value network of the established firms and introduces a different package of attributes from the one mainstream customers historically value."

The sharing economy is increasing its role in society because it is driven by technology, economic crisis, fewer jobs, the mind-set of the Z generation (Z gene), shifting consumption behavior, more serious attention to the environment, and decreasing trust in corporations. The sharing economy is a triangular relationship between the platform/media, the worker, and the customer.

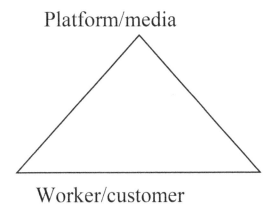

Uber, Gojek, Grab Bike, Airbnb, Lift, Netflix, Sporty, HomeExchange, TaskRabbit, and Skillshare are examples of the sharing economy. They don't just represent a new way of thinking or new services; they also provide a new way to use data effectively to provide services to people when and where they want. They have developed their platforms to allow service providers and users to connect to the benefit of both.

MUSIC/MEDIA	TRAVEL	FASHION	TRANSPORTATION	LABOR/SKILLS
• Netflix	• Airbnb	• Rent the Runway	• ZipCar	• TaskRabbit
• Spotify	• Home Exchange	• Fashionhire	• Lyft	• Mechanical Turk
• GameFly	• Vayable	• Bag Borrow or Steal	• Uber	• SkillShare

Source: http://www.gravitytank.com/pdfs/info_graphics/SharingEconomy_web.pdf.

The decline of the traditional economy is seen everywhere, which has led to many conflicts. Rejection of and protests against this business model have taken place in many countries such as France, where the presence of Uber in Paris has caused frequent protests by taxi drivers, some of which have turned violent. Such incidents have also happened in Indonesia, where conventional taxi drivers have launched huge demonstrations against online transportation providers such as Gojek and Grab Bike. Copenhagen taxi drivers have also protested against Uber, calling for equal terms and fair competition. The Danish Taxi Council supported the High Court ruling, stating it hoped UberPOP would now shut down in the city. In the Netherlands, there have also been protests by members of the taxi industry in response to Uber's operations (Listly, 2017).

Since their arrival in Spain, online rideshare applications (apps) have spurred a series of protests by members of taxi drivers' associations, In Sweden, as in many other countries, Uber's launch prompted debates on whether its services are covered by current legislation. The Swedish Transport Agency and the Swedish Taxi Association have both claimed that UberPOP should be categorized as a taxi service, making it an illegal operation in Sweden.

Demonstrasi menuntut pelarangan taksi online (Foto: Darren Whiteside/Reuters)

This phenomenon happens because the prices arising from the sharing economy are lower. Many people are interested in this business model because they can be their own boss and also manage their schedule to be more productive and creative. Another interesting thing about the sharing economy is that consumers/users can interact with producers everywhere with the help of technology. Many people claim that this kind of technology needs little or no government intervention. But is this true?

Some argue that novel regulation should catch up with this new business model because it will now exist forever. In reality, consumers will never feel the dangers of this business model, and in the end, a good product or service will determine its continuity.

No	Country	Ride Sharing Regulation	Home Sharing Regulation
1.	Bulgaria	Uber suspended services after being accused of unfair trade practices.	
2.	Denmark	Uber began operations in Denmark in 2014, but announced on March 28, 2017, that it is closing down its services in Denmark due to new taxi legislation. Uber was forced to pull out of the market because taxi meters have become a mandatory item for public transportation, seat sensors are needed for tax purposes, drivers must complete two weeks of training, and Uber would need to set up a transport office registering all rides.	a. Both hosted and un-hosted home sharing are legal in Copenhagen, but authorities have signaled stricter enforcement of short-term rental laws. b. The residence requirement restricts short-term rentals in the city to a maximum of seven weeks per year. c. An analysis by Inside Airbnb reported that more than 2,500 flats (16%) were rented out illegally (for more than seven weeks). d. There are no restrictions against Airbnb specifically, but this is currently under consideration (as of January 2017). The mayor has contended that an arrangement such as Amsterdam's may be suitable for Copenhagen, in order to avoid misuse and illegal hotels.
3.	Italy	Uber's services have been blocked in Italy because Uber's business practices were found to constitute unfair competition.	
4.	Hungary	A new law permits the Hungarian national communications authority to block internet access to "illegal dispatcher services," after allowing them to operate for two and a half years.	
5.	Finland	Uber's UberPOP service is illegal in Finland.	
6.	France	Uber's UberPOP service is illegal in France. UberPOP was banned in France on January 1, 2015, with the so-called Thévenoud Law taking effect. The law requires anyone carrying paying passengers to be licensed and to have appropriate insurance and 250 hours of professional training.	a. Un-hosted home sharing is legal for up to 120 days per year in Paris without city approval. b. Renting for more than 120 days per year and renting an apartment the host doesn't live in require approval from the city. c. Violators risk fines up to €28,500. d. In December 2015, officials found roughly 100 tourist apartments that were being rented out illegally during a sweep of 2,000 apartments in Marais.
7.	Spain	a. Uber's UberPOP service is illegal in Spain. b. In December 2014, a judge ordered Uber to cease all UberPOP operations in Spain, upholding	

(Continued)

No	Country	Ride Sharing Regulation	Home Sharing Regulation
		a complaint by the Madrid Taxi Association. c. The ruling was due to drivers lacking proper administrative authorization and agreed that their activities constituted unfair competition. d. Uber suspended its UberPOP operations in Spain on December 30, 2014, after receiving confirmation of the court ruling banning its services in the country.	
8.	Netherlands	Uber's UberPOP service is illegal in the Netherlands.	a. Home sharing is legal in Amsterdam with certain restrictions, and the city has recently made a new agreement with Airbnb, unique in Europe. b. In February 2014, the Amsterdam City Council approved a new policy for home sharing, creating a new category of accommodation called "private rental." Short-term rentals are allowed up to sixty nights per year as long as the number of guests does not exceed four people per reservation. c. In December 2016, Airbnb signed a deal with the city, agreeing to automatically remove listings rented out for sixty days, to install a day counter for hosts, and to install a neighbor tool and a 24/7 hotline where residents can report nuisances and noise complaints, among other measures. The agreement took effect January 2017 and expired December 2018.
9.	Germany	Uber's UberPOP service is illegal across Germany as it violates commercial passenger transport laws.	a. *Hosted* home sharing (spare room) is legal in Berlin. Un-hosted home sharing (entire home) is illegal without a city permit. Berliners can still rent out a spare room in their home if they are on site during the guests' stay, but not the entire home. b. On May 1, 2016, the Prohibition of Misuse of Residential Space Act took effect, banning all unregistered short-term rentals of the entire home. The law applies to those who rent out more than 50% of their property short term (less than two months) without a permit.
10.	Sweden	a. Uber's UberPOP service has been ruled illegal in Swedish courts, as	

No	Country	Ride Sharing Regulation	Home Sharing Regulation
		driving passengers for profit requires a taxi permit. b. Commercial passenger transport requires a taxi license in Sweden. c. In 2015, the Swedish government initiated an inquiry to review current carpooling rules and to investigate if taximeters should be mandatory in taxis, without dispensation, and if a new category of professional passenger transport should be established.	
11.	England	Uber is legally allowed to operate in London. Transportation for London (TfL) issued Uber a minicab (private hire) license in 2012.	a. Both hosted and un-hosted home sharing are legal in London, the latter up to ninety days per year. Under the Deregulation Act of 2015, short-term rentals of entire homes are allowed up to ninety days per calendar year in London, given the host is liable to pay council tax. b. A 2016 report by the Institute for Public Policy Research revealed that 4,938 of Airbnb's "entire home" London listings (23%) were rented out for more than ninety days per year.
12.	Iceland		a. Icelanders can legally rent out their homes through Airbnb for up to ninety days per year under a new law. b. Rentals exceeding the ninety-day or ISK 1 million limit require a state-issued license and are subject to business tax.
13.	New York		a. Short-term rental of most apartments in New York City has been illegal since 2010. Short-term rental (less than thirty days) of entire units in multiunit New York City buildings is banned. b. In October 2016, New York Governor Andrew Cuomo signed a bill imposing heavy fines on Airbnb hosts breaking state housing regulations for short-term rentals.
14.	Chicago		a. In Chicago, house sharing is legal, but a new ordinance has been passed to protect residents and neighborhoods from excessive home sharing. b. In June 2016, the Chicago City Council passed an ordinance requiring hosts to register their units with

(*Continued*)

No	Country	Ride Sharing Regulation	Home Sharing Regulation
			the city and to pay a 4% surcharge on short-term rentals.
			c. The ordinance requires companies, such as Airbnb, to establish 24/7 hotlines for residents, and lets some neighborhoods petition to have unwanted listings banned.
15.	Philadelphia		a. The city council of Philadelphia legalized short-term rentals through online marketplaces in June 2015, passing a string of new regulations.
			b. These regulations include a 180-day cap per unit per year, and a city license is required for rentals lasting more than ninety days per year.
			c. These regulations also include zoning adjustments that make most short-term rentals illegal in residential neighborhoods.
			d. Airbnb collects and handles the tax withholding during the rental reservation process.
16.	San Francisco		a. San Franciscans can legally rent out their entire home short term for up to ninety days per year.
			b. In November 2016, Airbnb agreed to provide the city with lists of addresses, hosts, and stays booked through its platform and to introduce a ninety-day cap on entire home listings, among other measures.
			c. Airbnb collects and remits a 14% Transient Occupancy Tax (hotel tax), and has done so since October 2014.
			d. Hosts are also required to pay a business personal property tax, or "furniture tax," on all furniture, fixtures, equipment, and appliances in their home.
17.	Norway	According to the Norwegian police, UberPOP is illegal.	
18.	Japan		a. Japan allows homeowners to rent out their property to paying guests for up to 180 days per year. They are subject to registering with local authorities who, in turn, have a license to implement their restrictions.
			b. Airbnb has said that Japan is one of its top ten markets worldwide, with 5 million people using its service in

(*Continued*)

No	Country	Ride Sharing Regulation	Home Sharing Regulation
			the country over the prior 12 months. The company claims its community "generated $8.3 billion of economic impact" in 2016. That figure is only likely to raise as the country's tourism industry gears up for a boom around the 2020 Olympic Games and the Rugby World Cup in 2019.
19.	China	Uber was bought out by Chinese competitor Didi Chuxing after allegedly losing billions of dollars in the Chinese market.	
20.	Taiwan	Services were suspended in Taiwan after Uber faced millions of dollars worth of fines. Uber has since reached an agreement to use rental car agencies on the ground under the Uber brand, but the service remains limited.	
21.	Australia, Northern Territory	The territory has completely banned Uber after refusing to change the law to accommodate the app's legality.	
22.	Canada, Vancouver	Ride sharing apps are illegal in Vancouver. The Liberal Party has promised to legalize Uber in the city. Transport Minister Todd Stone said: "We think we are striking a balance between what the vast majority of British Columbians want while respecting the industry that's been there for so many generations and get this right to protect the jobs that already exist."	

Sources: Tech Crunch (Japan passes a law legalizing Airbnb and other sharing economy rentals); Rhodes, 2017; Listly, 2017.

1.1 *Problem formulation*

Issues discussed in this paper are related to two main themes:

1. How do other countries regulate the sharing economy in ride sharing and accommodation sharing?
2. How should Indonesia regulate the sharing economy?

2 METHOD

The method is "the process, principles, and procedures by which we approach problems and seek answers. In social science, the term applies to how one conducts research" (Bogdan and Taylor, 2006, p. 46).

Research means searching again. *Search* refers to searching correct (scientific) knowledge because the result of the search will be used to answer certain problems (Amiruddin, 2004, p. 19). Another kind of research is normative legal research, a type of library research. Normative legal research is legal research using secondary data. Secondary data were sought on the sharing economy from Supreme Court decisions; however, secondary data sourced from these decisions also come from the study of legislation and research literature related to the research topic.

Secondary data derived from primary, secondary, and tertiary legal materials, and from journals, research, literature, electronic media, and tertiary legal materials included in encyclopedias, a legal dictionary, and an English–Indonesian dictionary.

The required data were collected by document study, which is studying and researching various literature sources that explain the research object to get a general description of matters related to the problem. Data analysis of the research result was performed qualitatively, which is collecting as much data as possible from literature study and field research. The collected data were selected; classified logically, systematically, and judicially; sorted between relevant and irrelevant data; and then arranged systematically before data analysis was performed. The collected data were classified by indexing, shorting, grouping, and filtering. Then, they were analyzed using content analysis, which interprets by viewing all of the data in depth (Strauss and Corbin Busir, 1999, p. 19).

3 DISCUSSION

The sharing economy has grown rapidly and challenged existing traditional business models and regulatory frameworks. Not all countries favor the development of the sharing economy; many nations have established boundaries for this new economic model through regulation. In the transport sector, sharing economy companies pose a challenge to governments in multiple ways, and responses should deal with employment issues, internal market regulations, the environment, taxation, consumer protection, etc. In the accommodation sector, the sharing economy faces issues of customer and host protection, safety regulations concerning fire and another disasters, unfair competition, etc. In Chicago, Airbnb has been accused of doing too little to protect residents from the proliferation of its business.

3.1 Regulation in many countries related to the sharing economy

From its beginnings as Air Bed and Breakfast, which hosted conference attendees on airbeds in the home of its founders, Airbnb has become a global entity with a market capitalization that exceeds those of traditional major hotel chains, while Uber has made headlines by attracting investments in the range of several billion dollars (Munkøe, 2017, pp. 28–44).

Uber is one of the most popular apps in the world. Gojek is also mentioned as the most popular app in Indonesia, but these apps are not without controversy. Multiple nations have already banned Uber outright, and local governments in West Java, Makasar, Solo, Pekanbaru, Batam, etc., have asked Gojek to temporarily stop or reduce its operations.

What follows is a list of countries that have banned or legalized ride sharing and home sharing electronic platforms.

3.2 Regulation of the sharing economy in Indonesia

Gojek has become a very popular app. This app has been downloaded 40 million times by netizens of Indonesia, a total of 50% of the ride sharing market in Indonesia. Not only that, Gojek's online food delivery feature, called Go-Food, is said to control 95% of the market share in Indonesia. Gojek's services include ride sharing apps titled Go-Ride and Go-Car, along with other features like Go-Food, Go-Tix, Go-Med, Go-Send, Go-Mart, Go-Box, and Go-Busway. There is also a Gojek derivative app called Go-Life, which includes Go-Massage, Go-Clean, Go-Auto, and Go-Glam. Gojek also features the Go-Pay digital wallet and Go-Points usage bonuses (Bohan, 2017).

According to Annie App, as compiled by TechinAsia's KompasTekno, the Grab app was downloaded 42.7 million times across Southeast Asia by the end of April 2017. Grab itself says the figure rose to 45 million downloads by May 2017. That number includes downloads by users in several countries in Southeast Asia, including Indonesia, Malaysia, and Singapore.

The size and scale of Uber, Gojek, Grab, Airbnb, and several other firms now rival or surpass those of some of the world's largest businesses in transportation, hospitality, and other sectors.

Many complaints have arisen against these business models: people are renting their apartments without complying with existing regulations for hotel accommodation; drivers are providing rides for a price without complying with taxi regulations. At the same time, new technologies have disturbed existing regulatory frameworks, triggering intense public debates about whether they are compatible with accepted social and legal norms (Bernstein, 2002). Legal and regulatory structures have historically been built around the idea of ownership, which is why the industry has mobilized, and several cases have been brought to court pleading for the prohibition of some of these business models (Goudin, 2016).

Competition is an important function of a healthy economy. So too is regulation because it provides the framework for a healthy business environment, sets standards, and ensures only fair advantages are gained in the marketplace. Companies shouldn't operate in a regulatory vacuum; neither should they risk degeneration at the hands of rules that are no longer fit for purpose (Fruman and Molfetas, 2016).

The government of Indonesia has passed the minister of transportation's Regulation No. 32 of 2016. This regulation applies to online and conventional four-wheeled vehicles. Under this legislation, online taxi service providers enter into a special lease, 1,000cc cars can be operated, local governments are entitled to set the upper and lower tax rates that apply to such businesses, and local governments are entitled to limit the number of taxis within the needs of their respective regions.

Online taxi companies must ensure that all vehicles and drivers are registered and that all drivers have passed the mandatory periodic test (KIR), have a pool, have a workshop, pay taxes in accordance with the Ministry of Finance, provide access to driver data (dashboard) for the Ministry of Transportation, and give sanctions in the form of strikes to blocking.

Supreme Court Decision No. 37 P/HUM/2017 Regarding the Material Test invalidated the minister of transportation's Regulation No. 26 of 2017. According to the Supreme Court's verdict, several articles of Regulation No. 26 contradict higher legislation and therefore have no binding legal force.

Fourteen points in Regulation No. 26 of 2017 are deemed contradictory to higher law, namely Law No. 20 of 2008 Regarding Micro, Small and Medium Enterprises and Law No. 22 of 2009 on Traffic and Road Transportation. The Supreme Court has instructed the Ministry of Transportation to revoke the articles related to these fourteen points.

Chairman of the Land Transport Organization (Organda) Andriyanto Djokosoetono has said the decision means that online taxi companies no longer can operate officially. Taxi fare apps should go back down starting November 2, 2017, and the KIR (testing of motor vehicles) is not needed anymore. Although the Supreme Court ruling already exists, there is still much debate in the community because transportation must contain the elements of safety, security, and comfort (Djoko Setijowarno, 2017).

Most assume that regulating novel technologies requires more information and longer decision-making processes than familiar, established technologies do[15]. Compounding the challenge, regulators in Indonesia must make four different types of decisions when confronting novel technologies or business practices. The regulator thus makes four related types of decisions (Cortez, 2014).

1. Timing: When should the agency intervene, if at all? Does waiting necessarily generate a better informational basis on which to regulate? What are the drawbacks of waiting?
2. Form: Should the agency regulate via rule, adjudication, guidance, or some alternate form? Given the costs and benefits of each, which best accommodates the uncertainties of the innovation? Does form even matter?
3. Durability: Should the agency's intervention be permanent, or temporary, or conditional? How long should it endure? And are there ways to better calibrate regulatory interventions to the innovation?
4. Enforcement: How rigorously should the agency monitor and sanction noncompliance? How much should the agency temper enforcement against novel products, firms, or industries?

This part considers how the regulator might respond to disruptive innovation along these four dimensions because the government should have the instruments to oversee the business of transportation anywhere to maintain the balance and arrangement of transportation nationally. The regulator in Indonesia should consider the form of intervention, the timing of intervention, how durable that intervention should be, and how rigorously the government should monitor and sanction noncompliance because the sharing economy in the end certainly has benefits and adverse effects. Therefore it is necessary for the government to provide fair and balanced rules.

4 CONCLUSION

1. The key to smooth operations in the sharing economy is getting the right balance between innovation and regulation. The obligation of the state and the government is to maintain a business-friendly climate and protect the balance of supply and demand, as well as to provide security and safety for the users of public transportation services. To protect consumers in the sharing economy, regulations governing online transactions should standardize problem solving, make a transparent database that consumers can access, and require the protection of personal data.
2. Regulators can choose to legalize the sharing economy or to ban it based on the experiences of other countries. States like Bulgaria, Denmark, Italy, Hungary, Finland, France, Spain, Netherlands, Germany, Sweden, Norway, Canada, and Australia have chosen to make ride

sharing illegal because they found that the sharing economy constitutes unfair competition. On the other hand, countries like England and Japan have chosen to legalize ride sharing. The cities of New York, Chicago, Philadelphia, and San Francisco have banned long-term home sharing rentals out of concern for consumer safety. Countries like Denmark, France, Netherlands, Germany, England, and Iceland and cities like Chicago have legalized home sharing because this economic model increases employment in the informal sector and promotes economic growth.

3. Related to tax payments, we can take an example from Denmark, through the installation of technology like the seat sensor, which connects between online vehicles and the state tax system. If such technology is too expensive for Indonesia, the tax obligation can be directed to the driver, because every citizen already has a Taxpayer Identification Number (NPWP). The types of taxes that apply to transportation business actors are in the form of corporate tax, personal income tax, transaction tax, and profit tax. The electronic system administrator also must be headquartered in Indonesia so that it can be registered as a taxpayer.

REFERENCES

Amiruddin, Zainal Asikin. (2004). *Pengantar Metode Penelitian Hukum*. Jakarta: Raja Grafindo Persada.

Bernstein, Gaia. (2002). "The Socio-legal Acceptance of New Technologies: A Close Look at Artificial Insemination." *Washington Law Review 77*, 1035–1036.

Bogdan, Robert and Steven J. Taylor. (2006). In Soekanto, Soerjono. *Pengantar Penelitian Hukum*. Jakarta: Universitas Indonesia Press, p. 46.

Bohan, Fatimah Kartini. (2017). Go-Jek Claims Defeat Grab and Uber in Indonesia. http://tekno.kompas.com/read/2017/06/13/18080087/gojek.klaim.kalahkan.grab.dan.uber.di.Indonesia.

Cortez, Nathan. (2014). "Regulating Disruptive Innovation." *Berkeley Technology Law Journal* 29. http://scholarship.law.berkeley.edu/bj.

Djoko Setijowarno. (2017). Supreme Court Revokes 14 Points Transportation Regulation of Taxi Online, Ministry of Transportation Will Obey. https://kumparan.com/nurul-hidayati/ma-cabut-bela san-pasal-permenhub-taksi-online-kemenhub-taat-azas.

Economy Co-responsibility Learning (ECORL). (2015). EC Project Number 2015-1-IT02-KA204-015467, Comparative Study on the Sharing Economy in EU and ECORL Consortium Countries. www.ecorl.it.

Fruman, Cecille and Aris Molfetas. (2016). Keeping Pace with Digital Disruption: Regulating the Sharing Economy. http://blogs.worldbank.org/psd/keeping-pace-digital-disruption-regulating-sharing-economy.

Goudin, P. (2016). *The Cost of Non-Europe in the Sharing Economy*. European Parliamentary Research Service.

Listly. (2017). Regulating the Sharing Economy. Teknologiradet. post@40teknologiradet.no.

Munkøe, Malthe Mikkel. (2017). "Regulating the European Sharing Economy: State of Play and Challenges." *Intereconomics: Review of European Economic Policy 52*(1), 38–44.

Rhodes, Anna. (2017). "Independent Travel." September 22 (Banned the Controversial Taxi App).

Strauss, A. and J Corbin Busir. (1999). *Qualitative Research: Grounded Theory Procedure and Technique*. London: Sage Publishing.

Wosskow, D. (2014). *Unlocking the Sharing Economy: An Independent Review*. London: UK Department for Business, Innovation and Skills.

Advancing Rule of Law in a Global Context – Susetyo, Rinwigati Waagstein & Budi Cahyono (eds)
© 2020 Taylor & Francis Group, London, ISBN 978-1-138-32782-5

Constitutional court dismissed the reconciliation effort

Junaedi
University of Canberra, Australia

1 INTRODUCTION

The transition from an authoritarian regime to a democratic government is often followed by a question concerning the settlement of human rights violations committed by the previous regime. This question is sometimes answered by a transitional government employing many mechanisms, and one such mechanism is the establishment of a truth and reconciliation commission (TRC) as created in South Africa (Van Zyl, 1999). The transitional government's establishment of the commission is an effort to provide the best middle way to punish the perpetrators of the previous regime on one hand and to provide forgiveness or amnesty for what happened in the past on the other. The option to establish the commission in many countries (Van Zyl, 1980) may not satisfy all factions within a nation, especially the victims or their families, but the choice should be determined as the past incidents are very complex in terms of their legal, social, and political dimensions.

The establishment of a TRC in the context of the transitional government is the way to turn from totalitarian governance toward democratic government (Geula, 2000, p. 57). Transitions such as these raise the question of responsibility for crimes against humanity committed by the previous regime. According to Phillips and Albon, this question contains two important issues: recognition (acknowledgment) and accountability (accountability) (Phillips and Albon, 1999). Confession contains two options: remember or forget. Accountability confronts us with a choice between prosecution and forgiveness.

In this controversy, the TRC is not just an alternative to ad hoc human rights courts, but also appears beside them. It is the key to a strong effort to use a human rights perspective and a humanist paradigm that promotes the interests of the victims on one hand and saving the lives of the general public on the other. The TRC is a tool for applying the concepts of restorative and reparative justice and constructive settlement (Drožđek, 2010, p. 5). This implies a concept of justice that comes out of a typical classical Aristotelian view (commutative justice that is contractual, distributive, corrective, and punitive) and from a Rawlsian-Habermasian mind-set that fosters fairness in equality (justice as fairness) that can only be applied in a more normal situation. The TRC also introduces the concept of progressive justice, which emphasizes punishment of crime (criminal justice), demolition of history (historical justice), preferential treatment and respect for the victim (reparatory justice), administrative justice, and the overhaul of the constitution (constitutional justice).

Following Huyse (1995), truth is both retribution and deterrence. On the retribution–reconciliation spectrum, the ideal response is selective punishment, a model that emphasizes formal, legal responsibility. Therefore, our type of transition is the replacement initiated by the people, which fits with the selective punishment model. Although referring to the handover of power from Suharto to Habibie, it appears that our type of transformation was initiated by the government, but the change was actually based on the insistence of the people, especially students.

The TRCs cannot and should not replace the function of the courts, because they are not part of the judiciary system, and also do not have the power to send someone to jail or to convict a person for a particular crime. They can do some important things that generally cannot be achieved through prosecution in criminal court. Cases handled by the TRCs are larger in scope than those tried before the criminal courts. Where serious human rights violations were widespread and systematic under the previous regime, a TRC as applied in South

Africa can investigate all cases or a large number of existing cases in a comprehensive manner (Ball et al., 2000). South Africa's TRC is also in a position to provide practical assistance to the victims specifically (Daly, 2002, p. 367), identifying individuals or families who are victims of any past crimes (Freeman and Quinn, 2003), so that they are legally entitled to receive reparations. The TRCs can also be used to try to answer questions such as how a human rights violation occurred (Bhargava, 2002, pp. 409–432), why it happened, and whether factors present in society and the state allowed the incident to occur (Gibson, 2006). These efforts can also highlight changes to make in order to ensure it does not happen again (Llewellyn and Howse, 1999).

2 DISCUSSION

2.1 *Law No. 27 of 2004 regarding the truth and reconciliation commission*

Indonesia has had a truth and reconciliation commission since the enacting of Law No. 27 of 2004 on the Truth and Reconciliation Commission (TRC Act). The general description of the TRC Act stated that human rights are basic rights bestowed by God on man that are universal and eternal. Therefore, they must be protected, respected, and maintained, and should not be ignored, reduced, or taken away by anyone.

Human rights violations that occurred before the enactment of Law No. 26 of 2000 on the Human Rights Court must be traced back to reveal the truth, to uphold justice, and to establish a culture of respect for human rights. Revealing the truth is also in the interest of the victims and their families as they are entitled to compensation, restitution, and rehabilitation.

To expose human rights violations, Indonesia's Truth and Reconciliation Commission was mandated by Article 47 paragraph (2) of the Human Rights Court Law and the provisions of the People's Consultative Assembly No. V/MPR/2000 on Stabilization of National Unity and Integrity.

The Truth and Reconciliation Commission is based in the capital of the Republic of Indonesia, and its working area covers the whole territory of the Republic of Indonesia. The Commission has the function of a public institution to disclose the truth about human rights violations and to implement reconciliation.

The Commission's tasks are as follows:

a. Receiving complaints or statements from perpetrators, victims, or victims' families who are the victims' heirs;
b. Investigating and clarifying gross violations of human rights;
c. Providing recommendations to the president concerning appeals for amnesty;
d. Providing recommendations to the president concerning compensation and rehabilitation; and
e. Providing annual and final reports to the president, with copies sent to the Supreme Court, concerning the accomplishment of its task and its authority in a case.

In performing these tasks, the Commission has the authority to:

a. Investigate under existing legislation;
b. Request information from victims, victims' families, perpetrators, and other parties, within and outside the country;
c. Obtain official documents from civil and military agencies as well as other institutions within and outside the country;
d. Coordinate with related agencies, both within and outside the country, to provide protection for victims, witnesses, individuals making statements, perpetrators, and evidence under existing legislation;
e. Call on concerned individuals to give information and testimony;
f. Decide compensation, restitution, and rehabilitation or refuse a request for compensation, restitution, rehabilitation, or amnesty, if the case has been registered to the Human Rights Court.

A TRC law mandated that the Commission should be established by six months after the legislation was promulgated (TRC Law, 2004). Therefore, in April 2005, President Susilo Bambang Yudhoyono, through Presidential Decree No. 7 of 2005, formed a selection committee to search for candidates to serve as the truth and reconciliation commissioner. The president also announced Presidential Regulation No. 27 of 2005 regarding the Procedure for Selection and Election of TRC Members.

The selection committee was to nominate forty-two candidates to the TRC. The entire list of selected candidates then was to be sent to the president to be narrowed down to twenty-one people before final approval of one candidate by the Representative Assembly.

At the end of March 2006, pros and cons emerged in the procedures and mechanisms of the Truth and Reconciliation Commission's work conducted by the advocacy team consisting of LBH Jakarta, Kontras, Elsam, National Solidarity, Impartial and Lembaga Pengabdian Hukum. The government planned to begin drafting the implementing rules of the TRC law, even though the president had not yet chosen members of the TRC from the forty-two candidates. The implementing rules were outlined in the Presidential Decree on Procedures and Working Mechanisms and the Presidential Regulation on the Organizational Structure of the TRC (Aru, 2006).

The initiative, as mentioned earlier, was considered by the advocacy team to be a government intervention that violated the independence of the TRC. Under Article 10 of the TRC Law, preparation of the code of conduct and the working mechanisms of the Commission fell under the authority of the Commission. The government was only authorized to support the work of the Commission through the Secretariat of the Commission, and its overstepping would potentially limit the work of the TRC.

The team also saw that the TRC Law still had many weaknesses and defects in communicating principles of justice, logical legal thinking, and human rights. For this reason, the team filed a judicial review application through the Constitutional Court. Judicial review of Law No. 27 of 2004 on the Truth and Reconciliation Commission was decided by the Constitutional Court on December 7, 2006, through Verdict No. 006/PUU-IV/2006. The court's consideration stated that Law No. 27 is contrary to the constitution and has no binding legal effect. With the cancellation of the TRC Law by the Constitutional Court, the selection process for the forty-two candidates for the TRC was terminated (Wisnu Nugroho, 2006).

2.2   *The court decision on the invalidity of the TRC law*

The Constitutional Court presented the fundamental decisions of lawmakers who determine the policy of reconciliation as a solution to the serious human rights violations coming before the Human Rights Court, not only as a political decision but also as a legal mechanism. This caused the assessment of such cases to be carried out based on the principles of law and the constitution, which includes exposing past abuses of power and human rights violations and implementing reconciliation in the common interest of the whole nation.

All parties had to agree that the mandate of Decree No. V/MPR of 2000 on Stabilization and National Unity, which continues into a package with an ad hoc human rights court, and is part of the implementation of the constitutional command to all state officials, was in line with the intent and purpose of the amendments to the Constitution of the Republic of Indonesia of 1945, especially Sections 28A–28J on human rights.

As a nation that has declared Pancasila its ideal philosophy and way of life, Indonesia has in its best interest to retrace gross human rights violations to uncover the truth, to bring justice, and to establish a culture of respect for human rights. These efforts should be made with the approach of understanding the conflict objectively, in order to achieve a safe and peaceful Indonesian state. Indonesia, as a member of the United Nations, has included human rights protections in its 1945 Constitution; then, in interpreting the 1945 Constitution, the Constitutional Court also considered UN documents on human rights within these paradigms. What follows is an examination of the Court's consideration of the constitutionality of particular articles outlined in the TRC Laws.

2.2.1 *Article 27 of Law KKR*

Article 27 specifies that compensation and rehabilitation as determined by Article 19 are granted when the application is granted. The explanation of the article specifies that when a perpetrator of gross human rights violations confesses a mistake, admits the truth of the facts, expresses remorse for his actions, and is willing to apologize to the victims or the families of the victims as their heirs, he can apply to the president for amnesty. Upon reasoned request, the president may accept the application, and the victims will be given compensation and rehabilitation. When the application is rejected, compensation and rehabilitation are not given by the state, and the matter is settled under the provisions of the Law on Human Rights Court.

This arrangement contains contradictions between one section and the other, especially the part that regulates:

a. The perpetrator has admitted the offense and the truth of the facts and has expressed remorse and willingness to apologize to the victim.
b. The perpetrator can apply to the president for amnesty.
c. Applications can be accepted or rejected.
d. Compensation and rehabilitation will be provided only if amnesty is granted by the president.
e. If amnesty is rejected, the matter will be submitted to an ad hoc human rights court.

One of the contradictions found in Article 27 of the TRC Law is holding an individual perpetrator criminally responsible, even in cases of human rights violations brought under the Human Rights Tribunal Act, when the perpetrator or victim and other witnesses are not easy to find again. Reconciliation between perpetrators and victims is referred to in the law as a precedent that becomes almost impossible to set under an individual approach to criminal responsibility. With this approach, which hinges on amnesty and just restitution, the compensation is given by the perpetrator or a third party.

If the goal is reconciliation, with no individual approach, then that becomes the point of departure for a serious human rights violation, and the victim receives a measure of reconciliation by accepting compensation and rehabilitation. The second approach, about restitution, compensation, rehabilitation, and amnesty has no relevance here because amnesty is the prerogative of the president.

The fact that serious human rights violations have taken place, which is really a state's obligation to avoid or prevent, has given birth to a legal obligation on the part of the state, and individual actors can be identified to provide restitution, compensation, and rehabilitation to the victims. Determination of amnesty is ruled legal protection and justice as guaranteed by the 1945 Constitution.

It is also the practice to provide reparation in proportion to the offense and the weight of loss, as published in the *Basic Principles and Guidelines on the Right to a Remedy and Reparation for Victims of Gross Violations of International Human Rights Law and Serious Violations of International Humanitarian Law*, which provides for adequate, effective, and prompt reparation for harm suffered and is intended to promote fairness in the treatment of serious human rights violations. It is such an interpretation that is used to look at Article 28 A, paragraph D of Article 28 (1), and Article 28, paragraphs (1), (4), and (5), so the petition on the grounds of Article 27 of the TRC Law is reasonable.

2.2.2 *Article 44 UU KKR*

Article 44 of the TRC Laws mentions: *"Pelanggaran Hak asasi Manusia yang beat yang telah diungkapkan dan diselesaikan oleh Komisi, perikarya tidak Dapat diajukan Lagi kepada Pengadilan Hak Asasi Manusia Ad Hoc."* ("The gross human rights violations having been expressed and resolved by the Commission, his case cannot be filed again to an ad hoc human rights court.") From the general explanation of the TRC Law, it can be concluded that the task is to uncover the truth, uphold justice, and establish a culture of respect for human rights. The TRC does not address the legal determinations of the process, but the measure of truth-seeking, restitution, and rehabilitation, as well as amnesty.

179

The question is whether the TRC is a substitute or surrogate court. A common explanation explicitly specifies that when serious human rights violations have been resolved by the TRC, an ad hoc human rights court has no authority to decide, unless the request was denied amnesty by the president. If an ad hoc human rights court has made a decision, the TRC is not authorized to decide. Although it is said that the TRC is only an alternative to the court and is not a law enforcement agency, it is clear that it is an alternative dispute resolution mechanism for deciding cases of human rights violations and, if successful, it will settle human rights disputes amicably.

The TRCs have been accepted in international practice, for example in South Africa, and also have been recognized in customary law. A closed legal process through an ad hoc human rights court when obtaining solutions in the TRC is a logical consequence of an alternative dispute resolution mechanism, so it should not be seen as a justification for impunity against perpetrators of gross human rights violations. In general, the legal mechanisms for addressing serious human rights violations before the Human Rights Tribunal Act experienced difficulty with delays resulting in the loss of the tools to be used as the basis of evidence in the individual criminal responsibility approach.

The TRC also has aimed to uphold justice as far as is possible in an alternative dispute resolution mechanism. Therefore, the court does not see enough basic and constitutional reason to grant it, especially because the provision only applies to serious human rights violations that occurred before the Human Rights Court Law.

2.2.3 *Article 1 Point 9 of the TRC Law*

Article 1 point 9 of the TRC Laws states, "*Amnesty Adalah pengampunan yang diberikan oleh Presiden kepada pelaku pelanggaran hawk asasi Manusia yang beat dengan memperhatikan pertimbangan Dewan Perwakilan Rakyat.*" ("Amnesty is a pardon granted by the President to the perpetrators of human rights violations by taking into consideration the opinion of the House of Representatives.") The meaning of gross human rights violations is specified in Article 1 Number 4 of the TRC Law as a "violation of human rights law as determined by the Human Rights Court, which in Article 7 states that the violation of human rights includes genocide crimes and crimes against humanity." The general explanation also mentions that the laws on the human rights tribunal refer to the Statute of Rome on the International Criminal Court, which has qualified genocide and crimes against humanity as the most serious crimes in the international community.

International practice as established in the General Comment of the UN Human Rights Commission generally considers that amnesty is not allowed for gross violations of human rights. Although the TRC was intended to create conditions conducive to the existence of peace and national reconciliation, it should attempt to determine the limitations of amnesty so that perpetrators cannot benefit from it. Amnesty cannot have all the legal consequences regarding victims' rights to reparation, and furthermore, amnesty cannot be granted to those who commit crimes violating human rights and international humanitarian law, which does not allow any other form of amnesty or immunity.

Although the General Comment has not been accepted as binding law, its substance appears in the 1945 Constitution: Article 28 G Paragraph (2) protects the right to be free of torture; Article 28 I Paragraph (1) protects the right to life and the right not to be tortured; Article 28 Paragraphs (4) and (5) confirm that the protection, promotion, and fulfilment of human rights is the responsibility of the state. Article 1 point 9 is just a notion or definition contained in the general provisions and not a norm linked with other chapters, so the petition is set aside and will be further considered in conjunction with the chapters related to amnesty.

Although the petition applied only to Article 27 of the TRC, because the overall implementation of the TRC Act depends on the article cited earlier, Article 27 is contrary to the 1945 Constitution and has no binding force. This occurs because Article 27 is closely related to Article 1 Paragraph 9, Article 6 Letter c, Article 7 Paragraph (1) Letter g, Article 25 Paragraph (1) Letter b, Article 25 Paragraphs (4), (5), and (6), Article 26, Article 28 Paragraph (1), and Article 29. In fact, Article 27 and those articles related to it determine whether the overall work provisions in the TRC Law are legally binding; if they are not, then the implication is that the entire chapter relating to amnesty is not legally binding.

Aforementioned deeds may not qualify as violations of procedural law, even though the petition concerns only Article 1 point 9, Article 27, and Article 44 of the TRC Law, because procedural law relating to judicial review of the Act of 1945 concerning the public interest has legal consequences *era omnes nature*, so it is not appropriate to see it as a thing that is known in civil law. Procedural law in district and religious courts in Indonesia prohibits hearing and deciding beyond what is required (*petite*) under Article 178 Paragraph (2) and (3) HIR and its equivalent in Article 189 Paragraph (2) and (3) RBG. It can thus be understood, because the initiative to preserve or not the rights which are owned by private individuals or the individual lies in the individual's will, or judgment lies in the will or judgment of such individuals, which can not exceded.

Nevertheless, developments did because and therefore, social interest, causing rules are not treated as well as longer absolute. Considerations of justice and decency has also become the reason, as shown, among others, in the judgment of the Supreme Court dated May 23, 1970, dated February 4, 1970, and on January 8, 1972, as well as other decisions are more, then where assigned that Article 178 Paragraph (2) and (3) HIR and Article 189 Paragraph (2) and (3) does not apply RBG absolutely because of the obligation of judges to be active and always have to try to make a decision that actually resolves the case.

Civil suits usually list all the applicants in a petition to the judge to reach the fairest decision (*ex aequo et bono*). Constitutional justice therefore implements procedural laws that relate to the public interest. Although the petitioner is an individual deemed to have legal standing, the legal consequences are broader than just the interests of the individual applicant. The Constitutional Court should consider the public interest and not dwell only on the application or petition that has been filed. This has also become a common practice of constitutional courts in other countries, for example, Article 45 of the Law on the Constitutional Court of South Korea (1978).

In this case, the provisions of the law cannot be implemented as a result of a decision on the petitioned article; a judgment of unconstitutionality can be imposed against the entirety of the law. See, for example, Decision on Case No. 001-0210022/PUU-I/2003 on Testing the Law of the Republic of Indonesia, and Law No. 20 of 2002 on Electricity. The following things are also found in the TRC Law;

1. The TRC is authorized to receive complaints, gather information and evidence of serious violation of human rights, call witnesses and then clarify with the perpetrators or victims, determine the serious category of human rights in a public court as mentioned in Article 18, conclude whether there is a violation of human rights, who are the victims and perpetrators, and also whether there is an apology. As mentioned in the General Introduction, the TRC's decision is final and binding. If the decision of the TRC provides compensation, restitution, or rehabilitation (Article 25 Paragraph (1) a), then the decision is final, and it does not have a holding capacity (binding force) if amnesty is rejected. The perpetrator and the victim or the government are also not bound by a decision that depends on the condition of the amnesty. Accordingly, the authority of the TRC is uncertain.

2. Article 28 Paragraph (1) states that if perpetrators and victims of serious human rights violations forgive each other and make peace, then the TRC can provide a recommendation to the president to grant amnesty. However, Article 29 Paragraph (1) states that amnesty recommendations shall be decided by the TRC, so there is no consistency in the TRC Law, which give rise to legal uncertainty.

3. If the perpetrator admits the truth of the facts, repents, and is ready to apologize to the victims, but the victims do not forgive, then the TRC decides amnesty independently and objectively. This situation does not provide an impetus for the revelation of truth.

4. If the perpetrators are not willing to admit the truth and their wrongdoing or do not regret it, then they lose the right to apply for amnesty and their case will be submitted to an ad hoc human rights court. In this case, a conflict of authority is possible between the

TRC and the House of Representatives, because Articles 42 and 43 of 2000 mandate that heavy human rights violations should be judged by an ad hoc human rights court but should also go before the parliament.

Based on Article 23 of the TRC Law, the TRC's finding of serious human rights violations is enough to bring the case to trial in front of an ad hoc human rights court without a decision by the House of Representatives.

Reconciliation offers alternative opportunities for offenders to admit their actions without dealing with the common law. Perpetrators have a chance to consider their actions. The TRC laws do not assure that perpetrators will choose the TRC to settle the case. Article 28 Paragraph (1) of the TRC Law states that when victims and perpetrators before a human rights court have mutually forgiven and reconciliation has been made, the Commission can recommend that the president grant amnesty.

Based on the provisions of Article 1 Paragraph 2, it can be concluded that three conditions must be met for reconciliation: (1) disclosure of the truth, (2) recognition, (3) forgiveness. Thus, if all three cannot be met, there certainly will not be any reconciliation. If a case does not reveal the truth, whether about the event, place, or time, reconciliation is not possible. The TRC Law does not contain provisions that directly state that the rejection of amnesty will lead to prosecution, but it does specify that perpetrators should be held legally responsible for their actions. From this description, it is clear that the TRC Law does not encourage offenders to settle their case through the TRC, as the TRC Law contains a lot of legal uncertainty. Meanwhile, if victims or their heirs are not willing to forgive, only then can they report offenders to law enforcement based on the evidence of confessions made by the offender.

Because this provision has raised the possibility of self-incrimination, it will be difficult to achieve reconciliation, which is the goal of the TRC. The laws do not expressly regulate whether a reconciliation process can occur without forgiveness by victims or their heirs. The provisions of Article 29 Paragraph (2) of the TRC can cause problems in cases where the victim took the initiative to report claims to the TRC. Such claims should have been reported from the beginning when the victim chose the TRC to resolve the case; the victim has the will to forgive the perpetrator. If the victim does not have the will to forgive the perpetrators, an alternative justice process is provided and not through the path of reconciliation. In other words, reconciliation requires a reciprocal willingness from the perpetrator and the victim.

5. When a complaint is accompanied by an application for compensation, restitution, rehabilitation, or amnesty, the Commission shall issue a decision within a period not later than ninety days from the date of the request receipt (Article 24 of the TRC Law). The question is whether matters should be decided by the Commission within ninety days, including decisions about the disclosure of "the truth about the serious human rights violations." (Article 1 Paragraph 3 and Article 5 of TRC Law). Article 25 Paragraph (1) states that the Commission's decision referred to in Article 24 may be to:

a. grant or refuse to provide compensation, restitution, and rehabilitation, or
b. provide recommendation regarding amnesty in the form of a legal opinion.

With the formulation of Article 25 Paragraph (1), the Commission has ninety days in which to decide applications for compensation, restitution, rehabilitation, or amnesty. It is equipped with the provisions of Article 25 Paragraphs (3), (4), (5), and (6), and Article 26, which sets a period for the decision-making process. The TRC Law does not specify a time limit for findings of serious human rights violations. With the time limit to decide upon compensation, restitution, rehabilitation, and amnesty within ninety days, if such a period has elapsed while the TRC is still in the process of investigation and clarification, the request for compensation, restitution, rehabilitation, and amnesty should be decided first.

A complaint or report may be submitted to the Commission, and the Commission shall conduct an investigation of the perpetrator and an examination and clarification of the events.

Article 24 contains provisions ruling that when the Commission has received complaints or reports of serious human rights violations that are accompanied by requests for amnesty, the word "with" means that the application for amnesty is submitted along with the complaint or report of serious violation of human rights.

The problem is that amnesty is only possible for persons who qualify as perpetrators based on the investigation that took place before the application was presented. In this situation, the perpetrators have applied for amnesty, and the president has the authority to decide. So the amnesty should be granted to a particular person based on an application submitted for amnesty. Offenders can be determined after the TRC has found serious human rights violations along with the culprit. Thus Article 24 may lead to confusion or legal uncertainty because of its time limit of ninety days. Amnesty may be requested, recommended, and given when it is known with certainty who the perpetrators are.

Revealing the perpetrators from the beginning is only possible if serious human rights violations have been disclosed as intended by Article 23 Letter a, or there is already a peace settlement between the victims and perpetrators, as stated in Article 28. On the other hand, Article 24 says that the process is based on Article 18 Paragraph (1) a, which addresses the authority of the sub-commission to actively investigate and clarify victims' complaints or reports. Article 23 Letter a grants authority to the TRC subcommittees, upon recognition of the perpetrators, to make a consideration for granting amnesty. Such a provision is judicially illogical if a request for compensation, restitution, rehabilitation, and amnesty is filed together with a complaint or a report that shall be terminated within a period not later than ninety days from the date of receipt of the application.

All these facts and circumstances contribute to the lack of legal certainty, both in the formulation of norms and their implementation. With regard to the considerations described earlier, the Court held that the principles and objectives of the TRC, as set out in Articles 2 and 3 of the law *a quo*, cannot be realized due to the lack of legal certainty. Therefore, the Court considers the law *a quo* as a whole against the 1945 Constitution and should declare that it has no binding legal effect. A declaration that the TRC Law does not have binding legal force as a whole does not mean the Court does not cover the settlement of serious human rights violations through reconciliation efforts.

Many ways can be taken to achieve reconciliation in the form of legal policies that are compatible with the 1945 Constitution and with international human rights instruments that apply universally, or in the form of political policies on rehabilitation and amnesty in general. Considering the application of Article 56 Paragraphs (2) and (3) and Article 57 Paragraphs (1) and (3) on Law No. 24 of 2003 concerning the Constitutional Court (LN Republic of Indonesia of 2003 No. 98, TLN Indonesia No. 4316), the Constitutional Court approved all requests from the applicants. The Court also decided that Law No. 27 of 2004 on the Truth and Reconciliation Commission is contrary to the 1945 Constitution, the implication being that it is not legally binding.

2.3 *Critics of the court decision*

Article 47 Law No. 26 of 2000 concerning the Human Rights Court (Human Rights Court Law) opened an opportunity to establish the Truth and Reconciliation Commission (TRC). This provision clearly states that the settlement of past gross human rights violations is to be completed by the TRC whenever possible. Two mechanisms are available for the settlement of past gross human rights violations: first, a judicial mechanism through the Human Rights Court, and second, an extrajudicial mechanism through the TRC.

Since its establishment, the Human Rights Court has decided two past gross human rights violations, which are the East Timor Referendum in 1999 and the Tanjung Priok incident in 1984. However, unfortunately, Law No. 27 of 2004 concerning the Truth and Reconciliation Commission (TRC Law), as the basis for the establishment of the TRC, had to be declared inapplicable before it was enacted, after the Constitutional Court handed down decision No. 006/PUU-IV/2006 dated December 7, 2006. In essence, the Constitutional Court decision declares that the TRC Law, especially Article 1 point 9, Article 27, and Article 44, are contrary to the 1945 Constitution.

Settlement of gross human rights violations that is more supportive of the victim through reconciliation is the best way to dismiss the notion that such settlements do not provide justice to the victim. Rambe, as one of the victims of the Tanjung Priok incident, admitted that justice for the victim is reached not solely by punishing the perpetrators, but also by considering the victim's condition after what the perpetrators have done. Moreover, some victims involved in *inshallah* (peace settlement in Islam) have asked the court not to punish the perpetrators. This relationship between the victim and the perpetrators seems to form solidarity in these case settlements, which Durkheim separates between repressive law and restitutive law. The request from the victims is what Durkheim called restitutive law, in which the settlement involves reparation, not expiation. This settlement mechanism involves a reconciliation process between the perpetrator and the victim so that justice is deeply felt by both parties.

Choosing reconciliation as one option for the settlement of past gross human right violations is a better alternative, considering the difficulties of finding proper and accurate evidence, which may have already been lost or destroyed.

Reconciliation has to be the basis for the settlement of past gross human rights violations through restorative justice mechanisms. Moreover, it should involve all victimized groups. In the Tanjung Priok incident, for example, at that time, the action conducted by the military forces was deemed correct since the mass demonstration was considered to endanger political and security stability. However, at present, such mass repressive actions conducted by the military forces are considered overreactive; therefore, different perspectives on that situation have to be considered in order to implement reconciliation. A general amnesty is needed so Indonesia can be a strong nation, both morally and legally, and so it can create a stable democracy.

In Chile, for example, the government appoints a truth and reconciliation commission, which is "the moral conscience of the nation," to investigate political murder and missing persons in the era of the military government. These violations have to be accounted for, but not by prosecuting the perpetrators instead of providing compensation to victims.

Furthermore, in Uruguay, truth and justice have become a subject of debate, the supporters of amnesty against those who seek disclosure of crimes. According to Senator Jorge Battle, "amnesty does not mean that the crimes did not take place; it means forgetting them" (Allier, 2006). Thus, truth is more important than justice, and revealing all the facts shall ensure that such violations will not happen again.

Democracy arose to reverse economic conditions, and it needs to be restructured to restore people's belief in the government, hoping that the new program shall be oriented to the future and also guarantee certainty. Compensation in the reconciliation process at least provides assistance to victims to overcome the material aspect of their loss. Moreover, such compensation at least offers official recognition of the suffering of the victims and prevent the state from making the same mistake in the future.

The House of Representatives of the Republic of Indonesia (DPR-RI) and the government should enact the Truth and Reconciliation Commission Law that clearly sets forth the settlement of gross human rights violations. This is an effort not only to settle the violations comprehensively but also to avoid the government's constant interruption of victims' demands. Settlement through the TRC is an effort not of impunity but of comprehensive settlement. Also, settlement through the TRC shall end violations on the legality principle, which is the most fundamental principle in criminal law, as well as resolve the state's burden in the past without tearing open old wounds and creating new ones.

In this matter, the government needs to amend the Human Rights Court Law, especially Article 43 regarding the establishment of an ad hoc human rights court and Article 47 regarding the establishment of the TRC. If the government would like to apply for settlement through reconciliation, then it should delete Article 43 and restate Article 47 on the establishment of the TRC as the settlement venue for cases of gross human rights violations. This is in accordance with Article 76 Paragraph (1) of Law No. 39 of 1999 concerning Human Rights (Human Rights Law), which establishes the Human Rights Commission to develop conditions conducive for the implementation of human rights protections that is in accordance with Pancasila, UUD 1945, the UN Charter, and the Universal Declaration of Human Rights. Moreover, the deletion of Article 43 is a concrete step to consistently implement Article 28 I Paragraph (1) UUD 1945 and Article 11 Paragraph (2) of the Universal Declaration of Human Rights, in which rights that cannot be prosecuted by the retroactive law are included as basic rights that cannot be restricted under any condition. Therefore, it is wiser to apply the principle that is not against UUD 1945 and the Universal Declaration of Human Rights.

The TRC Law has to be declared inapplicable by the Constitutional Court before it is enacted. Therefore, the new draft of the TRC Law should consider the stipulation in the constitution. To be accepted by all parties, the truth shall be formulated as the historical truth. This can then be followed by forgiveness for perpetrators and compensation for victims from the government. Franz Magnis Suseno argues, "We have to break the culture of revenge and retaliation since it shut down our spiritual culture. Therefore, reconciliation is important."

Marzuki Darusman states that the government and the DPR RI have to consider the establishment of a commission or an agency to implement the reconciliation process (Darusman, 1999, p. 169). Frans Hendra Winarta argues that Indonesia cannot move forward and establish a fair and prosperous nation if revenge is still sought. Moreover, Winarta contends that Indonesia should refer to the experiences of other nations in settling past gross human rights violations (*Suara Pembaruan*, 2001).

Meanwhile, Ma'mun Ibnu Ridwan quotes the opinion of Nucholish Madjid (a Muslim scholar) that wounds experienced by Indonesians are less severe compared to what has been experienced by the people in South Africa (*Pelita*, 2001, p. 4). So, if South Africa can implement national reconciliation, why cannot Indonesia do the same?

The Constitutional Court should agree with the government as well as the judge who provided a dissenting opinion that the establishment of the TRC shall be executed and be implemented through the following process:

1) Begin with the report/claim, then follow with the investigation and clarification of the gross human rights violation.
2) Issue a decision that grants or rejects the application for compensation, restitution, or rehabilitation, and provide recommendations concerning amnesty in the form of legal consideration.
3) In the event the application is approved. it becomes the obligation of the government and it shall be implemented within three years from the decision issued, and amnesty shall be decided within three days from when the TRC decision is submitted to the president.
4) Amnesty consideration shall be discussed by the DPR-RI as requested by the president.
5) In the event the president grants amnesty, then the case shall be settled and the perpetrators cannot be retried under either civil or criminal law because it is already the state's obligation to provide compensation and rehabilitation to the victims or their families.

If the president rejects amnesty, then the perpetrator shall be prosecuted through the Human Rights Court by the process stipulated in the Human Rights Court Law. Afterward, the president and the DPR-RI shall form an ad hoc human rights court. Although the implementation comprises five stages, this does not mean the TRC process is a judicial one. However, the decision shall be considered final as it grants or rejects the amnesty submission to the president because the TRC has no competency to settle gross human rights violations cases that have been registered in the Human Rights Court.

The establishment of the TRC is categorized as a *pro justitia* mechanism, which shall be a guarantee of legal certainty, especially because its establishment is based on the law. In the TRC mechanism, after the TRC carries out its investigation and determines there is a gross human rights violation, the Human Rights Commission shall make a recommendation to the attorney general, to be followed by a *pro justitia* investigation by the attorney general after the suspected perpetrator is named. Afterward, the DPR-RI shall make a recommendation to establish an ad hoc human rights court, then the court is established by presidential decree.

The DPR-RI has no authority to decide whether there is a gross human rights violation, but only to conclude the place and time of the gross human rights violation (Constitutional Court No. 18/PUU-V/2007).

The cancellation decision of the Truth and Reconciliation Commission by the Constitutional Court is less precise than the spirit and struggle of the victim who tries to stand on a formal legal foundation that allows the realization of accountability. Such a foundation should become the frame of consideration of the Constitutional Court's judges that the government has proven it has the political will to settle past gross human rights violations; such a mandate can also be seen as the nation's willingness to settle the past in order to establish a better life in the future.

The Constitutional Court bench does not consider carefully that the basis of the TRC in prevailing Indonesian law and regulation is really strong since it stems from the TAP MPR No. V of 2000 concerning Strengthening National Unity and Integrity, which recommends the establishment of the TRC as an extrajudicial agency with the mandate to reveal past abuses of power and human rights violations. The bench consideration seems to isolate the technical problem without taking into consideration the need to understand the tragedy in the past from a reparatory as well as a reconciliation point of view.

In its decision, the Constitutional Court bench associates the TRC with alternative dispute resolution, which is conceptually different. Alternative dispute resolution (ADR) is born from the contractual agreement between the parties in the event a dispute arises between them, while the TRC is the transitional mechanism for resolving violations conducted by the previous regime. The recommendation from the Constitutional Court bench to use reconciliation is also hard to implement without a valid legal mechanism, as described earlier.

3 CONCLUSION

Based on the aforementioned explanation, there are four important elements for the establishment of the TRC. First, it focuses on past gross human rights violations. Second, it aims to get a comprehensive description of the gross human rights violations in a certain period and does not focus only on a certain incident. Third, the TRC is established for a certain period with a certain goal, and it will be dissolved after issuing its final report. Last, the established TRC has to have the authority to access the information of every government institution so that no obstacles appear in accessing the information as well as summoning the specific parties involved.

Reconciliation is an effort to reach a peaceful solution by inviting all involved parties to settle past gross human rights violations, without any differences in motives, backgrounds, or goals. Such reconciliation should involve various settlement mechanisms to restore as well as to recognize the self-esteem of the disputing parties so that each party shall forget the past to face a better future.

Seven reasons are listed in the application for judicial review of the TRC Law:

1) The stipulation in Article 27 of the TRC Law causes compensation and rehabilitation to depend on the amnesty approval, not the substance of the case.
2) Amnesty in Article 27 of the TRC Law requires the existence of a perpetrator. As a consequence, without any perpetrator, then amnesty cannot be given, so that the victim will not have any guarantee of rehabilitation.
3) Such a stipulation puts the victim in an imbalanced legal position because they have to fulfill the hard requirement of amnesty approval to realize their rights.
4) Article 27 of the TRC Law creates an imbalanced position between the victim and the perpetrator. It discriminates against the rehabilitation rights of the victim, who has suffered as the result of the gross human rights violations.
5) Article 44 of the TRC Law positions the TRC as the same institution as the court that closed access to judicial process settlement.

6) Article 44 of the TRC Law does not allow settlement through an ad hoc human rights court; if the settlement has been conducted through the TRC, then it will be considered as abolishing the state's right to prosecute the perpetrator as stipulated in international law, either in practice or in international agreement.

7) Amnesty for perpetrators of gross human rights violations is contrary to international law; however, Article 1 point 9 of the TRC Law stipulates that amnesty can be given.

According to the Constitutional Court, Article 27 of the TRC Law contains contradictions, especially between the stipulation concerning the following: (1) the perpetrator admits his mistakes, affirms his regret, and is willing to ask the victim for forgiveness; (2) the perpetrator shall apply to the president for amnesty; (3) such an application can be approved or rejected; (4) compensation and/or rehabilitation can only be given if the amnesty is approved by the president; (5) if amnesty is rejected, the case shall be forwarded to an ad hoc human rights court.

Although the Constitutional Court granted judicial review only on Article 27 of the TRC Law, since the implementation of the TRC Law depends on such a stipulation, then Article 27 of the TRC Law goes against UUD 1945 and is not legally enforceable, the whole provision of the TRC Law becomes inapplicable. This is so because Article 27 of the TRC Law is related to Article 1 point 9, Article 6 point c, Article 7 Paragraph (1) point g, Article 25 Paragraph (1) point b, Article 25 Paragraphs (4), (5), and (6), Article 26, Article 28 Paragraph (1), and Article 29 of the TRC Law. Furthermore, Article 27 and its related articles will determine whether the overall stipulation of the TRC Law is applicable. Therefore, since Article 27 does not have legal binding force, the articles related to amnesty are not legally enforceable.

Although the TRC Law is not legally enforceable, the Constitutional Court does not choose other options to settle past gross human rights violations through reconciliation. This means such settlements can still be conducted under UUD 1945 and universal human rights instruments through political policy in the event of rehabilitation and amnesty in general. This dismisses the notion that settlements of past gross human rights violations cannot provide justice to the victim. As discussed, justice shall be given in two forms: restitutive justice and restorative justice, which mean justices is not only to punish the perpetrator but also has to consider the interest of the victim. Applying such reconciliation settlements will result in justice that will be deeply felt by the victim as well as the perpetrator.

Related to these conclusions, reconciliation shall be made in the form of legal policy (law), which is by UUD 1945 and universal human rights instruments through political policy in the event of rehabilitation and amnesty in general. Justice is not only to punish the perpetrator but also to consider the interest of the victim. Since the TRC Law has already been declared invalid by the Constitutional Court, the new draft of the TRC should not be contrary to UUD 1945 or universal human rights instruments. For the new TRC Law to be accepted by all parties, the historical truth must be revealed.

REFERENCES

Allen, Rob. (2005). "Rethinking Retribution: A Critique of Simple Justice." 32–39.
Allier, Eugenia. (2006). "The Peace Commission: A Consensus on the Recent Past in Uruguay?" *Revista Europea de Estudios Latinoamericanos y del Caribe/European Review of Latin American and Caribbean Studies* 87–96.
Aru. (2006). "LSM Minta Aturan Kerja Dibuat oleh KKR yang akan Terbentuk – Aturan Pelaksana UU KKR." www.hukumonline.com. March 27.
Ball, Patrick et al. (2000). "Making the Case: Investigating Large Scale Human Rights Violations Using Information Systems and." American Association for the Advancement of Science.
Bhargava, Anurima. (2002). "Defining Political Crimes: A Case Study of the South African Truth and Reconciliation Commission." *Columbia Law Review* 1304–1339.

Daly, Erin. (2002). "Reparations in South Africa: A Cautionary Tale." *University of Memphis Law Review 33*, 367.

Darusman, Marzuki. (1999). *Kohesi Nasional dan proses-proses Rekonsiliasi: Suatu Perspektif Hak-Hak Asasi manusia.* Komnas HAM, p. 169.

Droždek, Boris. (2010). "How Do We Salve Our Wounds? Intercultural Perspectives on Individual and Collective Strategies of Making Peace with Our Own Past." *Traumatology 16*(4), 5.

Durkheim, Emile. (2014). *The Division of Labour in Society.* Simon and Schuster, 2014. Also see Gibbs, Jack P. (). "A Formal Restatement of Durkheim's 'Division of Labor' Theory." *Sociological Theory 21*(2), 103–127.

Freeman, Mark and Joanna R. Quinn. (2003). "Lessons Learned: Practical Lessons Gleaned from inside the Truth Commissions of Guatemala and South Africa." *Human Rights Quarterly 25*(4), 1117–1149.

Geula, Marianne. (2000). "South Africa's Truth and Reconciliation Commission as an Alternate Means of Addressing Transitional Government Conflicts in a Divided Society." *Boston University International Law Journal 18*, 57.

Gibson, James L. (2006). "The Contributions of Truth to Reconciliation: Lessons from South Africa." *Journal of Conflict Resolution 50*(3), 409–432.

Huyse, Luc. (1995). "Justice after Transition: On the Choices Successor Elites Make in Dealing with the Past." *Law & Social Inquiry 20*(1), 51–78.

Kompas. "Presiden Bentuk Panitia Seleksi Calon Anggota KKR." April 2, 2005.

Llewellyn, Jennifer J. and Robert Howse. (1999). "Institutions for Restorative Justice: The South African Truth and Reconciliation Commission." *University of Toronto Law Journal*, 355–388.

Pelita. "Kabinet Baru dan Semangat Ishlah." August 1, 2001, p. 4.

Phillips Timothy and Mary Albon. (1999). When Prosecution Is Not Possible: Alternative Means of Seeking Accountability for War Crimes. In Belinda Cooper (ed.), *War Crimes: The Legacy of Nuremberg.*

Robert K. (1994). "Durkheim's Division of Labour in Society." *Sociological Forum 9*(1). Kluwer Academic Publishers–Plenum Publishers.

Suara Pembaruan. "Urgensi Pembentukan Komisi Kebenaran dan Rekonsiliasi." June 10, 2001.

Van Zyl, Paul. (1999). "Dilemmas of Transitional Justice: The Case of South Africa's Truth and Reconciliation Commission." *Columbia University Journal of International Affairs 52*, 647–668.

Wisnu Nugroho A. (2006). "Proses Seleksi 42 Calon Anggota KKR Dihentikan." www.kompas.com. December 18.

Advancing Rule of Law in a Global Context – Susetyo, Rinwigati Waagstein & Budi Cahyono (eds)
© *2020 Taylor & Francis Group, London, ISBN 978-1-138-32782-5*

Enhancing the ASEAN Way: Integrating international law into local initiatives in the ASEAN

Chad Patrick Osorio
University of the Philippines, The Philippines

ABSTRACT: The Association of Southeast Asian Nations (ASEAN) is a regional organization of countries aiming to promote amity and international cooperation between its member states. It seeks to build a mutually beneficial community supported by three pillars: political-security, economic, and sociocultural. The goal of the ASEAN is primarily economic in nature: improving the mobility of goods and services within the region as well as facilitating trade with external partners. The ASEAN was never meant to be a unitary political body similar to the United States or the European Union, hence the ASEAN Way: its emphasis on non-intervention in sovereign affairs and consensus-building. Because of this, the ASEAN remains a mere potential to be a global force for change, particularly in areas of pressing regional concern like international environmental law and transnational criminal law, among others. This paper forwards recommendations for transforming the ASEAN into more than just an economic community; rather, the ASEAN should be a leader of social change in the global setting. This paper focuses on cementing the ASEAN Way through more binding multilateral agreements, integrating various aspects of international law into meeting the United Nations' Sustainable Development Goals (SDGs). It does so through the perspective of citizen engagement, particularly youth empowerment, in order to rally regional political support for these initiatives. It concludes that for successful development in the region to take place, law and governance must be inclusive for all.

Keywords: ASEAN Way, Citizen Engagement, Inclusive Development, International Law, Sustainable Development Goals, Youth Empowerment

1   INTRODUCTION

The Association of Southeast Asian Nations (ASEAN) was established in 1967, a regional organization of countries aiming to promote amity and international cooperation between its member states. It seeks to build a mutually beneficial community supported by three pillars: political-security, economic, and sociocultural (Khoman, 2012). Indonesia, Malaysia, the Philippines, Singapore, and Thailand are the ASEAN's five founding members, later joined by Brunei Darussalam (1984), Viet Nam (1995), Lao PDR (1997), Myanmar (1997), and Cambodia (1999). The regional organization, in its 1967 Bangkok Declaration, sought to promote peace, progress, and prosperity in the region (ASEAN Declaration, 1967). Fifty years later, the ASEAN has moved toward loftier aspirations than just intercountry friendship: its new goal is to build a stronger, more stable economic cooperation, termed the ASEAN Economic Community (ASEAN Economic Community Blueprint, 2008). This initiative is considered the "largest integration effort attempted in the developing world" (Petri, Plummer, & Zhai, 2012, pp. 93–118), aiming for unitary production hubs coupled with a single market base, streamlining the movement of goods, services, and skilled labor within its collective territory, as well as facilitating investments and capitalism for its economies. The idea is to promote equitable and inclusive economic development among its member states and all relevant stakeholders.

Thus far, these efforts have yielded excellent results. Despite being an emerging entity, the ASEAN's collective economy is now considered the sixth largest in the world (Factsheet on ASEAN Community, 2017). As of the fourth quarter of 2015, its estimated worth lies at US$2.4 trillion, with an annual real growth rate of 5.3%. Economists predict a confident outlook (Lipton, 2015).

Because of its economic might and the possibilities that it offers, the regional organization possesses the excellent potential to be a platform through which areas of interstate cooperation can be launched, particularly regarding transnational concerns. These include government efforts to battle crimes beyond borders, as well as international environmental cooperation and climate action leadership, to name a few.

The reasons for cooperation beyond economics are varied. For one, many ASEAN countries share contiguous land boundaries and porous maritime borders. Geographically, the ASEAN countries are divided into two groups: Mainland Southeast Asia, composed of Cambodia, Laos PDR, Myanmar, Thailand, and Viet Nam; and Island Southeast Asia, comprising Brunei Darussalam, Indonesia, Malaysia, Singapore, and the Philippines (Boomgard, 2007). Because of this, there is a confluence of similar concerns that require cooperative action.

For example, various ASEAN member states share common environmental threats due to their geographical proximity. These include maritime pollution, affecting the common pelagic territories of countries such as Viet Nam, Indonesia, and the Philippines (Tahir & Pang, 2013), and transboundary haze, which similarly victimizes neighbors Indonesia, Malaysia, and Singapore (Jones, 2006). Another shared challenge in the region is the multilevel threat of climate change: in fact, ASEAN member states have been named as some of the countries that are the most vulnerable to the effects of global warming (Lee, 2013). Climate impacts vary, from severe periods of intense drought (Dai, 2013) to increasingly year alone, it attracted 7% of the total global foreign direct investments amounting to US$121 billion. With increasing relentless hydrometeorological disasters such as typhoons and flash floods (Abano, 2013).

Another area of common concern relates to terrorism, feared to be gaining a foothold in the region. The Islamic State (IS) has taken credit for attacks in the region, whether successful or not. Indonesia, Malaysia, Singapore, and the Philippines have recently been the object of these attacks. It is necessary to address this, especially since reports indicate that Southeast Asia is likely becoming a fertile source of new recruits for terror groups (Prameswaran, 2015).

Equally pressing and no less alarming are other transnational crimes, which are offenses whose elements and effects transcend national borders and the authority of a single sovereign state. ASEAN economic integration is a double-edged sword: the facilitation of mobility of people, goods, and services in the region also serves as a factor enabling crime. Eleven of these crimes are directly recognized under the ASEAN, from economic crimes like money laundering to crimes against national security like terrorism and piracy (Prameswaran, 2015).

To provide an exhaustive list of potential areas for international cooperation in the region requires a longer narrative, but many more exist than can be enumerated in this short paper. The ASEAN is not blind to these problems: in many cases, it actually and expressly states the need for immediate, compelling, and collective response. Unfortunately, the regional organization is also infamous for its ASEAN Way.

No formal definition for the ASEAN Way exists. Different scholars define the ASEAN Way differently. For example, Katsumata (2003) characterizes the ASEAN Way as the principles of noninterference in the internal affairs of other members, quiet diplomacy, the nonuse of force, and decision-making through consensus. Acharya, on the other hand, notes that the ASEAN Way has four key elements: informality, noninterference and nonconfrontation, consultation and consensus, and the preference of bilateral cooperation over multilateralism (Novikrisna, 2015). In any case, these definitions all similarly place sovereignty and consensus as utmost values in interstate cooperative action (Quah, 2015).

These values, however, have resulted in a lack of decisive action by the regional bloc and its individual member states, a problem that has long been noted and criticized:

> ASEAN's response [to these challenges] has primarily consisted of a rhetorical device. The members have adopted modest declarations that have failed to move beyond the proclamation of good intentions. They have focused on non-binding and unspecific measures without addressing the question of funding, setting target dates and establishing monitoring mechanisms to assess progress. Appropriate measures would have to deal with all these issues and would most likely need to be binding to lead to more tangible results. (Emmers, 2002)

This criticism succinctly summarizes the frustration faced by advocates for stronger bonds for cooperative action in the region. But why is this important?

2 DISCUSSION

2.1 *The interaction between international law and domestic law*

Multinational entities, including the United Nations, the European Union, and the ASEAN, to name a few, shape interstate cooperative action through the interplay of international and domestic laws. This they do through a number of means: from holding high-level meetings discussing global and regional agendas, to planning national strategies, to providing technical aid and assistance for project implementation at the level of the community (Elliot, 2003).

However influential these international institutions are, they still face numerous challenges, one of which is the translation of regional goals and agreements to domestic law and policy. At this point, it is important to emphasize the complementary relationship that the latter two share.

Virtually all transnational issues have roots in the community; they gain momentum only by replication and interconnections, making the issue a national concern, and thereafter an international one. Therefore, it is necessary to provide domestic laws that address these concerns at their inception; at the same time, these laws must be coordinated from a bird's-eye perspective in order for them to provide a coherent legal and policy framework. This is why effective international cooperation, in whatever field it may be, relies on the proper implementation of this twin system of national and international law. Concepts must be transformed into policy and policy into action. This overall system is governed by two separate but complementary legal regimes: international law, under which treaties fall, and state sovereignty governing domestic legislation.

The Vienna Convention on the Law of Treaties is applicable for the first legal regime. Article 2 (1) (a) of the said Convention defines a treaty as "an international agreement concluded between States in written form and governed by international law, whether embodied in a single instrument or in two or more related instruments and whatever its particular designation" (Aust, 2013).

Under the general principle of *pacta sunt servanda*, states are required to comply with the obligations set out by the international agreements to which they are parties; apart from this, however, they have the freedom to strategize the method and manner of fulfillment of such agreements (Mendez, 2013). Noncompliance with any provision therein is not excusable by the simple defense of sovereignty.

Unfortunately, despite this mandate to comply with the provisions of the treaties to which they are signatories, national legislative processes do not necessarily no automatically incorporate these provisions into the domestic legal regime. Oftentimes, it requires further state legislation to do so. In other words, the government must take an additional step – that is, through the proper legislative agencies, it must pass laws applicable within its sovereign territory, which specifically lends implementable form to the spirit of the treaties to which the government is a state party. Otherwise, unless legislated and rendered executable, the provisions of the treaty will be difficult to put into practice. The exceptions to this rule are treaties that contain provisions that are considered, under the domain of international law, as of a general or customary nature.

This means, then, that despite the tremendous influence that international and regional organizations hold, their impact is not maximized if their initiatives are not bolstered by national legislation and policy implementation. The extent of commitment provided by the domestic legal regimes dictates the level of effectiveness of international and multilateral agreements (Kheng-Lian, Robinson, & Lin-Heng, 2016). This concept encompasses various aspects of international cooperation, from environmental causes to trade agreements to law enforcement.

At the same time, at the domestic level, there must be unity of action among the different entities and institutions comprising the national government. While the legislative branch remains the primary actor in translating international law into domestic legislation, the executive still holds dominion over its effective implementation. At the same time, the judiciary must assist in these endeavors by taking proper consideration of the spirit of the applicable international law, and seek to enforce it to the best of its ability. Even in countries like Saudi Arabia, where the traditional system of checks and balances cherished by democratic governments has been replaced by a pure monarchy, agencies still exist that mimic the functions of government bodies abiding by the concept of separation of powers. This means that despite whatever instance, streamlined, cooperative, and collaborative intrastate action should be strived for.

To understand this more fully, let's take the example of an international regional organization like the ASEAN and one of its member countries, the Philippines. The ASEAN as a regional organization facilitates agreements between its member nations. The Philippines as a member of the ASEAN must uphold these agreements due to the principle of *pacta sunt servanda*; this applies both under international law and as mandated by the Philippine constitution. However, in many cases, this process is not automatic: the Philippines' bicameral congress, as its prime legislative body, must first concur with the treaty and then afterward pass measures to operationalize these provisions in the Philippine politico-legal context (Magallona, 2010). The two other branches and their instruments must follow the mandate of the law passed.

From this process, the inter-reliance of national and international law is clear. Agreements, declarations, and resolutions forged under international law must be cascaded to the community through domestic legislation, in order for them to be properly supported and implemented throughout the entire government bureaucracy. Unfortunately, this is the brunt of the criticism leveled against the ASEAN Way: because of the primacy the ASEAN Way places on the absolute sovereignty of its states, the ASEAN refuses to enter into legally binding agreements that member states view as an intrusion on their sovereignty. Because these agreements are not legally binding, there is no compelling force to issue domestic legislation supporting these initiatives.

This trend has been observed in various areas of cooperation in the ASEAN. One of them regards transnational criminal justice. The region's leaders have repeatedly claimed to understand the serious impact that transnational crimes pose to various aspects of national and regional well-being (ASEAN, 1999); because of this, they have promised stern measures to address these concerns (ASEAN, 1997). However, an analysis of the regional summits on the matter through the years has revealed that despite this acknowledgment, concrete actions were lacking, especially since there was little explicit and precise actual evaluation of transnational crimes and the dangers they present (Haacke & Williams, 2008).

The region's stance on transnational crimes therefore lacks teeth: policies are vague and implementation on the ground lacks coordinated international strategy. Domestic criminal laws do not reflect intercountry dialogue, rendering resolutions in the latter difficult to enforce. From this it is apparent that the transnational criminal law framework in the region is mere lip service: despite all these words, no actual or effective law enforcement or judicial management mechanism exists in the region to address this problem, leading to a greater threat of transnational crimes among its member states, particularly from organized syndicates (*Jakarta Post*, 2014).

This is a huge problem considering that organized transnational crime in the region is worth an estimated aggregate US$100 billion annually; the largest part is contributed by the illicit drug trade, with illegal exports of timber and wildlife the second largest part (Broadhurst, 2016). The United Nations Office on Drugs and Crime recognizes this and the potential harms that it may cause the region's integration efforts (United Nations, 2016). It urges ASEAN member countries to devise practical solutions to address the matter (Herman, 2014). In the recent period, this has been a topic touched on by the Philippines as the host country for the ASEAN's fiftieth year. It promised more teeth in its policies in battling crimes beyond borders (ABS-CBS, 2016). But until now, little improvement has been seen in the field.

Another topic worth touching on is the looming threat of terrorism in the region. While terrorism is considered a transnational crime under the ASEAN legal regime, it warrants a separate discussion because of its immediate and pressing nature. A historical analysis of the region's responses regarding cross-border terrorism shows remarkably vague plans of action for emergency situations at both the regional and national levels (ASEAN, 2001). True, there may be bilateral and multilateral agreements on the matter, and they have provided immensely important assistance to individual countries; however, collective strategic action within the ASEAN community remains lacking when it comes to antiterrorism measures (Haacke).

Criminal law is territorial in essence; with a few exceptions, it is applicable only within the confines of the sovereign country that promulgates it. It is therefore understandable that there is hesitancy in adopting legally binding criminal justice and enforcement treaties that seek to intrude in its conceptual territory. However, challenges with the application of the ASEAN Way are not limited to criminal law alone. Environmental law in the region also faces a similarly stubborn stance.

In fact, the ASEAN Way is critiqued as inadequate and inappropriate to handle worsening environmental challenges (Takahashi, 2000). The ASEAN initiatives are characterized as "demonstrably ineffective" in addressing regional environmental concerns (Elliot). This has been attributed to the regional body's staunch stand on sovereignty and the subsequent lack of binding international agreements: nonbinding plans tend to have weaker spheres of influence in domestic affairs as compared to those that are enforceable under international law (Takahashi). Because of the omission of national systems to comply with soft law environmental standards, initiatives often fall short of their goals (Elliot).

It is clear from multiple aspects of international law that the ASEAN Way as it currently stands poses barriers to addressing key regional concerns, including transnational crimes, terrorism, and matters related to the environment. Of course, on the other side of the coin are experts who defend the ASEAN Way, claiming that from the conceptual perspective, the ASEAN's emphasis on informality, consensus-building, and the low level of legalization of international agreements actually facilitates strategic action (Kahler, 2000).

While this may be true to a certain extent, scholars are also increasingly advocating for greater legalization in these key areas, similar to what has been done thus far for dispute resolutions for economic agreements between the ASEAN's member states (Alvarez, 2007). This entails changing the perspective of the ASEAN Way: while it should remain unchanging that sovereignty and consensus despite diversity should be the driving forces behind the region's unity, the perspective on international law and its accompaniment of legally binding agreements must change. The latter, especially, should be viewed not as an intrusion into the absolute power of the state, but rather as a substructure to support the foundation of stronger and more effective national and regional governance.

The next part of this paper discusses recommendations for how to integrate international law into local initiatives, but at the same time keep the spirit of the ASEAN Way alive.

2.2 *Proposed actions*

It is undeniable that multilateral cooperation in the ASEAN is increasing through the years. However, many of the resolutions and statements that the regional body issues provide mere lip service to the causes that they seek to serve (Varkkey, 2012). These aspirations of cooperative action are considered nonbinding under international law and are not as easily translatable to domestic legislation as compared to legally binding treaties and agreements (Takashi). While there may be a lot of projects, programs, and initiatives at the regional level, they do not necessarily cascade to their intended beneficiaries. In many cases, domestic policies and action plans fail to reflect the intent and to realize the goals of the international instruments from whence they originate (Elliot).

This is why it is important to increase the legalization of certain aspects of the ASEAN: its member states must be encouraged to enter into more legally binding agreements, particularly regarding areas of pressing concern such as environmental law and transnational criminal justice (Tiquio & Marmier, 2017).

A system of incentives and sanctions for state actors can unarguable that there are certain aspects of international cooperation that should not be limited to soft law. In fact, improving compliance with these initiatives and legally binding frameworks in various aspects of international law can strengthen this mechanism in order to provide support to national legislation and implementation (Takahashi).

In line with this, the mind-set that legally binding agreements are intrusions on the sovereignty of the state should be changed. Instead, the exercise of such agreements should be viewed as a tacit recognition of sovereignty; it should be regarded as the authority of the state to choose to uphold or to waive the rights that accompany such sovereignty (Saguisagv. Executive Secretary, 2016).

It must also be emphasized that it is important to integrate different aspects of the law with each other. For example, binding environmental treaties could adopt provisions related to economics and trade: this assures a mechanism for incentives and sanctions, leading to more impact. Provisions may include climate financing, transactional payments for ecosystem services, and green employment, among others (ASEAN, 2009). Studies show that this system works: debt-for-sustainable-development swap agreements help stimulate trade, with the resulting economic gains as the proverbial carrot (ASEAN, 2007). On the other hand, acting as the stick are sanctions like trade restrictions and tariffs, which apply to entities not compliant with the environmental agreements their respective countries have signed (Urpelainen, 2017).

This amalgamation of international trade law and international environmental law upholds the preservation of resources and the protection of the environment across the region; more so when it provides a clear mechanism of responsibilities and accountabilities in both international statements of consensus and domestic legal frameworks.

A number of other examples of equal importance appear in integrating fields of law. A heavy criticism against the exercise of transnational criminal policies is its lack of focus on human rights (Osorio, 2016); in the same way, antiterrorism measures take little notice of potential applications of provisions subsumed under international humanitarian law (Osorio, 2017). Allowing intersections between these fields of international law enriches their practice and implementation on the domestic front. It also creates avenues by which experts from different fields can collaborate to solve pressing regional problems, except this time with the added value of a varied perspective.

It is equally important that people are educated about international law and how it applies to everyday life. This is to remove the practice and philosophy of international law from its ivory towers and lend citizens the mind-set that in this day and age, it is the individual who is at the center of international law. This kind of thinking has the potential to empower citizens to take more active roles in policy development and decision-making in the community, and to lend their voices to discussions on matters of national and international importance.

Conceded, this is easier said than done, considering that an estimated 8% of the entire ASEAN citizenry does not know how to read or write (ASEAN, 2015). Per member country, this varies wildly, from Cambodia's 78.1% literacy rate to Brunei Darussalam's 97.6%. It remains a challenge to reach out and inform people of their rights, obligations, and roles under both national and international law. This is why members of the legal profession are key in spearheading this montage of education and development, primarily by holding legal education campaigns in communities translated into non-legalese language. These education campaigns should apply both to hard law, as in binding treaties like the Paris Agreement, and to soft law, including meeting the United Nations' Sustainable Development Goals. By doing so, advocates gain indispensable partners in grassroots implementation, as well as political support in properly forwarding the application of international law on the domestic front.

Interestingly, gaining communities as partners can lead to citizen advocacies and public clamor, both instrumental elements in driving social change (Johnson, 2013). This political capital can not only translate to proper policy implementation at the national level but may also boost the accession, ratification, and adoption of international treaties and agreements that serve the best interests of these communities. Citizen engagement, inclusive in nature, is therefore important at all levels: from local initiatives all the way to regional and international policy setting, law, and governance.

An important consideration is the role of the youth in forwarding the application of international law in various areas of development. The ASEAN's young population is projected to be the driving force of the region's future, and it is important to consider this impact early on. This is why structural changes must be applied to the education sector, in order for students and educators to integrate legal education and basic concepts of international law into academic discussion. This knowledge of the law enables the ASEAN youth to clearly see the consequences of their actions in the global context, and to realize that they can do much to contribute to cascading principles of international law into their own spheres of social influence.

3 CONCLUSION

The ASEAN Way has been much maligned due to its emphasis on sovereignty and not entering into more binding multilateral agreements. This tendency for low legalization in the region leads to diminished potency of transnational policies to be implemented in local communities. This is why regional initiatives in various areas of cooperation fall short of their goals, or otherwise violate important tenets of international law.

It is necessary to change this perspective. While maintaining the ASEAN Way of primacy on consensus and sovereignty, entering legally binding agreements under international law should be viewed not as the shedding of a piece of the sovereignty of the state party, but rather as the affirmation of its utmost authority to either recognize or deny the rights associated with its exercise. A simple change in mind-set may lead to the more effective and integrative application of treaty provisions in domestic legal frameworks.

At the same time, citizen engagement also plays a huge role in integrating international law into local initiatives. Not only can citizens provide political support for their governments to accede, ratify, or adopt legally binding treaties, they are also partners for change at the grassroots. An important demographic which needs to be engaged is the youth sector, an emerging social, political, and economic force in global and regional governance. To reach the youth, it is necessary that legal education, which includes discussions on principles of international law, be forwarded through formal and informal channels.

Through these proposals, it is clear that the ASEAN Way of consensus-building and respect of sovereignty need not be changed. They are guiding principles that allow cooperation and facilitate amity among member states. It is only the mind-set in relation to international law that must be revised. Having done so, the barriers that the ASEAN Way has seemingly erected can now dissolve, hopefully paving the way for a more peaceful and more prosperous ASEAN community.

Advancing Rule of Law in a Global Context – Susetyo, Rinwigati Waagstein & Budi Cahyono (eds)
© 2020 Taylor & Francis Group, London, ISBN 978-1-138-32782-5

Enriching legal studies with socio-legal research

Sulistyowati Irianto
Faculty of Law, University of Indonesia, Indonesia

ABSTRACT: The aim of this paper is to describe Indonesian law studies that should prepare to go beyond its old tradition toward quickly growing knowledge and science, in line with fast-changing society. Future knowledge will be multi-, inter-, and cross-disciplinary in character, as a result of collaboration among scholars of different fields who focus on certain issues (including law) and who seek to meet society's needs through academic novelty. Legal science is obliged to explain injustice among disadvantaged groups in society. Legal research should be open to all possibilities for enrichment in its methodology, with methods transplanted from other social sciences, humanity studies, even natural sciences. Legal science will not lose its paradigmatic character when its research uses a combination of a doctrinal approach and fieldwork with various methods. Socio-legal research whose home is at faculties of law offers such enrichment to legal science.

Keywords: Legal Research, Legal Science, Multidisciplinary and Interdisciplinary Approach to Law, Socio-legal Research

1 INTRODUCTION

1.1 *The need of cross-disciplinary approach to law*

The failure of many legal policies, including the well-known "Law and Development Movement" and "The Rule of Law Orthodoxy Movement" in distributing justice has shown that, to a certain extent, on both a theoretical and a practical basis, the mainstream of law studies has limitations (Carothers, 2006; Golub, 2006). Many problems of injustice cannot be explained by a solely textual approach. Breakthroughs are urgently needed in many aspects of legal academic work and legal practice in order to make the law work for everyone (Commission on Legal Empowerment of the Poor, 2008, 2009). This new idea leads to methodological consequences in legal studies. This paper proposes a multidisciplinary or interdisciplinary approach in legal studies to get a better understanding of the relation between law, society, and culture, which has been named *socio-legal studies* and *socio-legal research*. Cross-disciplinary research methods, which are transplanted to legal research, are essential to find such in-depth, basic, and enlightening explanations of the lawmaking process and whether it works in society.

It is a fact that the law has many faces. Therefore, legal scholars have no single agreement on its definition. In general, the law has been defined as a set of rules of conduct that regulates and coerces society; the law also regulates how to settle disputes (Otto, 2007, pp. 14–15). Following the (obsolete) paradigm of legal centralism, the law is always perceived as the state's law, the only legal reference in society and correlated with the state's legal institutions. This perspective has been criticized for a long time by many scholars, led by Griffiths (1986). Law is not made in a vacuum, neither is it capable of being studied in one. This is especially true for state law, which intermingles with other legal systems in ways that are characteristic of legal pluralism. Legal anthropologists perceive the law from a wider perspective that includes but is not limited to state law and the laws and system of norms that work outside the state's system altogether with all of its process and actors within it.

The coexistence of state law and other laws (*adat* law, religious law, or systems or norms) in certain social arenas is a fact, and the study of legal pluralism has become a tremendous school of thought developed by legal scholars and other social scientists from universities, research centers and academic forums, and epistemic communities (Wiber, 2005). Nowadays, globalization provides almost unlimited connection between countries, and international law, mainly in the humanitarian fields, makes the legal pluralism scheme more complex.

The coexistence of different laws in one social arena requires reshaping, as several legal systems seem to lose their respective demarcations when a case reflects an intersection of different laws. State law has never been clear-cut, and it is a completely separate entity from other laws. Scholars have suggested that globalization results in globalized law. This can be seen in the extension of international (humanitarian) law into national law; *adat* law and religious law are also affected in some ways (Benda-Beckmann, Benda-Beckmann, and Griffiths, 2005). Vice versa, some values and norms of *adat* law are also adopted in new arrangements of global law, such as methods of non-adjudicative or alternative dispute resolution that stem from explicitly customary principles. However, the notion of law should not be limited to state, international, and transnational law, but should refer to all those objectified cognitive and normative conceptions for which validity for a certain social formation is authoritatively asserted. Law manifests in many forms and is comprised of a variety of social phenomena (Benda-Beckmann and Benda-Beckmann, 2006, p. ix).

The most important thing to glean from this definition is that the law does not consist only of normative conception, i.e., what is forbidden and what is permissible, but it also consists of cognitive conception. On a normative basis, stealing, murder, and corruption are forbidden by state, religious, or customary law. However, in the cognitive context, the definition of murder and corruption may vary depending on the political and cultural context. The Maduranese or the Bugis who feel their pride has been disgraced and accordingly commit what is called *carok* and *Siri* do not feel that they are committing any crime, especially murder. The same logic applies to corruption. Any law would forbid the crime of corruption due to its destructive nature. However, the cognition of corruption may vary in society. Does corruption conducted in a group for a "good purpose" have the same definition as that meant by the legislation?

2 DISCUSSION

2.1 *Socio-legal studies*

The study of law in developing countries needs to go beyond the doctrinal approach. The analytical approach is required to know the content of the legislation. However, the status quo textual approach in legal studies does not give an adequate understanding of how the law works on a daily basis, or about the relation between law and society and "the effectiveness of law and its relation with its ecological context" (Otto, 2007, p. 11). The need is especially to get a better understanding of how law works in society, how people respond to the law, and how certain enigmas in law cut across social, political, cultural, and economic issues (Commission on Legal Empowerment of the Poor, 2008, 2009). Law is not isolated from the society in which it exists. Law should be challenged in legal practices. In a way, legal research should welcome other methodological approaches derived from other disciplines like anthropology, sociology, history, psychology, feminist studies, or even natural science. Based on such reasoning, the study of law should take a multidisciplinary and interdisciplinary approach.

It is important to clarify, especially for Indonesian legal scholars, that socio-legal studies should be distinguished from at least three other disciplines: these are sociology of law, sociological jurisprudence, and anthropology of law. Sociology of law is most highly developed in Western Europe and in the school of thought of law and society in America that has adopted a disciplinary bond with the social sciences (Banakar and Travers, 2005). Socio-legal studies is different from sociology of law, in which the intellectual root comes from sociology, and its purpose is to construct a theoretical understanding of the legal system. Legal sociologists have conducted sociology of law by placing law in a wide social structure. Law as a social regulation mechanism, as a profession, and as a discipline, has become the concern of sociology of law (Cotterell, 1996, p. 6). It mainly concerns how the law works in the daily experience of society members (Wignosoebroto, 2002). The law, as meant here, is the social norm that has been formalized in the form of legislation (state law). The scope of this study is mainly to examine the function of the law in society by observing the structure of law and its enforcement. Important concepts to study in this vein include social control, socialization of law, stratification, and law and social changes (Wignosoebroto, 2002, pp. 3–16). Since this mainly refers to sociology, a sociological methodology is used that is characterized by sociology's tradition of quantitative research.

While the prefix "socio" in "socio-legal studies" does not refer to sociology or social science, it does refer to the intersection of legal issues with nonlegal issues. The "socio" in "socio-legal studies" represents law's interface with the context within which it exists. According to Wheeler and Thomas (cited in Banakar and Travers, 2005), socio-legal studies is an alternative approach to challenge the doctrinal studies of law. This is why when socio-legal researchers use social theory for analysis, they do not mean to call attention to sociology or other social sciences, but to focus on law and the studies of law. Besides, law, the prescription of law, and the definition of law are not assumed or taken for granted, but they are analyzed problematically in terms of their emergence, articulation, and purpose (Banakar and Travers, 2005). Socio-legal studies is a study of law that uses a combination of legal research methods and social sciences methods in a wider context. As a "new" school of thought, socio-legal studies, through various recently published books and journals, has described legal theories, methods, and topics that are more grounded and that have garnered much attention from legal scholars.

Sociological jurisprudence is a mainstream legal theory founded by Roscoe Pound and developed in America since the 1930s. The term "sociological" refers to realism in the field of study of law (Holmes, cited in Langone, 2016; Wignosoebroto, 2002, pp. 8–16), which argues that even though law is produced logically through a process that is imperatively accountable, the life of law has not been logical; rather, it is a sociopsychological experience. A presiding judge must be proactive in making a verdict to settle a dispute by taking social facts into consideration. Thus, the verdict will fulfill society's sense of justice. From this thought, a well-known doctrine emerged in sociological jurisprudence that the law is a tool of social engineering.

One more well-known discipline is anthropology of law or legal anthropology, which might be misperceived as socio-legal studies. Anthropology of law studies how law as part of culture works in daily life. It focuses on how actors respond to a state law or create a justice mechanism through which they can do so. The work of the law is elaborated through its relation with another aspect of culture, such as the economy, social and power relations, and religion. Ethnography, the main method in anthropology, has been adopted in legal anthropology as ethnography of law (Latour, 2009).

It happens in many parts of the world that faculties of law house socio-legal scholars. They are mostly legal scholars who are experts in major legal studies like constitutional law, state law, penal law, civil law, and labor law. However, at the same time, they are paying great attention to socio-legal issues and researching and publishing on issues that cut across law and society, work later known as socio-legal research. They are called "double box legal scholars" because they are experts in major subjects of law, but also in related area studies (like Asian, Southeast Asian, Latin American, or African studies) or in specific issues in area studies. They perhaps have limited contact with sociology experts because these particular studies are not developed in departments of sociology or other social science departments (Banakar and Travers, 2005).

Banakar and Travers explain that in England, the development of socio-legal studies is mainly driven by the needs of faculties of law to nourish and develop interdisciplinary studies of law. Socio-legal studies is perceived as a discipline, subdiscipline, or methodological approach that has emerged in its relation or its opposition to the law. This study has never been developed by any social scientist or sociologist. This can be seen in sociology syllabi or in the traditions developed in sociology departments that rarely pay interest to the theoretical as well as the practical (Banakar and Travers, 2005, pp. 1–26).

Considering the wide scope of methodology that might be used as an entry point into socio-legal studies, it is therefore not acceptable to reduce socio-legal research to empirical legal research, or to a scope of legal research that is usually associated with field studies to discover how the law works and operates in society. The method of socio-legal studies is beyond that. However, socio-legal experts must pose a firm understanding of legislation, instrument, and substantive law that is related to their field of studies, then critically analyze it.

The proximity of socio-legal studies to social science is apparent; it is on the same page of methodology. Method and research techniques from social science are studied and used to collect data. The method in sociology and anthropology, the so-called mother of social science, is developed by socio-legal researchers. Under the sociological or anthropological approach, the substance of the law is more deeply elaborated. At this moment, along with the latest approaches such as discourse analysis, cultural studies, feminism, and postmodernism, post-colonialism has gained a place in socio-legal research. The issues studied also vary and include the lawmaking process, courtroom studies, non-ligation dispute settlement mechanisms, corruption, environmental law and natural resources, the legal issues of labor, gender justice, and many more.

Even with the characteristic differences between sociology of law, sociological jurisprudence, anthropology of law, and socio-legal studies, all of these schools of thought commonly understand socio-legal studies as an alternative legal study. The common understanding is that socio-legal studies perceives the law in a wide social context, with all of the methodological consequences of that approach. This method emphasizes the importance of studying law without perceiving law as an independent study isolated from culture (way of thought, a system of knowledge) and power relations among legal drafters, law enforcers, parties, and society at large.

2.2 Socio-legal research

Socio-legal research utilizes a combination of doctrinal and empirical approaches. The doctrinal approach is strengthened with broader tools of analysis derived from other disciplines. Under the framework of multidisciplinary and interdisciplinary studies of law, legal studies will not lose its paradigmatic character.

A socio-legal research method can be identified through the two following aspects. First, the socio-legal researcher conducts textual studies. Therefore, socio-legal studies deals with the heart of legal studies, as it analyzes the range of legal instruments from the constitution in the highest structure of law to village regulation in the lowest one. The textual and documentary approach is used to analyze and to criticize certain article(s). In this regard, the article should be carefully read to identify the key words and to grasp its meaning. As learned from feminist jurisprudence, legal questions concerning women can be applied as a tool of analysis (Dalton, 1988; Juergens, 1991; Whitman, 1991). Certain articles raise these questions: What identities or images of women, including their sexuality, capacities, roles, and value, are projected by law? Does the law reflect women's realities and experiences, including their violations, and, if so, which women? Is the issue addressed by the law? Should it be the issue given women's experiences? What aspects of women's lives are affected? Given the experiences and realities of women, does the law protect or benefit them, and, if so, which women? The analysis lies on elaboration of the word's meaning and its implication for the subject of law (including disadvantaged groups and women).

Studies of judicial decisions or verdicts are highly important (Hammerslev, 2005). Judicial decisions or courtroom studies provide very rich legal materials. Based on its extensive experience, for instance, the Center of Women and Gender Studies of the Universitas Indonesia has developed courtroom studies (Irianto et al. 2004; Irianto and Cahyadi, 2008; Irianto and Nurcahyo, 2006). The method applied is called courtroom studies based on the textual judicial decision that is combined with fieldwork by gathering data from observation and interviews of the parties involved in the case. One of the important points to examine is whether the judicial decision has been a breakthrough for the fulfillment of justice. The technique used here involves reconstructing the case. Who are the disputing parties? What is the legal problem (scarce resources under dispute)? What is the plea of the parties? What is the judicial consideration in the verdict?

Second, socio-legal studies has developed various "new" methods from legal research methods and from social science, such as qualitative socio-legal research (Ziegert, 2005) and socio-legal ethnography (Flood, 2005). Thomas Scheffer employs actor-network theory to describe the performance of judges and lawyers using the micro-histories of legal discourse (Scheffer, 2005). Banakar and Seneviratne conduct studies that focus on the use of text and discourse analysis to examine the performance of ombudsmen (Banakar and Travers, 2005). Reza Banakar develops a case study to research legal culture (Banakar and Travers, 2005). Sally Engle Merry, in one beautiful ethnographical work, describes the ethnography of an international tribunal in which the problems of social justice, human rights, and women are promoted. It is an elaboration of various treaties, documents, policies, and declarations that produces what is called transnational consensus-building (Engle Merry, 2005).

Generally, legal anthropologists have developed ethnography of law to study the dispute settlement process among community members, which is common in their daily life. Samia Bano (2005) uses ethnography to study the use of unofficial dispute settlement forums such as Shariah councils among immigrant South Asian Muslim women who live in England. Anne Griffiths (2005) uses field research of the Bakwena society of Africa to explain the experience of doing law in that society and to respond to the concept of Western law. Both Bano and Griffiths use the feminist qualitative approach to law.

A researcher who studies legal pluralism developed modern ethnography of law according to global issues that make legal pluralism come closer to perceiving the phenomenon of variety of law (Benda-Beckman et al., 2005). Therefore it is important to study the actors that have formed the melting pot of the legal system and that cause the law to move, for instance, those who conduct interreligious marriage (Glick-Schiller, 2005), immigrants (Nuijten, 2005; Zips, 2005), or an epistemic community (Wiber, 2005). Methods developed by such an interdisciplinary approach may explain the wide scope of this new legal phenomenon and its relation to power and to the social, economic, and cultural contexts in which such law exists.

This development of legal studies in the perspective of socio-legal studies has become important. This is because many Indonesian legal scholars are still searching for a "pure" legal research method as a mono-disciplinary approach that is uncontaminated by social science. This misunderstanding has happened among legal scholars at large. In their eyes, the social science method is only identified with a quantitative approach that deals with variables, measurement, hypothesis, experiment through statistical measurement, and the problem of sampling and representation with tight procedures, which is more in the scope of the positivistic paradigm. Such a perception may, of course, be misleading because the research method in social science is only noticed partially while the quantitative approach under the positivistic paradigm includes other paradigms: interpretivism (phenomenology, constructivism) and criticism with countless research techniques (Neuman, 1997; Sarantakos, 1997).

It mostly happens that a scholar believes that social science or humanity studies, including legal studies, will be considered scientific if they perform as natural sciences. In this regard, the law phenomenon must be materialized, observed, and measured. Objectivity and neutrality are preserved. Researchers and objects of study may be placed apart. The application of this principle in natural science is a must and, like objectivity, precision, or accuracy, it can be achieved. Natural science in the laboratory has a fixed character.

When the natural science principle is transplanted to social and humanities studies, it becomes problematic. Can humans be perceived as other natural materials? Humanity is the creator of meaning, and humans are born as creatures of will, free, and dignified, not bordered by principles beyond themselves such as in other natural materials. Humanity lives in the space of interpretation; therefore, the concepts of self, reality, and knowledge (including legal knowledge) are also the result of interpretation. Is it possible that the social science scholar creates distance from his subject of study, which is also human? Can the legal phenomenon be perceived as natural materials? Can law be materialized, measured with certainty? How about the concept of *ubi societas ibi ius*, which states that wherever there is society, there will be law? It is a well-known adage that shows that the existence of humans is closely related to law. It is barely possible to separate law from society and culture. It may be hard to measure how a human being thinks about or interprets the law.

Indeed, the quantitative approach is acceptable to gain a general description or to map causal relations among variables. Nevertheless, this approach cannot be used to gain in-depth understanding of the legal phenomenon being studied. This approach has failed in giving an in-depth explanation of how humans practice law in daily life.

The study of law in Indonesia is highly influenced by the "black letter" legal tradition. The legal text is treated as a given natural material, and law is isolated from society and context. Even though handbooks of modern constitutional theory may always contain "new" legacies such as critical legal theory or feminist jurisprudence, it seems that mainstream legal thought other than legal positivism only attracts the interest of a few legal scholars.

Let us examine another paradigm within social and humanities studies that may help explain the legal phenomenon and its relation with human beings. The interpretive and critical paradigms have surprisingly become the dominant theoretical and methodological approaches in social science and the humanities, and, in fact, they have become home to a pure legal research method that is based on textual analysis.

An interpretive paradigm that is close to the hermeneutics tradition emphasizes the examination of text, including legislative text. The researcher tries to find the meaning embedded in the text. When conducting studies of a text, the researcher attempts to understand and enter the inner part of a certain view, which represent a holistic view. Then he or she builds an in-depth understanding of how each part of the text is connected and forms a reconstructed explanation. Such meaning is rarely simple or obvious. The researcher may only understand the meaning if he/she carries out a detailed study of a certain text, contemplating numerous messages in a text and searching for the relation between parts of the text (Neuman, 1997, p. 68).

The critical paradigm that includes other critical theory such as feminist jurisprudence, in my opinion, has helped the mainstream legal research method with the question of disadvantaged groups of people and women in analyzing certain legislation. Word by word, sentence by sentence, the legislative text is carefully observed and analyzed with a critical question: Is the law made as a formalization of common will and society's interests? Or is the law drafted as a tool to define the power of the ruling, and, if so, whose power? By what means may power be exerted through the drafting of legislation? Furthermore, the critical paradigm argues that law may be used as a tool to engineer and improve a certain condition.

3 CONCLUSION

The needs of an alternative legal approach may be seen in the roots of law school in Indonesia since the beginning of higher legal education. Law school (*Rechtshogeschool*) was established in Batavia in 1924, and, ever since, the foundation of socio-legal discourse has been laid. Paul Scholten, a former judge and lawyer and one of the prominent founders of *Rechtshogeschool*, contends that the science of law has sought meaning for its existence (*het betaine*). However, such meaning may not be perceived without connecting law, history, and society. The purity of law has been preserved by the legal scholar, while, in fact, legal materials consist of impurity; therefore, to force such methods will produce only bloodless phantoms (Scholten, 2005, p. 13) or skeletons without muscle (Hoebel, cited in Ihromi, 2001, p. 194).

Scholten's argument initially stems from his criticism of the Kelsenian legacy that perceives law as a natural material. The law is treated a soulless material isolated from society and historical context (Cotterell, 1986; Scholten, 2005). There is a distance between the object of study and the researcher, and such distance is preserved in the name of objectivity and neutrality.

According to Scholten, the law consists not only of legislation and regulation but also verdicts, customs, agreements, wills, and torts conducted by people (Scholten, 2005, p. 4). Law is not a natural material. In my point of view, even legislation and regulation are the product of political bargaining, and it is hard to believe that law may be isolated from any political interest or power relation.

This is consistent with Cotterell in explaining the weakness of legal positivism. Treating legal data as mere legal text does not represent the dynamic of the legal phenomenon. It also does not represent the reality of regulation and its continuous changes that come from the complex interaction of individuals and groups in society. Legal positivism identifies legal data as far as possible, seeing the backstage process of drafting legislation and without considering the attitudes and values of the drafters. As long as the law is found, it is not necessary to understand the definitions of justice and injustice, policy, efficiency, morality, and politics of law (Cotterell, 1986, pp. 10–11). The cultural perspective is mainly discussed by a legal anthropologist. Law is closely related to culture; the notion of interpreting the law as mere "black letter" legislation is not realistic since law can be perceived as a living anthropological document.

Thus, presenting alternative studies of law will enrich the doctrinal study of law in developing countries, including Indonesia. Legal scholars who have studied law using a multidisciplinary or interdisciplinary approach have developed disciplines in philosophy of law, sociology of law, anthropology of law, *adat* law, feminist jurisprudence, and legal studies. Their studies cut across what we know today as socio-legal studies. They have done critical analysis of legal texts (documents) while presenting legal practices in a complicated constellation of power relations in society. They have combined doctrinal and empirical law studies. In conducting such empirical studies, they are free to borrow research methods from a wide range of social sciences and humanities studies, which are continuously changing and leaving classical methods behind. In today's world, the need for socio-legal studies has grown as modern knowledge and science head in the direction of cross-disciplinary studies.

Indonesia has four times throughout its history adopted a new constitution, i.e. the 1945 Constitution (effective from August 18, 1945), the 1949 Federal Constitution (effective 1949–1950), the Provisional Constitution of 1950 (effective 1950–1959), and the application of the 1945 Constitution with its explanatory notes adopted by presidential decree on July 5, 1959, which was continuously effective until the first constitutional amendment of 1999. After the constitutional crisis in 1998, the process of change was conducted not by replacement but by addendum, according to the consensus made before the agenda of constitutional change was agreed to by all political parties in the People's Assembly in 1999. Under the new approach, the constitutional change could smoothly fulfill the need for continuity and change within the dynamic competition between the conservatives and the reformists during the critical time of change in 1998–1999.

We could say that the Indonesian constitutional changes of 1999, 2000, 2001, and 2002 (the first four stages of integral changes) were incremental in a formal sense. However, the contents of change are nothing less than a big bang. The Indonesian constitution of 1945 is the shortest constitution in the world, containing only 1,393 words comprising 71 ruling verses. After the Reformation, with only four amendments, the content of the constitution increased by nearly 300%. By the adoption of the fourth amendment in 2002, the contents included 199 ruling verses. Of the 199 verses of the new constitution, only 25 came from the original document. The other 174 verses were new rules adopted into the 1945 Constitution. The total words written in the new version of the 1945 Constitution grew from 1,393 in the original to 4,559 in the new version (a 327% increase). Therefore, the Indonesian experiences between 1999 and 2002 can be understood as an incremental big-bang constitutional change.

3.1 *Respecting, fulfilling, and promoting human rights*

The new contents of the 1945 Constitution cover a wide range of aspects and almost all parts of the constitution received new articles. One of these is the article concerning human rights, which were adopted in Articles 28A through 28J. Originally, the 1945 Constitution was designed without special respect for human rights, which were associated with liberal and individualistic discourses of Western colonial power. Besides, the framers of the constitution objected to the adoption of the articles on human rights due to the integralistic notion of state shared among the founding leaders of Indonesia. Soepomo claimed that Indonesia could not adopt the idea of separation of power or Montesquieu's doctrine of trial politics and that the articles on human rights were not suitable to be outlined in the constitution.

By the second constitutional amendment of 2000, almost all international instruments on human rights were adopted into Articles 28A through 28J, consisting of twenty-six verses on human rights. By its adoption, the 1945 Constitution had become one of the most modern humanistic constitutions in the world. Under the new 1945 Constitution, human rights are fully respected, promoted, and must be fulfilled accordingly by all state and non-state actors.

3.2 *The emergence of the green constitution today and the blue constitution tomorrow*

Today, many constitutions in the world have adopted a pro-environment policy. This is the new phenomenon of the green constitution. Two of the greenest constitutions in the world are the French Constitution of 2006, and the Ecuador Constitution of 2008. In 2006, the Preamble of the French Constitution, which originated from the 1789 Declaration of Man and of Citizens, was added along with the adoption of the Charter for Environment of 2004. Then in 2008, the new constitution of Ecuador was adopted in which the idea of nature's fundamental rights are set forth explicitly, just as important as the articles on human rights. These two greenest constitutions in the world reflect the new human awareness of the importance of constitutional protection of the living environment and nature's rights for the sustainable future development of mankind.

In Indonesia, the second amendment, in 2000, of the 1945 Constitution also adopted a pro-environment article. Article 28H (1) of the amended constitution[1] reads: "Every person shall have the right to live in physical and spiritual prosperity, to have a home and to enjoy a good and healthy environment, and shall have the right to obtain medical care." Article 33 (4)[2] reads: "The organization of national economy shall be conducted based on economic democracy upholding the principles of togetherness, the efficiency with justice, sustainability, environmentalism, self-sufficiency, and keeping a balance in the progress and unity of the national economy." Under the green policy, the government and the parliament may not make any law or regulation, or any administrative decisions, contrary to the green constitutional principles and policies.

In the future, the phenomenon of the green constitution must also be followed by the idea of the blue constitution related to space and the virtual world. Article 33(3) of Indonesia's 1945 Constitution only states that "the land, the waters and the natural resources within shall be under the powers of the State and shall be used to the greatest benefit of the people." It does not say anything about resources above the land and waters, as if they do not belong to the State of Indonesia. Compare this to the US Constitution, which asserts ownership over the territory and property in one integrated clause; the articles of property and territory in Indonesia's 1945 Constitution are separated into Chapter IXA on territory and Chapter XIV on national economy and social welfare. In Article 10, however, "the President holds the supreme power of the Army, the Navy and the Air Force," meaning that the president has the supreme power to protect that country and its interests, along with its territory and property in the air, in space, and in the virtual world above the land and waters, by use of the air force.

The rapid development of information and communication technology has made mankind dependent on the virtual world. The constitution must cover any policies regarding the air, the virtual world, and even outer space for the national interest, and for peace and human prosperity. Therefore, constitutional law experts from all over the world have to take part in developing a blue constitution.

3.3 *Moving from a merely political constitution to the new perspectives of economic, social, and cultural constitutions*

The Constitution of the United States of America is political by nature. As a modern political constitution, its contents are merely about politics and relationships between functions and institutions of power, and political relations between state institutions and citizens. That's why C. F. Strong used the term "political constitution" in his book *Modern Political*

---

1. Second Amendment of 2000.
2. Fourth Amendment of 2002.

Constitutions (1966).[3] Most constitutions in the world follow the political tradition of the US Constitution that other aspects, such as economic, social, and cultural policies, are not addressed in the constitution. Among the reasons why the US Constitution does not contain economic subjects is that the drafters of the US Constitution were fully occupied with only political consideration for the establishment of the independent federal state of America, and they left economic affairs to be regulated in and by the marketplace.

Before the establishment of the federation, America developed as an industrious society in which economic affairs were handled independently under free market capitalism. Therefore, the framers of the US Constitution did not think it important to include any articles of economic policy. Only later in US history did the economic aspects of the US Constitution emerge from the interpretation of the Supreme Court, as James Buchanan discussed in his *Economic Interpretation of the American Constitution*.

Today, many countries include articles of economic policy in their constitutions. Not only communist countries have a tradition of covering economic policy in their constitutions, but some noncommunist countries have such a tradition as well, as exemplified in the Irish Constitution, India's constitution, the Indonesian 1945 Constitution, and many others. Alongside the growing role of the market economy in today's globalizing era, the role of the economic constitution is also growing, to control the free market. The Indonesian experience is one such example. Chapter XIV of the 1945 Constitution contains special articles about the national economy and social welfare policies. Therefore, I call the 1945 Constitution a political constitution as well as an economic constitution that makes the national economic system "a constitutional market economy," i.e. a free market limited by the constitution as the highest policy norm.[4]

Besides the growing concern about economic constitutions, scholars have to look further at the ideas of social[5] and cultural[6] constitutions. Constitutions today are used not only for the organization of state power but also for the organization of civil societies, corporate constitutions, and even villages' constitutions, such as in American tribal villages. Therefore the study of the constitution expands from the perspectives of the political constitution to include economic, social, and even cultural constitutions. Scholars engaged in constitutional studies could devote more attention to the wider aspects of constitutions, not limited to the conventional meaning of the political constitution.

3.4 *Constitutional law and ethics*

Today, a constitution cannot be regarded as the source only of constitution law but also of constitutional ethics. The role of ethics in public offices has developed rapidly since the last decade of the twentieth century. Even the General Assembly of the United Nations in 1997 recommended that its member countries develop ethics infrastructures for public offices. By doing so, the world is expected not only to be dependent on the role of law but also to develop an effective ethics for public offices. We need legal and constitutional government as well as good government. Since its Reformation, Indonesia has adopted a system of ethics for public offices. For the enforcement of judicial ethics, we have established the Judicial Commission set forth in Article 24B, Chapter IX of the 1945 Constitution. The People's Assembly's Decision No. VI/2001 on Ethics for the Nation's Public Life has supplemented the guiding principles of ethics laid out in the 1945 Constitution.

Based on the Indonesian Constitution and the ruling of the People's Assembly, almost all state institutions have now been equipped by law with the code of ethics and a special committee to enforce it. The balanced roles of constitutional law and constitutional ethics are expected to overcome the weakness of the system of the rule of law by

3. Strong, C. F. (1966). *Modern Political Constitutions*. London: Sidgwick & Jackson.
4. Asshiddiqie, Jimly. (2010, 2012). *Konstitusi Ekonomi, Penerbit Kompas*. Jakarta.
5. Asshiddiqie, Jimly. (2014). *Gagasan Konstitusi Sosial: Institusionalisasi dan Konstitusionalisasi Kehidupan Sosial Masyarakat Madani, LP3ES*. Jakarta.
6. Asshiddiqie, Jimly. (2017). *Konstitusi Kebudayaan dan Kebudayaan Konstitusi, Intrans*. Malang.

the application of the system of rule of ethics at the same time. Not all problems can be overcome through a legal approach. Some problems related to public offices are more effectively addressed by the rule of ethics. Besides, the effectiveness of the legal norm is also dependent on the effectiveness of the ethical norm in practice. Law is like a ship, and ethics is like the sea. The law will never sail and reach the island of justice when there is not enough water in the sea. Therefore, the future of constitutional law needs support from the concept of constitutional ethics; every constitutional lawyer should take part in developing the study of constitutional ethics.

For instance, in the field of ethics for election management bodies and for members of parliament, the code of ethics is enforced through a court of ethics. In the House of Representatives, Indonesia has established "Mahkamah Kehormatan," or the court of honor. For election management bodies, the Honorary Council of Electoral Management Bodies enforces the code of ethics through an adjudication process, like in an ordinary court of law. By application of this system, Indonesia has begun to introduce a new system of courts of ethics, besides the conventional system of the courts of law and justice. Along with the notion of constitutional law and the principles of the rule of law, we have to develop also the notion of constitutional ethics and the principles of rule of ethics for the future.

3.5   *The new* Quadro Politica

Another important issue is a new form of separation of power within a constitutional state. A new phenomenon of power relations has arisen in the world with the emerging hegemonic role of electronic and social media. In the middle of the twentieth century, the Montesquieu doctrine of trial politics of 1689 has been added with the new branch of power, i.e. the media, as the fourth estate of democracy. The executive branch, legislative branch, judicial branch, and media have become four branches of power or four estates of democracy. It changed from a trial politics to a quadru politics.

The actual form of quadru politics now consists of the new domain of powers, i.e. (i) the state, (ii) civil society, (iii) business corporations, and (iv) the media. In the Indonesian experience, there is a new phenomenon of businessmen developing television broadcast companies, then establishing a political party or managing several political parties through financial support, and finally running for the presidency. If no regulation limits these potential conflicts of interest and separates the four power domains, democracy will be hijacked.

The new four estates of democracy should therefore be separated from each other to avoid any potential conflicts of interest. The new dominant factors in democratic politics are corporate, political, and economic capitalism and electronic and social media hegemony. How to control them for the benefit of the people as a whole has become an important issue. Without separation of power between the new four estates, there will be no more democracy based on the future standard of quality. Democracy and the democratic system will be hijacked or ploughed under by corporate capitalism. Therefore, we have to separate the new four estates and to promote the principles of government of the people, by the people, and for the people, and continue to be with the people.

3.6   *The impact of globalization on the new relation between international public law and domestic constitutional law*

In the globalizing world, constitutional law as a subject of study has been changing significantly from positive and domestically oriented to a general science of law applicable everywhere in the world. The phenomenon of the European Union and its constitutional treaty, for instance, has blurred the distinction between domestic constitutional and international public law in Europe. We are now on the move from exclusively domestic-oriented constitutionalism to inclusive regional and international constitutionalism.

Within this new world, the study of comparative constitutional law becomes more and more important. At present, the judicial interpretation of the constitution through comparative reference to other countries' constitutions has become a common cause. Even in Germany's Constitutional Court, EU law is now treated as on the same level with the German Constitution; German law may not be contrary to the German Constitution or EU law. The structure of the European Union is also new to the conventional political history of our world. The organizational structure of the European Union is just like that of a state consisting of executive, legislative, and judicial branches of power. The European Union also has its monetary system, just like an independent state. Therefore, in Europe, there is no longer a clear distinction between international public law and domestic constitutional Law. A professor of constitutional law in Europe today is just like an international law professor.

Because of this phenomenon, today, we are moving toward universalization of constitutional values around the world. All modern constitutions share common values, some of which some are borrowed or transplanted from other countries or other international best practices, making them all look similar in substance. However, we have to look into the problems of interpretation and implementation of these constitutions in the cultural contexts of the respective countries concerned. Universalization of constitutional values is not identical with internationalization or globalization of values. Universal values may come from outside as well as from our respective cultural histories. Therefore, the issues of constitutional culture should become pivotal in constitutional studies today. It is now time to pay more attention to the issues of "cultural constitutions and the constitutional culture" of each country so that the institutionalization of constitutional rules will not conflict with their living cultural traditions.

Indonesia has its own long historical experiences with so many pluralistic cultural traditions throughout the country. But most of the ideas adopted into the constitution are transplanted or borrowed from other countries. So it is the task of scholars to build an intellectual bridge between modern constitutional state institutions and the cultural living traditions of the people. We have to avoid cultural divide or discrepancy between the political institutions and the living traditions of the people. The state and its power is nothing less than the power of the people, for the people, and by the people themselves. Even today, I always adapt that famous quotation from Abraham Lincoln: not only the government of the people, by the people, and for the people, but also with the people. The government is always of the people, by the people, for the people, and with the people.

3.7   *From universalism to multiversalism, toward a cosmopolitan legal pluralism*

Another issue following the trend of universalization of legal and constitutional values and the need for a cultural reading of the law and the constitution is the issue of global pluralism. We live now in a globalizing and borderless world, where a single act or actor is potentially regulated by different, multiple legal or quasi-legal regimes. We have to live in new complex relationships among international, regional, national, and subnational legal systems, where non-state actors such as industry-setting bodies, nongovernmental organizations, religious institutions, ethnic groups, and others exert a significant normative pull. We cannot depend anymore on the old perspectives of sovereign territorialism, or on a substantive universalist approach that requires people to be conceptualized as fundamentally identical to be brought within the same normative system. We are now moving from universalism toward multiversalism, which Paul S. Berman calls "cosmopolitan legal pluralism"[7] as a useful approach to the design of procedural mechanisms, institutions, and discursive practices.

A cosmopolitan pluralist approach manages multiplicity without attempting to erase the reality of that multiplicity. The key solution for legal scholars to the problem of legal pluralism is not pluralism itself, but a global, comparative study of laws and constitutions. Legal scholars in every country have to pay more attention to this need. Scholars of countries with

---

7. Berman, Paul S. (2012). *Global Legal Pluralism: A Jurisprudence of Law beyond Borders*. Cambridge University Press.

common law traditions have to know civil law traditions. Lawyers of one country should know and understand well the legal systems of other countries with intensive interactions between people or businesses. For example, Indonesia, as a nation that has very close relations with Western countries as well as with China, India, Japan, Korea, and the Middle East, has to pay more attention and promote comparative studies of the legal systems of those countries, including comparative studies of the legal systems of every ASEAN member country, such as Malaysia, Singapore, Thailand, Vietnam, the Philippines, etc.

We have to develop new perspectives in the study of the constitution. Besides continuing to develop (i) the positive constitutional law approach, we have to promote also in a more active way the studies of (ii) comparative constitutions, (iii) the science of constitutional law, (iv) constitutional ethics, (v) cultural constitutions, (vi) economic constitutions, and (vii) green and blue constitutions. So many aspects of the inner structures of constitutional law are also changing rapidly, such as the structures and domains of power, that the principle of separation of power among the executive, legislature, and judiciary is no longer rigid in its implementation. Press media are emerging rapidly and changing the modes of communication and the information system to such an extent that they have become the fourth estate of today's democracy. But the four estates no longer consist of the executive branch, the legislature, and the judiciary, but the state, civil society, corporate business, and the media, because of the pivotal roles of the rapidly growing market economy and civil societies globally. To guarantee the quality of today's democracy, we need to separate the new four branches of powers, i.e. the state, civil society, market, and media, so that they will not present conflicts of interest.

REFERENCES

Banakar, Reza and Max Travers. (2005). Law, Sociology and Method. In Reza Banakar and Max Travers (eds.), *Theory and Method in Socio-legal Research*. Oxford: Hart Publishing, pp. 1–26.

Bano, Samia. (2005). Standpoint, Difference, and Feminist Research. In Reza Banakar and Max Travers (eds.), *Theory and Method in Socio-legal Research*. Oxford: Hart Publishing, pp. 91–112.

Benda-Beckmann, Franz. (1990). Changing Legal Pluralism in Indonesia. Sixth International Symposium Commission on Folk Law and Legal Pluralism. Ottawa, Canada.

Benda-Beckmann, Franz. (2006). "The Multiple Edges of Law: Dealing with Legal Pluralism in Development Practice." *World Bank Legal Review: Law, Equity, and Development 2*, 51–86.

Benda-Beckmann, Franz & Keebet Benda-Beckmann. (2006). "The Dynamics of Change and Continuity in Plural Legal Orders." *Journal of Legal Pluralism and Unofficial Law*. Special double issue no. 53–54, pp. 1–44.

Benda-Beckmann Franz, Keebet Benda-Beckmann, and Anne Griffiths (eds.). (2005). Introduction. In Franz Benda-Beckmann, Keebet Benda-Beckmann, and Anne Griffiths (eds.), *Mobile People, Mobile Law: Expanding Legal Relations in a Contracting World*. Farnham: Ashgate.

Carothers, Thomas. (2006). The Problem of Knowledge. In Thomas Carothers (ed.), *Promoting Rule of Law Abroad: In Search of Knowledge*. Washington, DC: Carnegie Endowment for International Peace, pp. 15–30.

Carothers, Thomas. (2006). The Rule-of-Law Revival. In Thomas Carothers (ed.), *Promoting Rule of Law Abroad: In Search of Knowledge*. Washington, DC: Carnegie Endowment for International Peace, pp. 3–14.

Chai-Itthipornwong, Nilubol. (2006). Iriai-ken Court Cases in Japan: Lessons for Traditional Local Community Right's Court Cases in Thailand. In *The Work of 2003/2004 API Fellows, Power, Purpose, Process, and Practice in Asia*. Japan: Asian Public Intellectuals Program, pp. 151–160.

Channell, Wade. (2006). Lessons Not Learned about Legal Reform. In Thomas Carothers (ed.), *Promoting Rule of Law Abroad: In Search of Knowledge*. Washington, DC: Carnegie Endowment for International Peace, pp. 137–160.

Commission on Legal Empowerment of the Poor. (2008). *Making the Law Work for Everyone: Report of the CLEP, vol. 1*. New York: UNDP.

Commission on Legal Empowerment of the Poor. (2009). *Making the Law Work for Everyone: Report of the CLEP, vol. 2*. New York: UNDP.

Dalton, Clare. (1988). "Where We Stand: Observation on the Situation of Feminist Legal Thought." *Berkeley Women's Law Journal 3*, 1–13.

Engle Merry, Sally. (1988). "Legal Pluralism." *Law and Society Review 22*, 869–896.

Engle Merry, Sally. (2005). Human Rights and Global Legal Pluralism: Reciprocity and Disjuncture. In Franz Benda-Beckmann, Keebet von Benda-Beckmann, and Anne Griffiths (eds.), *Mobile People, Mobile Law: Expanding Legal Relations in a Contracting World*. Farnham: Ashgate.

Flood, John. (2005). Socio-legal Ethnography. In Reza Banakar and Max Travers (eds.), *Theory and Method in Socio-legal Research*. Oxford: Hart Publishing, pp. 27–32.

Glick-Schiller, Nina. (2005). Transborder Citizenship: An Outcome of Legal Pluralism within Transnational Social Fields. In Franz Benda-Beckmann, Keebet Benda-Beckmann, and Anne Griffiths (eds.), *Mobile People, Mobile Law: Expanding Legal Relations in a Contracting World*. Farnham: Ashgate, pp. 27–50.

Golub, Stephen. (2003). *Beyond the Rule of Law Orthodoxy: The Legal Empowerment Alternative*. Rule of Law Series, Democracy and Rule of Law Project, Number 41, October 2003. Washington, DC: Carnegie Endowment for International Peace.

Golub, Stephen. (2006). A House without a Foundation. In Thomas Carothers (ed.), *Promoting Rule of Law Abroad: In Search of Knowledge*. Washington, DC: Carnegie Endowment for International Peace, pp. 105–136.

Golub, Stephen. (2006). The Legal Empowerment Alternative. In Thomas Carothers (ed.), *Promoting Rule of Law Abroad: In Search of Knowledge*. Washington, DC: Carnegie Endowment for International Peace, pp. 161–190.

Griffiths, Anne. (2005). Using Ethnography as a Tool in Legal Research. In Reza Banakar and Max Travers (eds.), *Theory and Method in Socio-legal Research*. Oxford: Hart Publishing, pp. 113–132.

Griffiths, John. (1986). "What Is Legal Pluralism?" *Journal of Legal Pluralism and Unofficial Law 18*(24), 1–65.

Hammerslev, Ole. (2005). How to Study Danish Judges. In Reza Banakar and Max Travers (eds.), *Theory and Method in Socio-legal Research*. Oxford: Hart Publishing, pp. 195–202.

Harding, Sandra. (1987). "The Instability of the Analytical Categories of Feminist Theory." *Signs: Journal of Women and Culture and Society 11*(4), 645–665.

Ihromi, Tapi Omas (ed.). (1993). *Beberapa Catatan mengenai Metode Kasus Sengketa yang Digunakan dalam Antropologi Hukum (Some Notes on Trouble Case Methods Used in Legal Anthropology)*. In Tapi Omas Ihromi (ed.), *Antropologi Hukum, Sebuah Bunga Rampai (Legal Anthropology in Editorial)*. Jakarta: Yayasan Obor Indonesia, pp. 194–213.

Irianto, Sulistyowati and Antonius Cahyadi. (2008). *Runtuhnya Sekat Perdata dan Pidana. Studi Peradilan Kasus Kekerasan Terhadap Perempuan (Permeable Wall of Criminal and Civil Court: Courtroom Study on Cases of Violence against Women)*. Jakarta: Yayasan Obor Indonesia.

Irianto, Sulistyowati and Lidwina Nurcahyo. (2006). *Perempuan di Persidangan. Pemantauan Peradilan Berperspektif Perempuan (Women in the Courtroom: A Court Watch on Ten Law Cases in Jakarta and Tangerang West Java)* Jakarta: Yayasan Obor Indonesia.

Juergens, Ann. (1991). "Feminist Jurisprudence: Why Law Must Consider Women's Perspectives." *William Mitchell Magazine 10*(2), October.

Langone, Richard. (2016). "The Science of Sociological Jurisprudence as a Methodology for Legal Analysis." *Touro Law 17*(4), article 5.

Latour, Bruno. (2009). *The Making of Law: An Ethnography of the Conseil d'Etat*. Cambridge: Polity Press.

MacNaughton, Andrew. (2002). Humbling the Litigious: Foreign Nationals Exploring Rights in a Japanese Company. In Rajendra Pradhan (ed.), *Legal Pluralism and Unofficial Law in Social*, Economic *and Political* Development, *Papers of the XIIIth International Congress*. April 7–10. Chiang Mai, Thailand.

Morgan, Jenny. (1988). "Feminist Theory as Legal Theory." *Melbourne University Law Review 16*, 743–759.

Moore, Henrietta. (1988). *Feminism and Anthropology*. Minneapolis: University of Minnesota Press.

Moore, Sally Falk. (1983). Law and Social Change: The Semi-autonomous Social Field as an Appropriate Subject of Study. In Sally Falk Moore, *Law as Process. An Anthropological Approach*. London: Routledge & Kegan Paul, pp. 54–81.

Moore, Sally Falk. (1994). The Ethnography of the Present and the Analysis of Process. In Robert Borofsky (ed.), *Assessing Cultural Anthropology*. McGraw Hill, pp. 362–376.

Nader, Laura and Harry Todd. (1978). Introduction. In *The Disputing Process: Law in Ten Societies*. New York: Columbia University Press, pp. 1–40.

Neuman, W. Lawrence. (1997). *Social Research Methods: Qualitative and Quantitative Approaches*, 3rd edition. Boston: Allyn and Bacon.

Otto, Jan Michael. (2007). *Some Introductory Remarks on Law, Governance and Development*. Leiden: Van Vollenhoven Institute, Faculty of Law, Leiden University.

Sarantakos, Sotirios. (1997). *Social Research*. Melbourne: Macmillan Education Australia.

Scheffer, Thomas. (2005). Courses of Mobilization: Writing Systematic Micro-histories of Legal Discourse. In Reza Banakar and Max Travers (eds.), *Theory and Method in Socio-legal Research*. Oxford: Hart Publishing, pp. 75–90.

Scholten, Paul. (2005). *Cetakan kedua, Struktur Ilmu Hukum. Cet.2, terjemahan Arief Sidharta*. Bandung: Alumni.

Seneviratene, Mary. (2005). Researching Ombudsmen. In Reza Banakar and Max Travers (eds.), *Theory and Method in Socio-legal Research*. Oxford: Hart Publishing, pp. 161–174.

Shapland, Joanna. (2008). Contested Ideas of Community and Justice. In Joanna Shapland, *Justice, Community and Civil Society. A Contested Terrain*. Devon, UK: Willan Publishing.

Steny, Bernardinus. (2005). *Free and Prior Informed Consent dalam Pergulatan Hukum Lokal* (*Free and Prior Informed Consent in Conflicted Local Law*). Jakarta: Huma.

Stephenson, Matthew. (2006). A Trojan Horse in China? In Thomas Carothers (ed.), *Promoting Rule of Law Abroad: In Search of Knowledge*. Washington, DC: Carnegie Endowment for International Peace, pp. 191–216.

Tanner, Nancy. (1993). Matrifocality in Indonesia and Africa and among Black Americans. In M. Z. Rosaldo and Louise Lamphere (eds.), *Woman, Culture and Society*. Stanford, CA: Stanford University Press, pp. 129–156.

Tong, R. P. (1998). *Feminist Thought: A More Comprehensive Introduction*, 2nd edition. Boulder, CO: Westview Press.

Whitman, Christina Brooks. (1991). "Review Essay: Feminist Jurisprudence." *Feminist Studies 17*, 493–507.

Wignosoebroto, Soetandyo. (2002). *Optik Sosiologi Hukum dalam Mempelajari Hukum, dalam Paradigma* (*Optic on Sociology of Law in Studying Law in Paradigm*). In Soetandyo Wignosoebroto, *Metode dan Dinamika Masalahnya*. Jakarta: Huma, pp. 3–16.

Ziegert, Klaus A. (2005). Systems Theory and Qualitative Socio-legal Research. In Reza Banakar and Max Travers (eds.), *Theory and Method in Socio-legal Research*. Oxford: Hart Publishing, pp. 49–68.

Zips, Werner. (2005). "Global Fire": Repatriation and Reparations from a Rastafari (Re)Migrant's Perspective. In Franz Benda-Beckmann, Keebet Benda-Beckmann, and Anne Griffiths (eds.), *Mobile People, Mobile Law: Expanding Legal Relations in a Contracting World*. Farnham: Ashgate, pp. 69–90.

Advancing Rule of Law in a Global Context – Susetyo, Rinwigati Waagstein & Budi Cahyono (eds)
© 2020 Taylor & Francis Group, London, ISBN 978-1-138-32782-5

Counterterrorism, deradicalization, and victimization in the aftermath of terrorism in Indonesia

Heru Susetyo

Faculty of Law, University of Indonesia, Depok, Indonesia

ABSTRACT: In the aftermath of the Bali bombing on October 12, 2002, which is considered the biggest terrorist attack ever against the country, the Indonesian government launched the so-called war against terrorism. Laws against terrorism were soon enacted and applied retroactively. The Special Police to Combat Terrorism – namely, the Detachment of 88 – was formed. A special agency to coordinate counterterrorism measures was also established, called the National Agency for Terrorism Countermeasures (Badan Nasional Penanggulangan Terorisme). Since the early 2010s, the Agency has conducted a special deradicalization program to combat terrorism while the Detachment of 88 has launched massive manhunts for suspected terrorists all over Indonesia, supported by the Indonesian military. The deradicalization program targets former terrorism suspects and inmates. It also targets schools, universities, and other academic institutions suspected of housing agents of radical ideology dissemination in Indonesia. Both the deradicalization program and the massive manhunt have created problems. On one hand, the Agency has claimed that it has prevented terrorism, but on the other hand, it has victimized innocent people such as the family members of terrorism suspects, and it has violated the civil rights of former inmates. In addition, because of its targeting of specific academic institutions, the Agency has been accused of arbitrarily labeling people as terrorists or future terrorists, which is a clear violation of human rights. This research, therefore, is a study of counterterrorism, including the deradicalization program in Indonesia. It tries to describe the dynamics of the program between the needs of law enforcement and its impact on human rights and civil liberty.

Keywords: Deradicalization, Human Rights, Indonesia, Terrorism, Victimization

1 INTRODUCTION

"Where is your father?" a neighbor in Lamongan asked Zahra when her family returned to Lamongan. "He is looking for Allah's mercy," Zahra replied innocently. Zahra copied her mother's answer anytime someone asked her about it. "No, your father is a bomber. He is now in prison in Bali!" the neighbor rudely replied.[1]

Zahra was just a little girl when her father, Fadlulloh, was taken to prison in Bali by Indonesian antiterrorism police following the Bali bombing on October 12, 2002. Her father was arrested in East Kalimantan Province on January 13, 2003. Fadlulloh left a wife, Titin, and two little children. Shortly after the arrest, Titin returned to her home in Lamongan with Zahra and her little sister, who was still just a few months old. However, returning home was another challenge for Titin. She not only had to survive as a single parent and teach at the Women's Islamic Boarding School (Pondok Pesantren Muslimah) in her village, but she also had to withstand verbal abuse

---

1. Taken from a book about a Bali bomber suspect, Noor Huda Ismail, *Temanku Teroris, Saat Dua Santri Ngruki Menempuh Jalan Berbeda* (Bandung: Mizan Press, 2010), p. 357.
   Note: Paper presented for ICLAVE 2017 in Depok City, November 1–2, 2017.

from her neighbor, who kept asking about her husband. These kinds of abuses deeply affected Zahra. One day when she reached home, she asked Titin: "Ummi, where is Abi, actually?" Titin replied, "Be patient, Zahra, you will meet Abi someday."

Another incident took place following the arrest of suspects in Kebumen, Central Java, Indonesia, in May 2013. A suspect named Bayu was killed in an antiterrorism police raid. Bayu used to live in Bugel Kulon Village, Kendal Regency, Central Java. When the police took his dead body to be buried in the village, villagers strongly rejected it and said that Bayu was not an inhabitant of the village, even though his wife and two kids lived in the village. Bayu married his wife in 2008, but he never reported it to the village chief. He never talked to his neighbors and always went away from the village for a few months at a time. Therefore, many villagers did not recognize Bayu. Eventually, the villagers strongly rejected his dead body for burial in the village by stating that they were worried that their village would acquire bad influence and a bad name (as a terrorist village, for instance).

These cases are real stories about the life of the family members of terrorism suspects in Indonesia in the aftermath of terrorist offenses involving their family member/s. There are many stories related to victimization of the family members of terrorism suspects in Indonesia. Sometimes suspects' families, neighborhoods, former schools, former workplaces, and even racial groups are subject to condemnation.

Indonesia is the third largest democracy in the world after India and the United States, the world's largest archipelago, and also the world's most populous Muslim country with around 205 million Muslims in 2013.

Indonesia also has experienced numerous terrorist attacks in recent years. Some of these attacks are the Bali bombings in 2002 and 2005, the bombing of the Philippine ambassador's residence in Jakarta in 2000, the bombing of the Australian Embassy in Jakarta in 2004, the bombing of Jakarta churches on Christmas Eve 2000, the J. W. Marriot Hotel bombings in Jakarta in 2005 and 2009, the shooting of civilians in Aceh in 2010, the robbery of CIMB Niaga Bank in Medan in September 2010, a suicide bombing inside the headquarters of the Cirebon City District Police in 2011, grenade attacks on various police stations in Solo City in 2012, a suicide bombing inside the headquarters of the Poso Regency Police, and many more.

Since the first Bali bombing in 2002, the Indonesian government has applied harsher policies to combat terrorism. The government was strongly pressed by internal and external parties to immediately prepare specific regulations on combating terrorism. In 2003, the Indonesian government enacted Law No. 15 on the Eradication of Terrorist Acts. This law provides for the protection of victims and their heirs due to criminal acts of terrorism (Mansur et al., 2007).

Another program to combat terrorism is called "deradicalization." A special agency for eradicating terrorism, the National Agency for Terrorism Countermeasures (Badan Nasional Penanggulangan Terorisme) (BNPT) was founded to focus on its implementation. Among its jobs is to conduct deradicalization activities within Indonesian educational institutions.

2 PROBLEM IDENTIFICATION

This paper scrutinizes counterterrorism programs in Indonesia in the aftermath of the Bali bombing in 2002, with the objective to examine whether these programs are in conjunction with the grounds and corridors of law enforcement or, on the contrary, they are actualizing victimization and human rights violations in the name of law enforcement.

3 TERRORISM AND ANTITERRORISM LAW IN INDONESIA

Defining *terrorism* is quite challenging, since there is no universally accepted definition of the term. Walter Laquer states that no single definition will cover the whole pattern of terrorism in the world.

The League of Nations in 1937 defined *terrorism* as follows (Upendra, 2004, p. 6):

Article 1: Terrorism is criminal acts directed against a state and intended or calculated to create a state of terror in the minds of particular persons, or a group of person or the general public.

Another definition of *terrorism* is provided by the US Department of Justice: "the unlawful use of force or violence against persons or property to intimidate or coerce a government, the civilian population, or any segment thereof, in furtherance of political or social objectives" (Graeme, 2004, p. 6).

Article 2 (1) of the Draft Comprehensive Convention on International Terrorism states that:

Any person commits an offence within the meaning of this Convention if that person by any means, unlawfully and intentionally causes:

1. Death or serious bodily injury to any person
2. Serious damage to public or private property including a place of public use, a state or government facility, a public transportation system, an infrastructure facility of the environment.
3. Damage to property, places, facilities, or systems resulting or likely to result in: major economic loss, when the purpose of the conduct, by its nature or context, is to intimidate a population or to compel a government or an international agreement to do or abstain from doing any act.

Terrorism is a highly subjective and contested term. States and scholars may have failed to offer a universally agreed definition, but they do agree to its four important elements: *1. Terrorism is an act of violence. 2. It is deliberately undertaken. 3. Its primary targets are unarmed civilians. 4. Its immediate motive is to create fear.*

Since states have a monopoly over the use of violence, even though with some legally specified preconditions, state and even scholarly definitions of *terrorism* identify only non-state actors as perpetrators.

In the Indonesian context, *terrorism* had no legal definition until the enactment of special laws on terrorism following the Bali bombing, namely PERPU No. 1/2002, UU No. 15/2003, and UU No. 16/2003.

The "interim law" No. 1 of 2002 on the Eradication of the Crime of Terrorism was issued by former President Megawati Soekarnoputri under her emergency "legislative" powers. The Indonesian president can only issue regulations; however, Article 22 of the Constitution permits the president to issue Government Regulations in Lieu of Law (Peraturan Pemerintah sebagai Pengganti Undang-undang) (PERPU), herein referred to as "Interim Laws" (Simon Butt). As required by the Constitution, the DPR adopted Interim Law No. 1 of 2002 at its next sitting – on April 4, 2003 – through Law No. 15 of 2003. This statute simply confirms the Interim Law, so that Interim Law No. 1 – or, at any rate, its substantive provisions – remains Indonesia's antiterrorism law.

The law does not define *terrorism* per se. Rather, Article 1 (1) simply states that the crime of terrorism is an act that fulfills the elements of a crime under this Interim Law. These elements are set out in Articles 6 to 23. Article 6 provides a generally worded description of terrorism: "any person who by intentionally using violence or threats of violence, creates a widespread atmosphere of terror/fear or causes mass casualties, by taking the liberty or lives and property of other people, or causing damage or destruction to strategic vital objects, the environment, public facilities or international facilities, faces the death penalty, or life imprisonment, or between 4 and 20 years' imprisonment." The language used in Article 7 is almost identical to that of Article 6, except that it, first, refers to an act that is intended to create terror or mass casualties, but that does not actually cause either; and, second, the maximum penalty available under Article 7 is life imprisonment.

Following Law No. 15 of 2003, the government and the parliament (House of Representatives) enacted Law No. 16 of 2003, which stipulates that PERPU No. 1 of 2002 on the Eradication of Terrorism was applied retroactively to the 2002 Bali bombing.

Another issue emerged when the Constitutional Court (through Decision No. 013/PUU-I/2003) annulled Anti-terrorism Law No. 16/2003 that stipulates the implementation of Anti-terrorism Law No. 15/2003 to investigate the 2002 Bali bombing. Thus, the unfinished investigation and trials of the suspects of the 2002 Bali bombing did not have legal ground anymore.

The latest regulation concerning terrorism was enacted in 2013 – namely, Law No. 9 of 2013 on Prevention and Eradication of Financing of Terrorism. "Financing of terrorism" means (Article 1 (1)): "All measures intended to supply, collect, provide or lend funds, either directly or indirectly, known to be used for terrorist activities, terrorist organization or terrorists."

4 COUNTERTERRORISM PROGRAM IN INDONESIA

Indirectly influenced by the 9/11 World Trade Center (WTC) bombing in New York and directly influenced by the 2002 Bali bombing, the Indonesian government started a "war against terrorism" in 2002 (Joko Widodo, 2015). A special police unit designated to combat terrorism, Special Detachment 88, was formed. Many terrorism suspects were immediately hunted, arrested, tortured, jailed, and convicted. Since then, terrorism issues have been widely exposed by the media.

However, media exposés, to some extent, did not bring a positive impact for the families of terrorism suspects. In many cases, the media investigated where the person suspected as a terrorist used to live and exposed his family, even the face of his children. Family members were subsequently socially isolated (Nugroho, 2011).

As part of its efforts to counter violent extremist narratives, Indonesia continued to amplify the voices of victims of terrorism, as well as former terrorists who have renounced violence. Numerous nongovernmental and religious organizations sponsored workshops and conferences, emphasizing the need to respect diversity and foster greater tolerance. Indonesia also invited religious leaders, in coordination with civil society and faith-based organizations, to be part of outreach efforts to violent extremists.

Although domestic counterterrorism efforts are civilian-led, the Indonesian military maintains counterterrorism units that can be mobilized to support domestic operations if needed. These units train regularly with law enforcement agencies to ensure greater capability and coordination for potential domestic counterterrorism operations. However, the Indonesian military and its counterterrorism units are primarily responsible for external terrorist threats to the archipelago and in certain other specific situations.

The latest military efforts to combat terrorism took place in Poso, Central Sulawesi, on March 30–31, 2015. However, since the Indonesian military (Tentara Nasional Indonesia) (TNI) has no legal mandate to enforce domestic security, it pretends as if it is only conducting military training. Yet the "military training" has led to massive manhunts to target terrorism suspects.

In 2014, Poso, a town on the northeastern coast of Central Sulawesi, Indonesia, remained the epicenter of terrorism in Indonesia, followed by Bima in West Nusa Tenggara. The Mujahidin Indonesia Timur (Mujahidin of Eastern Indonesia) (MIT), led by Santoso, was allegedly behind a series of terrorist attacks targeting the police. In general, the Indonesian police remained key targets of terrorist attacks throughout 2014, reflecting the continuity in the shift from the "far enemy" to the "near enemy" – a trend that has manifested more intensively since the dismantling of the Aceh terrorist training camp in 2010. At the same time, the police's antiterror unit – Detachment 88 – foiled several terrorist plots during raids conducted across the country and captured at least fifty-four terrorist suspects – a majority of whom were linked to the MIT (Araianti).

In 2014, Detachment 88 successfully foiled several plots and arrested dozens, including those involved in the attacks. It also seized up to ninety-three motorcycles that were allegedly stolen by MIT members in Central Sulawesi. In early August 2014, the Indonesian government banned the Islamic State of Iraq and Greater Syria (ISIS). However, despite the ban, the existing legal regime in Indonesia does not give authorities the power to arrest supporters of ISIS unless there is evidence of their involvement in terrorist acts. According to official estimates, the number of Indonesians joining ISIS soared from 56 in midyear to more than 110 by the end of 2014. A number of individuals who were openly displaying ISIS paraphernalia in their vehicles or house windows were arrested but eventually released.

4.1 Deradicalization program

Broadly speaking, the word *radicalization* can be used to describe a process whereby individuals (and even groups) develop, over time, a mind-set that can increase the risk that he or she will engage in violent extremism or terrorism. It, therefore, follows that the word *deradicalization* should only be used to refer to the methods and techniques used to undermine and reverse the completed radicalization process, thereby reducing the potential risk to society from terrorism. However, confusion can arise as the term is also erroneously used as a broad catch-all to encompass other, different but related methods and techniques aimed at reducing society's risk from terrorism, including *counter-radicalization* (the term used to describe methods to stop or control radicalization as it is occurring) and *anti-radicalization* (the term used to describe methods to deter and prevent radicalization from occurring in the first place) (Clutterbuck, Lindsay).

Through Presidential Decree No. 46/2010, the government created the BNPT. The mandates of the BNPT are as follows:

1. Creating policies, strategies, and national programs in terrorism countermeasures, namely in prevention, protection, deradicalization, action, and preparation of national preparedness;
2. Coordinating related government agencies in executing policies in terrorism countermeasures; and
3. Creating related task forces.

Instead of repressive measures, counterterrorism programs enforce a deradicalization program. According to Agus Surya Bakti (2014), the objectives of deradicalization programs are: (1) counterterrorism; (2) preventing radicalism; (3) preventing provocation, spreading hatred, and interreligious hostilities; (4) protecting the community from indoctrination; (5) raising people's knowledge of terrorism; and (6) learning to understand different ideologies/school of thoughts.

At the end of 2013, the BNPT developed a deradicalization blueprint. As envisioned, deradicalization would include the efforts of the Indonesian government in coordination with civil society organizations and selected academic institutions. The BNPT opened six additional branches of the Terrorism Prevention Communication Forum, which is now present in the capital cities of twenty-one of Indonesia's thirty-four provinces. Members of each forum include civic and religious leaders who coordinate outreach, facilitate communication among key stakeholders at the local level, and work closely with communities and families on reintegration programs for released terrorism prisoners.

In dealing with radicalism, the deradicalization program initiated by the BNPT has been included in Indonesia's counterterrorism strategy, particularly after the 2002 Bali bombing. At least 700 suspected Jamaah Islamiyah (JI) members have been arrested, and some of them are involved in the program (Mahardika Satria Hadi, 2011). However, the success of a deradicalization program is unlikely due to several implemental hindrances such as rejection of such programs by some Muslim communities. Muslim communities argue that such programs are targeted to suppress Muslims and eradicate the principles of Sharia law.

The BNPT and the police have started to use "soft" strategies and put more concern on how to stop the spread of the group's radical ideology. The "soft" approach is based on trust between terrorist prisoners and converted JI terrorists who have renounced radical ideology. The most famous success story is that of Nasir Abbas, a former Afghan militant who trained the Bali bombers. Since his 2004 release from prison, he has been involved in the police's deradicalization program and helped to track down and arrest several of his former companions. Nasir Abbas has traveled to several Indonesian prisons to visit his former colleagues serving imprisonment for terrorist offenses and convinced them to stop the violence (Harding, 2010).

Law enforcement, particularly Detachment 88, has aggressively and successfully pursued terrorists and disrupted their networks. In 2013 alone, at least seventy-five suspects were arrested in more than forty separate raids. The arrests of dozens of suspected terrorists indicates that Detachment 88 remains proactive and mostly successful in cracking down on terrorism in the country. However, the arrests, detentions, and other repressive measures conducted by Detachment 88 and other law enforcement officials have led to human rights violations.

Since its inception soon after the 2002 Bali bombing and its achievement of full operation in 2003, Detachment 88 has frequently arrested or killed the wrong suspects and tortured and abused others. And not just the suspects themselves have been abused; in many cases, the family members or surrounding relatives of the suspects were also subjected to degrading treatment and punishment by Detachment 88.

In Poso Regency, Central Sulawesi, fourteen people were wrongfully arrested from December 20–27, 2012, by the police. They were arrested based on the wrongful allegation of killing four police officers a week before. In the detention center, they were kept incommunicado with no access to their families and legal representatives. Within seven days, they were abused and tortured by the officials. Most of them got serious injuries. In January 2013, in Bima and Dompu Regency, Sumbawa Island, Nusa Tenggara Barat Province, Detachment 88 wrongfully killed innocent civilians suspected of terrorism in a massive raid.

The continuing human rights violations and abuses committed by Detachment 88 led to public protests and anger, particularly from Muslim organizations. Muslim groups have called on the National Police to dissolve Detachment 88 in the wake of the allegations. Muhammadiyah chairman Din Syamsuddin, who is also deputy chairman of the Indonesian Ulema Council (MUI), said the police force must change its current approach to combating terrorism as it would prove counterproductive to terrorism eradication efforts. Din and representatives of other Islamic groups, including Nadhlatul Ulama (NU), DDII, and Persatuan Islam (Persis), reported alleged human rights abuses perpetrated by Detachment 88 members against suspected terrorists. The groups claimed to have presented to the police video footage depicting men in Detachment 88 uniforms intimidating and torturing what appears to be a suspected terrorist.

In 2013, a viral video depicted a man who was tortured by alleged Detachment 88 officers. The man's legs and hands were tied when the officers verbally abused and shot him. "You are going to die, now *istighfar* [go ask for God's mercy]." The video insulted Islam and could arouse sympathy among Indonesian Muslims, who largely oppose terrorism, for the militants. "Detachment 88 should be evaluated, or dissolved if necessary. It could be replaced by another institution that promotes a different approach, to combat terrorism together because terrorism is our common enemy," Din Syamsuddin said.

Siyono's case is another problem in counterterrorism. Siyono was arrested after prayers at a *masjid* near his residence in Klaten, Central Java, in March 2016. He was arrested in front of his parents, but the officers did not explain the reason for the arrest and said that Siyono had a debt problem. Two days after the arrest, police officers searched Siyono's house without explaining his condition to his wife. The next day, police contacted Siyono's family, delivering news that Siyono had died without explaining the cause of death. Police then told Siyono's family to pick up the body in Jakarta. At the same time, Siyono's parents were intimidated by the Klaten Precinct Police and the head of the village. According to Satrio, they were asked to sign a statement saying that they would not file a lawsuit for Siyono's death. The National Police said that Siyono died on the way to a hospital after being involved in a brawl with an antiterror member escorting him.

Human rights activists treat the case of Siyono as a criminal investigation after an autopsy confirmed that torture was the cause of his death. Counterterrorism is indeed a priority for Indonesia as for any other country; however, the mechanisms used should be in accordance with the law and the national ideology of Pancasila. Dahnil Anzar Simanjuntak, the chairman of Pemuda Muhammadiyah, expressed his concern that counterterrorism efforts in Indonesia neglected human rights principles and such violations had not been addressed by the government. Siyono was one among 121 other victims of Detachment 88's counterterrorism operations that neglected human rights principles, were allegedly conducted without warrants, or involved torture, which the autopsy results had proven in Siyono's case.

Hence, the counterterrorism program in Indonesia, partly conducted by Detachment 88 and coordinated by the BNPT, is subject to controversy. The country definitely needs the counterterrorism program to combat the numerous terrorist attacks in Indonesia. On one hand, the counterterrorism program must meet human rights standards. Do not let counterterrorism lead to unlawful victimization or another form of "terrorism" by state apparatuses (usually called "state terrorism").

The deradicalization program is rather problematic. Some Muslim organizations reject this since the program seems to target "Muslim terrorists" or ex-terrorist inmates. Anti-radicalization programs are also targeted mostly to the Muslim community. Terrorists are not only Muslims and radical ideologies could originate from various sources and ideologies, not just from the misinterpretation of Islamic teaching.

An ICG report mentioned that deradicalization programs are important, but they will inevitably be based on trial and error; no single intervention can produce rejection of violence among a disparate group of people who have joined radical movements for many different reasons. Within Jamaah Islamiyah alone are the ideologues, the thugs, the utopians, the followers, the inadvertent accomplices, and local recruits from Poso motivated by very different factors than those who graduate from JI-affiliated schools in Central Java.

Much more thought needs to be given to how to evaluate the "success" of deradicalization programs, because there is a danger that many people deemed deradicalized are those who were never the real problem, or that the reasons individuals renounce violence have nothing to do with police programs. Even if we could measure the number of people deradicalized according to specific criteria, that figure would only have meaning if we had some sense of the number of new recruits and knew that the balance was going in the right direction.

Part of the reasons why counterterrorism programs in Indonesia frequently lead to human rights abuses is the flawed Indonesia laws on antiterrorism. For instance, the definition of *terrorism* in PERPU No. 1/2002 (Interim Law) is still too broad. According to Indonesia's experiences in dealing with a repressive regime, this weakness can be used to oppress democratic society. Investigators may arrest any person strongly suspected of committing a criminal act of terrorism based on adequate preliminary evidence as defined in Article 26 (2) for a maximum period of seven times twenty-four hours.

In Article 28 of the said law, for the purpose of investigation and prosecution, the investigator is given authority to detain the accused for a maximum of six months. Article 25 (2) of the said law expanded the criminal procedures. Intelligence reports may be used as legal evidence and investigators can examine personal emails and tap telephone or other communication conversations for a period of up to one year. Thus, with these extensive powers, many arbitrary arrests are happening and it is common for the law enforcement apparatus to use extraordinary action.

Further, when a suspected person is arrested or detained, it is difficult to get access to her/him. The suspect is usually made incommunicado with no access to lawyers or legal aid. In most cases, when someone is arrested, no immediate information is given to her/his family. Instead, the family often has to go to many police stations over several days to find their family member.

These kinds of limitations and the lack of judicial oversight have given broad chances for investigators to commit torture or other cruel, inhuman, or degrading treatment. Unfortunately, no serious responses have come from the authorities, including the court, whenever there is a report on torture. Some of the defendants who call for the court's attention regarding the use of torture get no adequate answer.

Therefore, counterterrorism laws, policies, and practices have impacted certain political rights such as freedom of expression, freedom from fear, and right to liberty and security of person. People are intimidated by the excessive power of judicial and nonjudicial institutions because there is no adequate guarantee of a fair trial or protection from wrongful arrest.

Indonesia has enacted numerous national laws and ratified major international human rights conventions. In 1993, the country founded the National Human Rights Commission. In 1999, the Law of Human Rights No. 39 was enacted. In 2000, the People's Consultative Assembly (Parliament) amended and added ten articles on human rights to Article 28 of the 1945 Constitution (Second Amendment).

Prior to the Second Amendment, Indonesia ratified the Convention on the Elimination of All Forms of Discrimination Against Women (CEDAW) through Law No. 7 of 1984, the Convention on the Rights of the Child in 1990, and the Convention Against Torture, Inhuman and Degrading Treatment and Punishment through Law No. 5 of 1998.

Two major human rights conventions, the International Covenant on Civil and Political Rights and the International Covenant on Economic, Social and Cultural Rights, were also ratified by Indonesia in 2005 through Law No. 11 and Law No. 12, respectively.

These laws and human rights conventions explicitly mention the prohibition of torture, abuse, and other forms of inhuman and degrading treatment and punishment of criminal suspects as well as their families and surrounding relatives.

6 CONCLUSION

As a country well known for numerous terrorist attacks, Indonesia must establish a counterterrorism program to maintain order, security, and peace. Repressive measures taken by law enforcement officials are understandable to some extent. Terrorists have instigated and committed violence ubiquitously, so countermeasures must be in line with various methods of crimes.

Nevertheless, even in counterterrorism programs, human rights should be respected, promoted, and protected. Law enforcement officials cannot simply neglect human rights in the name of the "war against terrorism." The principles of presumption of innocence, fair arrest, detention, and trial, and also freedom from torture and inhuman and degrading treatment and punishment must also be taken into account.

Counterterrorism measures must still meet human rights standards as stipulated in Indonesian laws. Moreover, Indonesia ratified the Convention Against Torture, Inhuman and Degrading Treatment and Punishment in 1998 and the International Convention on Civil and Political Rights in 2005.

Acts of torture, wrongful arrest, detention, and killing must be avoided at all costs. The government also needs to amend the flawed antiterrorism laws – namely, PERPU No. 1 of 2002 and Law No. 15 of 2003 – to avoid future human rights violations. Counterterrorism measures must be reviewed and redefined. Also, it is important for Indonesia to take notice and apply the ICJ Declaration on Upholding Human Rights and the Rule of Law in Combating Terrorism (the Berlin Declaration).

Last but not least, the deradicalization program introduced in the aftermath of the 2002 Bali bombing, must not jeopardize and victimize innocent people. Unfair labeling of specific communities just because of their appearances and their sociocultural association to a group of terrorists must be hindered. Moreover, justice, rule of law, and certainty of law must prevail.

REFERENCES

Becker, H. S. (1991). *Outsiders: Studies in the Sociology of Deviance*. The Free Press.
Djelantik, Sukawarsini. (2010). *Tinjauan Psiko-Politis, Peran Media, Kemiskinan dan Keamanan Nasional*. Jakarta: Yayasan Obor.
Hasani, Ismail and Bonar Tigor Naipospos. (2012). *Dari Radikalisme Menuju Terorisme*. Jakarta: Setara Institute.
International Crisis Group. (2007). *Deradicalisation and Indonesian Prisons*. Asia Report N° 142. November 19.
Ismail, Noor Huda. (2010). *Temanku Teroris? Saat Dua Santri Ngruki Menempuh Jalan Berbeda*. Bandung: Penerbit Mizan.

Istiqomah, Milda. (2011). Deradicalization Program in Indonesian Prisons: Reformation of the Correctional Institution, Proceedings of the 1st Australian Counter Terrorism Conference, Edith Cowan University, Perth. Western Australia, December 5 –7.

Khan, S. (2003). *Children Forgotten in This War against Terrorism*. Retrieved July 25, 2011 from www.globeandemail.com/servlet/ArticleNews/TPStory/LAC/20030728/COSHEEMA28/.

Lutz, James M. and Brenda J. Lutz. (2004). *Global Terrorism*. London: Routledge.

Mansur, et.al. (2007). *Urgensi Perlindungan Korban Kejahatan*. Jakarta: Rajawali Press, pp. 130–131.

Majid, Munir. (2012). *9/11 and the Attack on Muslims*. Kuala Lumpur: MPH Group Publishing.

Mustofa. (2002). "Memahami Terorisme: Suatu Perspektif Kriminologi." *Jurnal Kriminologi Indonesia FISIP UI 2*(3), 30.

Norris, Pippa, Montague Kern, Marion Just, et al. (2003). *Framing Terrorism: The News Media, the Government and the Public*. London: Routledge.

Purwanto, Wawan. (2012). *Satu Dasawarsa Terorisme di Indonesia*. Jakarta: CMB Press.

Saikia, Jaideep and Ekaterina Stepanova. (2009). *Terrorism: Patterns of Internalization*. New Delhi: Sage Publications.

Samudra, Imam. (2004). *Aku Melawan Teroris!* Solo: Jazera Publishing.

Silke, A. (2003). *Terrorists, Victims, and Society: Psychological Perspective on Terrorism*. John Wiley & Sons, p. 192.

Sharma, Mukul. (2010). *Human Rights in a Globalized World. An Indian Diary*. New Delhi: Sage Publications.

Solahudin. (2011). *NII Sampai JI: Salafy Jihadisme di Indonesia*. Depok: Komunitas Bambu.

Thackrah, John Richard. (2004). *Dictionary of Terrorism*, 2nd ed. London: Routledge.

Equality before the law for women in Indonesia: An analysis of international law and its implementation at the national, regional, and sector level

Widya Naseya Tuslian[1]
Faculty of Law, Universitas Indonesia, Indonesia

ABSTRACT: Equality before the law is one of the core values of the Convention on the Elimination of All Forms of Discrimination Against Women (CEDAW). Achieving equality for either sex is fundamental in promoting human rights across the nations. Moreover, this principle is mandated by Article 15 of CEDAW, which stipulates that state parties shall integrate gender equality before the law in civil matters, legal capacity, entering into contracts, etc. Any state parties bound by this article will be obligated to make sure that their legal system and policies will guarantee an equal position amongst women. Indonesia has been a part of this convention and is expected to be committed to its provisions. However, to date, the laws that impede women to achieve an equal position before the law have remained widespread. The initial commitment to harmonize the legal system with the concept of equality has been done, but there is no sustainable attempt to keep up the work. Up until now, laws and policies that institutionalize discrimination still exist at the national, local, and sectoral levels. Taking this perspective into account, this paper explores the extent to which equality before the law has been achieved in Indonesia after the ratification of CEDAW. In doing so, this paper describes and examines the concept of equality before the law in CEDAW, and the enforcement of Indonesian rules and regulations in court practices and policies.

Keywords: CEDAW, Equality before the Law, Gender, Human Rights, Women

1 INTRODUCTION

Equality before the law is a core concept in human rights law. Achieving an equal position between women and men under the law and eliminating all forms of discrimination on the basis of sex is fundamental in promoting human rights. The Convention on the Elimination of all Forms of Discrimination against Women (CEDAW) is a landmark instrument for the achievement of gender equality in most aspects of life. Although CEDAW is a non-sanctioned law, it articulates important principles of gender equality and state responsibility and clarifies how these principles apply to a state party (Engle Merry, 2007, p. 75). CEDAW not only proscribes discrimination but also advocates positive steps, such as the elimination of sex-role stereotypes in the media and educational materials and the creation of "temporary special measures" to benefit women. This international convention is thus expected to redress existing regulations, conditions, and systems which do not yet support women's position under the

---

1. Widya Tuslian is a researcher and a junior lecturer at the Department of Law and Society of the Faculty of Law, University of Indonesia. She finished her studies on the Faculty of Law, University of Indonesia, for her bachelor's degree (LL.B), and the Faculty of Law, Maastricht University, the Netherlands, for her master's degree (LL.M). She is currently doing research on the mechanism of handling cases of violence against women in Southeast Asia.

law. It also raises awareness among all elements of the nation, including state actors, wider society, and women themselves, of the importance of gender equality protection. However, in many developing countries, including those that have adopted CEDAW, gender equality is not yet a reality.

Indonesia, the world's fourth largest country in terms of inhabitants and home to the world's largest Muslim population, is a case in point. The country has reinforced its commitment to development by including human rights protections in its 1945 Constitution, particularly since its amendment in the early period of reformation (*reformasi*). Like many countries, Indonesia wants to show its support for human rights by ratifying international law instruments that promote human rights, including international law with a gender perspective, such as CEDAW. Indonesia ratified CEDAW on September 13, 1984, through Law No. 7 of 1984. The adoption of CEDAW is the cornerstone of efforts toward women rights promotion in Indonesia. The ratification of this convention also asserts the commitment of the Government of Indonesia to respect nondiscrimination principles in the formulation and implementation of public policies and services. The harmonization of national laws and regulations with CEDAW is ongoing at the national and regional levels.

The harmonization of national laws, regulations, and government policies with the gender equality concept can be identified at various levels in Indonesia, focusing on different areas, such as education, health care, and political participation. Important measures to note are the presidential instruction on gender mainstreaming in national development, the establishment of a ministry for women's empowerment and child protection with five main work units (gender mainstreaming, the improvement of women's lives, women's protection, child protection, and community empowerment), and the creation of a "Forum for the Advancement of Women's Economic Productivity."

Despite these measures, gender equality has not been fully achieved in Indonesia as of yet. According to the report by the National Commission on Violence against Women, which was established in 1998 to promote the elimination of violence against women and protect victims of violence, there were 54,425 reported cases of violence against women in 2008. Husbands or personal relationships account for more than 90% of violence. In 2015, the number of reported cases saw a dramatic increase, amounting to 321,752 cases separated into three spheres (personal, community, and politics). Moreover, according to the United Nations' global database on violence against women, the number of child marriages accounts for 17% of the total number of marriages, while 49% of women have become victims of female genital mutilation in Indonesia. A patriarchal mind-set persists, illustrated by widespread practices of polygamy, illegal and early marriage, and female circumcision.

Taking this perspective into account, this paper attempts to address the research question: To what extent is equality before the law reflected in regulations and implementation in prevailing conditions in Indonesia after the ratification of CEDAW? This paper also tries to explain why there still is a major gap between the ideal of equality before the law between men and women as set in CEDAW and other international law instruments on one hand and legal practice on the other. It examines the national laws and regulations at various levels as well as their implementation. It shows that the lack of gender equality is not caused only by existing loopholes in national laws and regulations but also by problems in the implementation of such legislation by law enforcement agencies and the courts, which often form obstacles to rather than facilitate achieving justice for women who become the victims of gender-based violence.

This paper starts with a discussion of the concept of equality before the law in CEDAW and briefly describes the process of ratifying CEDAW in Indonesia. The next section describes provisions in various laws and regulations at different levels that contain loopholes as far as the promotion of gender equality is concerned. Laws and regulations that are the focus of this paper include Law No. 1 of 1974 (Marriage Law) at the national level, a set of gender-biased bylaws at the regional level, and Regulation of the Minister of Health No. 1636/MENKES/PER/XI/2010 regarding female circumcision at the sectoral level. The last section addresses the problems of implementation and enforcement of national laws and regulations in the protection of gender equality.

2 DISCUSSION

2.1 *Equality before the law in CEDAW and its ratification by Indonesia*

International and regional human rights treaties and declarations, as well as most national constitutions, contain guarantees or safeguards relating to sex and/or gender equality before the law and an obligation. This is to ensure that everyone benefits from the equal protection of the law. The right to equality before the law and access to justice for women is essential to the realization of all the rights protected under CEDAW. The concept of equality before the law is the tenet of gender equity. Equality entails that persons should in principle be treated equally if they can be considered to be in an equal situation (Westendorp, 2012, p. 68). Equality before the law is actually multidimensional. It encompasses justifiability of regulations without any distinction, accessibility to judiciary forum, good-quality and accountable justice systems, and professional state apparatus to implement the law.

Article 15 of CEDAW provides that women and men must have equality before the law and benefit from the equal protection of the law. The equality before the law principle acknowledged by CEDAW is further emphasized in Article 1, which prohibits state discrimination against women. Discrimination is defined as: "any distinction, exclusion or restriction made on the basis of sex which has the effect or purpose of impairing or nullifying the recognition, enjoyment or exercise by women, irrespective of their marital status, on a basis of equality of men and women, of human rights and fundamental freedoms in the political, economic, social, cultural, civil or any other." This provision means that women and men should be given an equal position to act upon herself or himself and participate equally in terms of political, economic, legal, and cultural life.

CEDAW assigns a special role to state parties to take part in the realization of gender equality. Article 2 stipulates that state parties must take all appropriate measures to guarantee the substantive equality of men and women in all areas of life, including through the establishment of competent national tribunals and other public institutions to ensure the effective protection of women against any act of discrimination. The content and scope of this article are further detailed in the Committee on the Elimination of Discrimination against Women's General Recommendation No. 28 on the core obligations of state parties under Article 2 of CEDAW. Article 3 of CEDAW mentions the need for appropriate measures to ensure that women can exercise and enjoy their human rights and fundamental freedoms on a basis of equality with men. Therefore, state parties have further treaty-based obligations to ensure that all women have access to education and information about their rights and remedies and how to access these, and to competent, gender-sensitive dispute resolution systems, as well as equal access to effective and timely remedies (Committee on the Elimination of Discrimination against Women, 2013, p. 4).

Indonesia has ratified several international human rights instruments that directly or indirectly relate to the obligations to respect, protect, and fulfill gender equality. The ratified instruments include CEDAW, the Convention on the Rights of the Child (CRC), the Convention against Torture and Other Cruel, Inhuman and Degrading Treatment or Punishment (CAT), and several key International Labour Organization (ILO) instruments, including ILO Convention No. 138 on the Elimination of Child Labour, ILO Convention No. 105 on the Abolition of Forced Labour, and ILO Convention No. 182 on the Elimination of the Worst Forms of Child Labour. However, to date, some human rights instruments that guarantee the protection and fulfilment of the rights of women and children, such as the Optional Protocol to CEDAW, the Optional Protocol on Children in Armed Conflict, and the Optional Protocol on the Sale of Children and Child Prostitution, have yet to be ratified by the government of Indonesia.

The Indonesian government is working to harmonize the principle of human rights, particularly equality before the law, by making amendments to the 1945 Constitution and enacting certain laws. The fourth amendment of the 1945 Constitution on 2002 on Article 28 D asserts that every person shall have the right of recognition, guarantees, protection and certainty before a just law, and of equal treatment before the law. All laws and regulations in Indonesia should be in compliance with this principle as the 1945 Constitution is the highest level of law in Indonesia which may not be contradicted by lower levels of law. However, to date, a substantial gap remains between provisions stipulated in the 1945 Constitution and actual implementation on the ground. Violence and unequal treatment before the law of marginalized groups, especially women, remain persistent.

2.2 Laws and regulations in Indonesia

2.2.1 Marriage law

Indonesia's Marriage Law was enacted in an attempt to unify the pluralist system of marriage legislation that was created by the colonial government. Unification was aimed at creating uniformity in marriage laws, which would serve several development goals, such as economic growth, while at the same time supporting demographic policies to reduce population growth and control marriage practices. However, in fact, unification never took place as there is an additional set of rules, referred to as the Islamic Law Compilation, which only applies to Muslims, who form 87.2% of Indonesia's population of more than 250 million people. Indeed, it is hardly possible to come up with one law governing in a uniform manner in all aspects of marriage that is applied across the whole population. Instead, the current system is still based on legal pluralism and continues to be the object of political struggle (Bedner, 2010, p. 176). To date, the Indonesian legislators have failed to make a revision to this law, although this has been an agenda of the national legislation program (*Prolegnas*) since 2006 (Amnesty International, 2012, p. 9). This condition is ironic, as many of the provisions of the Marriage Law conceive loopholes due to the background of this law legislation process, which was based on the patriarchal culture and on the partial interpretation of sources of Islamic law.

The Marriage Law was enacted in 1974, long before the ratification of CEDAW in 1984, and even longer before the acknowledgment of an extensive list of human rights by an amendment of the 1945 Constitution in 2000. Not surprisingly, many provisions of this prevailing law conflict with CEDAW.

Not all provisions of the Marriage Law conflict with CEDAW. A sense of "equality" is reflected in Article 31 of the Marriage Law:

(1) The rights and the position of the wife are equal with the rights and the position of the husband both in family and social life.
(2) Either party to the marriage has legal capacity.
   The aforementioned provision seems to promote the concept of gender equality by asserting an equal position between women and men. However, this notion seems to contradict with the following section of Article 31 of the Marriage Law:
(3) The husband is the head of the family while the wife is the head of the household.
   This section clearly stereotypes the role of women and men in the family, positioning men to have a role enabling them to participate in the public arena, while women have to manage the household and restrict themselves to the domestic area.

The stereotypical separation of tasks between women and men in marriage life is further confirmed by Article 34:

(1) The husband has a responsibility to protect his wife and provide her with all the necessities of life in a household in accordance with his capabilities.
(2) The Wife has the responsibility and shall be taking care of the household to the best of her ability.

The article clearly reflects the distribution of function between men and women on the basis of a patriarchal culture and conservative religious view without having further consideration for the concept of equality referred to in Article 31.

The Marriage Law also discriminates between women and men by setting different marriage ages, allowing for underage marriage of women. This is stipulated in Article 7:

> A marriage is only allowed if the male candidate has reached the age of 19 while the female candidate has reached the age of 16.

Articles 3 and 4 of the Marriage Law contain other provisions, which spark long-standing protests from feminists, nongovernmental organizations, and human rights organizations. Article 3 authorizes a man to have more than one wife:

(1) In principle, a man may be married to one woman only, women shall be married to one man.
(2) The court may permit a husband to have more than one wife if the party concerned so wishes.

According to Article 4, the court can grant permission to a husband who is willing to have more than one wife if his first wife cannot perform the conjugal duties, his wife suffers from physical infirmity or an incurable disease, or his wife cannot bear children.

This provision clearly stigmatizes women and their role in marriage as an object of reproduction. It encourages the stigmatization of women who cannot bear children as impaired in the marriage. This provision is not too flexible for a contrary interpretation in a situation where a man cannot produce children due to infertility or disease.

In conclusion, the Indonesian Marriage Law codifies the cultural and social constructions of women pertaining to patriarchal culture. This law also perpetuates gender discrimination and hinders women to achieve equality before the law as it limits the function of married women to the domestic arena.

2.2.2 *Regional regulations*

Gender equality lacks implementation not only at the national level but also at the regional level. An increasing number of regional regulations pose a threat to the equal treatment of men and women in Indonesia, as research conducted by the Indonesian National Commission on Violence against Women shows. According to the Commission, as many as 154 local bylaws enacted at the provincial government level (19 district/municipal-level policies, 134 policies, and 1 village-level policy) between 1999 and 2009 have become instrumental in the institutionalization of discrimination, either in intent or as an impact. These discriminatory local bylaws have been enacted in 69 districts/municipalities in 21 provinces (National Commission on Violence against Women, 2010, p. 3). More than half of them (80 policies) were enacted almost simultaneously between 2003 and 2005. Many districts in 6 provinces (West Java, West Sumatra, South Kalimantan, South Sulawesi, West Nusa Tenggara, and East Java) have been enforcing the discriminatory local bylaws. Only 39 local bylaws (14 at the provincial level, 22 at the district/municipal level, and 3 at the village level) aim to fulfill the right of victims to recover. Of the 154 local bylaws, 63 directly discriminate against women by restricting their right to freedom of expression (21 policies regulating dress codes) by criminalizing them, thus impairing their rights to legal protection and legal certainty (37 policies on the prohibition of prostitution) by nullifying their rights to legal protection and certainty (1 policy on the prohibition of *Khal wat*), and by excluding their rights to protection (4 policies on migrant workers).

One regulation that sparks controversy and that is still in force today is Regulation of the Municipality of Tangerang No. 8 of 2005 on the Prohibition of Prostitution, which contains a very dubious provision about what constitutes prostitution. Worse, Article 4(1) of the Regulation limits the assumption of innocence:

> Anyone whose attitude is suspicious giving rise to a presumption that they are prostitutes are banned from public roads, fields, lodging houses, hostel, hotels, dormitories, houses / rented, coffee shops, entertainment places, theater, corners or halls of the road or other places in the region.

This equivocal provision leads to arbitrary interpretation, wrongful arrest, and imprisonment of innocent women suspected of prostitution by the Municipal Police (Sat Pol PP) for the simple reason that they were still "outside the house" late at night. Women's organizations have undertaken many attempts to have the regulation annulled, but to no avail. In 2006, several women supported by representatives of the Anti-discriminative Regional Regulations Advocacy Team (TAKDIR) filed for judicial review to the Supreme Court of Indonesia, requesting that the regulation be declared unconstitutional. However, the Supreme Court rejected the request, arguing that the municipal government and council had followed correct procedures in the regulation's formation and that the municipality had the right to produce such a regulation. The Supreme Court did not agree with the applicants' argument that the regulation was in conflict with the Indonesian Criminal Code, Law No. 8 of 1981 on Criminal Procedure, Law No. 7 of 1984 on the Ratification of the Convention on the Elimination of all Forms of Discrimination Against Women, Law No. 39 of 1999 on Human Rights, Law No. 12 of 2005 on Civil Rights, Law No. 10 of 2004 on Lawmaking, and Law No. 32 of 2004 on Regional Government.

Interestingly, many other regional regulations criminalize private behavior of women, such as Regulation of the Province of Aceh No. 14 of 2003 on *Khal wat*, which means women and men who have no family relation being together without the presence of other people, and other regulations that regulate how women should dress. Just as in the case of Regulation of the Municipality of Tangerang No. 8 of 2005, these regulations continue to be in force, despite the tireless efforts of human rights groups to have them annulled.

The aforementioned examples illustrate that regional legislation does not yet promote equality before the law. Various regulations are prone to ambiguous interpretation and are more repressive to women, even though the government usually argues that these regulations are gender neutral. Moreover, the regional laws usually standardize minor stigmatization of women, leading to further drawbacks to equality between men and women, especially with regards to women's position in public areas.

2.2.3 *Sectoral legislation*

A further limitation to women's equality before the law can be found at the sectoral level. Article 133 of Law No. 36 of 2009 on Health stipulates that every child has the right to be protected from any form of discrimination and violence that can harm their health. However, just a year later, the Indonesian minister of health issued Regulation No. 1636/MENKES/PER/XI/2010 on Female Circumcision. This regulation authorizes medical professionals (e.g., doctors, midwives, and nurses) to perform female genital circumcision. The regulation defines female circumcision as: "'the act of scratching the skin covering the front of the clitoris, without hurting the clitoris.' The procedure includes 'a scratch on the skin covering the front of the clitoris (frenulum clitoris) using the head of a single-use sterile needle.'" (Amnesty International Indonesia, 2012, p. 6).

Regulation of the Minister of Health No. 1636/MENKES/PER/XI/2010 should not contradict higher legislation, including Law No. 36 of 2009 on Health. However, to date, the minister of health has not repealed the regulation and there has been no attempt on the part of the government of Indonesia to limit the discriminative practice of female circumcision, which is based on the stigmatization of women's bodies. The government of Indonesia thus fails to promote equality before the law.

2.2.4 *Implementation of law in Indonesia*

Written laws and regulations play a crucial role in the promotion of gender equality in Indonesia, considering that the country has a civil law system using written laws and regulations as the utmost instrument to achieve justice. However, a set of written laws and regulations that preserve an equal position between men and women is not enough. These regulations should also be implemented, which requires an effective law enforcement mechanism. The success of such a mechanism is highly dependent on the social context in which the legislation is implemented.

The victims of gender-based violence in Indonesia have to face a range of obstacles in law and practice when trying to seek justice (Amnesty International Indonesia, 2012, p. 21). The first obstacle is the obscure provisions in Indonesian criminal law in relation to domestic violence. Concepts such as "rape" and "sexual violence" are applied inconsistently in the law and are defined narrowly. Article 285 of the Indonesian Criminal Code defines rape as: "any person using force or threat of force on a woman to have sexual intercourse with him out of marriage."

This article limits "rape" to any act of sexual violence *outside* marriage, which means that any act of rape in this provision cannot be categorized as rape when this happens between married people. Meanwhile, Law No. 23 of 2004 regarding the Elimination of Household Violence does not categorize rape as a crime. This means that rape inside of marriage has yet to be acknowledged in the Indonesian law system.

The next obstacle limiting access to justice for victims of sexual violence is law enforcement agencies. A report by Amnesty International, which conducted an investigation in collaboration with nongovernmental organizations in Aceh, illustrates this point. The report tells the story of Sari, a fourteen-year-old girl who was accused by the police of committing adultery when she went to report that she had been raped.

> Sari went to the police station in Aceh to report that she had been raped by a 25 year old married man. However, police officials initially did not believe her. Instead they alleged that she had sex with a married man because they liked each other. Police officials accused her of breaching the legal provision on adultery. (Amnesty International, 2010, p. 21)

This case shows that the police, as a state apparatus with the authority to investigate criminal offenses, is prejudiced against women who become victims of sexual violence. This is an alarming case because of the incapability of state actors to implement the law owing to their lack of gender equality perspective.

The prevalence of gender stereotyping, which is still a mainstream point of view of the state apparatus, jurists, and state actors, is portrayed in their treatment and application of the law in cases involving marginalized women, their policies, and their public statements. And many of those in public authority still have the mind-set that undervalues women, such as the West Aceh district head who said in 2010 that if women don't dress according to shariah law, they are asking to get raped (Amnesty International, 2012, p. 21). The regulations and their implementation by state actors that attempt to control women's bodies are perpetuating the impunity of the criminals because women victims are considered the responsible party in the sexual violence they experience.

Even the courts, which are expected to be the last resort when women become victims of sexual violence, fail to give a sense of justice. For instance, in cases of sexual violence inside marriage, courts will draw inferences about the credibility, character, or predisposition to the sexual availability of a victim based on her prior or subsequent sexual conduct (Amnesty International, 2012, p. 23). The judge can only impose a criminal sentence if there are two elements of proof. According to the Code of Criminal Procedure, these can be testimony from a witness (including the victim), the defendant, an expert, a letter, or a sign. In practice, most cases require evidence of semen through medical records (*visum et repertum*), which makes it practically impossible for female victims of rape and other forms of sexual violence to obtain justice through the courts (National Commission on Violence against Women, 2008, p. 2). As sexual violence inside marriage occurs in the private arena, it is really hard to obtain testimony from the victim and witnesses. Victims will also have to deal with a lot of pressure, as jurists and the state apparatus in Indonesia are not accustomed to cases of rape inside marriage. In addition, victims usually have a financial dependency on the abuser, while at the same time they have to deal with social, cultural, and religious values (Saraswati, 2013, p. 1).

Decisions of the courts are often gender-biased due to the fact that most judges are not yet familiar with a gender equality perspective. This particularly applies to judges in religious courts, which often handle marital and family matters. With more than 200,000 cases tried in religious courts each year, the following are examples of gender bias in religious court decisions:

1. Women survivors of domestic violence are denied their right to divorce and instructed to reconcile with their husbands.
2. Women who initiate divorce are not granted any property or financial compensation and lose child custody.
3. Women are awarded inadequate alimony payments in cases of divorce.
4. Charges of marital rape are dismissed. (Asia Foundation, 2013)

According to a report by the National Commission on Violence against Women, which was established in 1998 to promote the elimination of violence against women and protect victims of violence, there were 54,425 reported cases of violence against women in 2008 (Japan International Cooperation Agency, 2011, p. 3). Husbands or personal relationships account for more than 90% of violence. In 2015, the number of reported cases saw a dramatic increase, amounting to 321,752 cases, which are separated into three spheres (personal, community, and politics). Moreover, according to the UN women's global database on violence against women, the number of child marriages accounts for 17% of the total number of marriages, while 49% of women have become the victim of female genital mutilation in Indonesia. A patriarchal mind-set persists, illustrated by widespread practices of polygamy, illegal and early marriage, and female circumcision.

3 CONCLUSION

In conclusion, we need to recognize the fact that Indonesia has ratified CEDAW and several other international instruments that promote gender equality, it has amended 1945 Constitution to better protect gender equality, and it has made efforts to harmonize national legislation with the concept of equality. Still, the equal position of men and women has yet to become a reality in most parts of the country. Many laws and regulations at the national, regional, and sectors levels still institutionalize local cultural and religious stereotypes that undervalue women. In terms of implementation of such legislation, the country is still facing many problems, especially in law enforcement agencies and courts, which still reflect the absence of an appreciation toward gender equality and the lack of will to address obstacles that perpetuate this socioeconomic-political problem.

REFERENCES

Amnesty International. (2010). *Left without Choice: Barrier to Reproductive Health in Indonesia*. London: Amnesty International.
Amnesty International (2012). *Indonesia Briefing to the UN Committee on the Elimination of Discrimination Against Women*. London: Amnesty International.
Asia Foundation (2013). *Religious Courts: Improving Women's Access to Justice in Indonesia*.
Bedner, A. W. & S. van Huis. (2010). "A Plurality of Marriage Law." *Utrecht Law Review* 6.
Engle Merry, S. (2006). *Human Rights & Gender Violence: Translating International Law to Local Justice*. London.
Ina Noor Inayati. (2016). *Soepra Jurnal Hukum Kesehatan (Soepra Health Law Journal)* 2(1).
Japan International Cooperation Agency. (2011). *Country Gender Profile: Final Report*.
National Commission on Violence against Women. (2008). *Indonesia's Compliance with the Convention Against Torture and Other Cruel, Inhuman and Degrading Treatment or Punishment: Issues for Discussion with the Committee Against Torture*.
National Commission on Violence against Women. (2010). *Executive Summary: In the Name of Regional Autonomy*.
Saraswati, R. (2013). *Justice and Identities of Women: A Case of Indonesian Women Victims of Domestic Violence Who Have Access to Family Court*. University of Wollongong.
UNICEF (2016). *Child Marriage in Indonesia Progress on Pause: An Analysis of Child Marriage in Indonesia*.

Westendorp, I. (ed.). (2012). *The Women's Convention Turned 30: Achievements, Setbacks and Prospects.* Cambridge – Antwerp – Portland: Intersentia.

World Health Organization. (2008). *Eliminating Female Genital Mutilation: An Interagency Statement.* Geneva: World Health Organization.

INTERNET SOURCES

http://evaw-global-database.unwomen.org/en/countries/asia/indonesia

www.hukumonline.com/berita/baca/hol16520/ma-tolak-permohonan-uji-materiil-perda-pelacuran-tangerang

www.indexmundi.com/indonesia/demographics_profile.html

www.komnasperempuan.go.id/siaran-pers-komnas-perempuan-catatan-tahunan-catahu-2016-7-maret-2016/

http://print.kompas.com/baca/2015/03/23/Perda-Diskriminatif-Bertentangan-dengan-Konstitusi

www.upr-info.org/sites/default/files/document/indonesia/session_13_-_may_2012/js9_upr_idn_s13_2012_jointsubmission9_e.pdf

Advancing Rule of Law in a Global Context – Susetyo, Rinwigati Waagstein & Budi Cahyono (eds)
© 2020 Taylor & Francis Group, London, ISBN 978-1-138-32782-5

Contemporary issues on constitutional law

Jimly Asshiddiqie
Faculty of Law, University of Indonesia, Depok, Indonesia

1 INTRODUCTION

Indonesia has four times throughout history adopted a new constitution, i.e. the 1945 Constitution (effective from August 18th, 1945), 1949 Federal Constitution (effective from 1949-1950), Provisional Constitution of 1950 (effective from 1950-1959), and the application of 1945 Constitution added with its explanatory notes adopted by the Presidential Decree of July the 5th, 1959 which has been continuously effective until the first constitutional amendment of 1999. After the constitutional crisis in 1998, the process of change was conducted not by replacement but by an addendum, according to the consensus made before the agenda of constitutional change was agreed upon by all political parties in the People's Assembly in 1999. By the new approach, the constitutional change could smoothly fulfil the need for continuity and change within the dynamic competition between the conservatives and the reformists during the critical time of change in 1998-1999.[1]

We could say that the Indonesian constitutional changes of 1999, 2000, 2001, and 2002 (the first four stages of integral changes) was incremental in its formal sense. However, the contents of change are nothing less than a big-bang change. Indonesian Constitution of 1945 is the shortest constitution in the world, containing only 1.393 words consisting of 71 ruling verses. After the Reformation, by only four amendments, the content of the constitution has changed to nearly 300%. By the adoption of the fourth amendment in 2002, the contents had 199 ruling verses. From the 199 verses of the new constitution, there are only 25 verses which are originated from the original document. The other 174 verses are all new rules adopted into the new 1945 constitution. The total words written in the new version of the 1945 Constitution grow from 1.393 of the original to 4.559 words in the new version (327%). Therefore, I suggest that the Indonesian experiences between 1999 – 2002 can be understood as an incremental big-bang constitutional change.

2 DISCUSSION

2.1 *Respecting, fulfilling, and the promotion of human rights*

The new contents of the 1945 Constitution cover a wide range of aspects and almost all parts of the constitution added with new articles. One of them is the articles of human rights, which were avoided to adopt in 1945, were adopted in articles 28A until 28J. Originally, the 1945 Constitution was designed without special respect for human rights which were associated with liberal and individualistic discourses of western colonial power. Besides, the framers of the constitution objected the idea to adopt the articles of human rights, because of the idea of the integralistic notion of state shared among the founding leaders of Indonesia. Soepomo claimed that Indonesia could not adopt the idea of separation of power of Montesquieu's doctrine of 'trial political' that the articles of human rights were not suitable to be outlined in the constitution.

Now, by the second constitutional amendment of 2000, almost all International instruments of human rights were adopted into the articles 28A until 28J consisting of 26 verses of human rights. By the adoption, the 1945 constitution become one of the most modern humanistic constitution in the world. By the new 1945 Constitution, human rights are fully respected, promoted, and must be fulfilled accordingly by all the state actors as well as non-state actors.

2.2 *The emergence of green constitution today and blue constitution tomorrow*

Today, there are many constitutions in the world have adopted a pro-environment constitutional policy. This is the new phenomenon of the green constitution in the world. Two of the greenest constitutions in the world are French Constitution of 2006, and Equador Constitution of 2008. In 2006, the Preamble of the French Constitution which was originated from the Declaration of Man and of Citizens of 1789 was added with the adoption of the Charter for Environment of 2004. Then in 2008, the new Constitution of Equador was adopted in which the idea of nature's fundamental rights are set-forth explicitly, just as importance as the articles of human rights. These two greenest constitutions in the world reflect the new human awareness of the importance of constitutional protection of the living environment and nature's right for the sustainable future development of mankind.

The Indonesian 1945 Constitution of the second amendment in 2000 has also adopted the article of environment. Article 28H (1) of the Constitution reads: *"Every person shall have the right to live in physical and spiritual prosperity, to have a home and to enjoy a good and healthy environment, and shall have the right to obtain medical care."* Article 33 (4) reads: *"The organization of national economy shall be conducted based on economic democracy upholding the principles of togetherness, the efficiency with justice, sustainability, environmentalism, self-sufficiency, and keeping a balance in the progress and unity of the national economy."* By the constitutionalization of the green policy, the government and the parliament may not make any law and regulation, nor administrative decisions, contrary to the principles of the green constitutional principles and policies.

In the future, the phenomenon of the green constitution must also be followed by the idea of the blue constitution related to the space and the virtual world for peace. Article 33 (3) of the 1945 Constitution only states that *"The land, the waters and the natural resources within shall be under the powers of the State and shall be used to the greatest benefit of the people"*. It does not say anything about resources above the land and waters, as if they do not belong to the State of Indonesia. Compare to the US Constitution which rules the territory and property in one integrated clause, the articles of property and territory in 1945 Constitution are separated in different Chapter IXA on the Territory, and Chapter XIV on National Economy and Social Welfare. Luckily that in Article 10 of the Constitution, "The President holds the supreme power of the Army, the Navy and the Air Force", meaning that the President has the supreme power to protect that country and the nation's interest, its territory and property in the air, in the space and the virtual world above the land and waters, by the use of the Air Force.

Today, the rapid development of the use of information and communication technology has made the world and mankind depend upon the virtual world. The constitution must cover any policies regarding the air, the virtual world, and even the outer space for the national interest, and peace and human prosperity. Therefore, the constitutional law experts from all over the world have to take part in developing the study for the emerging need for a blue constitution, besides today's application of the green constitution.

2.3 *Moving from a mere political constitution to the new perspectives of economic, social, and cultural constitution*

The Constitution of the United States of America is a political constitution in nature. As a modern political constitution, the contents are merely about politics and political relationship between functions and institutions of power, and political relation between the state institutions and the citizens. That's why C.F. Strong used the term of the political constitution in his book, "Modern Political Constitution" (1966) (Strong: 1966). Most constitutions in the world follow the political tradition of the United States Constitution that other aspects, such as economic, social, and cultural policies are not regarded as that important to be formulated in the constitution. Among the reason why the US Constitution does not contain economic subject is that the drafting of US Constitution was fully occupied with only political consideration for the establishment of the independent federal state of America, and left the economic affairs be regulated in and by the market place.

Before the establishment of the federation, American society has been developed as an industrious society that economic affairs have been run independently in the free market capitalism. Therefore, the framers of the constitution did not think of the importance to formulate any articles of economic policy in the constitution. Just later in the history that the economic aspects of the constitution come out from the interpretation by the Supreme Court as it was discussed by James Buchanan in his Economic Interpretation of the American Constitution.

Today, many countries formulated articles of economic policy in the constitution. Not only communist countries have the tradition of formulating economic policy in the constitution, but also non-communist countries, such as the Irish Constitution, India's Constitution, the Indonesian 1945 Constitution, and many others. Among the growing roles of market economy in today's globalization era, the role of the economic constitution is also growing, to control the free market. Therefore, the Indonesian experience is one of the examples. Chapter XIV of the 1945 Constitution has special articles about the national economy and social welfare policies. Therefore, I call the 1945 Constitution as a political constitution as well as an economic constitution that make the national economic system, "a constitutional market economy", i.e. a free market limited by the constitution as the highest policy norm (Jimly: 2010).

Besides, the growing concern on the economics of the constitution, scholars have to look further at the ideas of the social constitution (Jimly: 2014) and the constitution of cultures too (Jimly: 2017). Constitution today is not only used for the organization of state power, but also related to the organized civil societies, corporate's constitutions, and even the villages' constitution, such as in American tribal villages. Therefore the study of the constitution grows from the perspectives of the political constitution to the economic constitution, social constitution, and even cultural constitution. I hope that scholars of the constitutional study could develop more attention to the wider aspects of the constitution, not only limited to the conventional meaning of the political constitution.

2.4 *Constitutional law and constitutional ethics*

Today, the constitution cannot be regarded only as of the source of constitution law, but also the source of constitutional ethics. The role of ethics for public offices developed rapidly since the last decade of the 20th century. Even the General Assembly of the United Nations in 1997 recommended its member countries to develop ethics infrastructure for public offices. By doing so, the world is expected not only dependent upon the role of law but should also develop the effective role of ethics for public offices. We need legal and constitutional government as well as good government. Since the Reformation, Indonesia has also adopted the system of ethics for public offices. For the enforcement of judicial ethics, we have established the Judicial Commission set-forth in the Article of 24B, Chapter IX of the 1945 Constitution. The People's Assembly's Decision No. VI/2001 on Ethics for the Nation's Public Life has supplemented the guiding principles of ethics in the 1945 Constitution.

Today, based on the Constitution and the ruling of the People's Assembly, almost all state institutions have been equipped by law with the code of ethics and a special committee to enforce the code. The balanced roles of the notion of constitutional law and constitutional ethics are expected to overcome the weakness of the system of the the rule of law by the application of the system of rule of ethics at the same time. Not all problems should be overcome through legal approach. Some problems related to public offices are considered more effective to be addressed by the rule of ethics. Besides, the effectiveness of the legal norm is also dependent upon the effectiveness of ethical norm in practice. Law is like a ship, and ethics is the sea. The law will never sail and reach the island of justice when there is no enough water in the sea. Therefore, the future of constitutional law needs support from the idea of constitutional ethics, that every constitutional lawyer should take part in developing the study of constitutional ethics in the future.

For instance, in the field of ethics for election management bodies and ethics for members of parliament, the code of ethics is enforced through and by a system of the court of ethics. In the House of Representatives, we have established "Mahkamah Kehormatan" or court of honour. While for the election management bodies, we established the Honorary Council of Electoral Management Bodies which perform the enforcement of the code of ethics through the adjudicating process, like in the ordinary court of law. By application of the system, Indonesia has begun to introduce the new system of the court of ethics, besides the conventional system of the court of law and justice. Besides the notion of constitutional law and the principles of the rule of law, we have to develop also the notion of constitutional ethics and the principles of rule of ethics for the future.

2.5 *The new Quadro Politica*

Another important issue is about the new form of separation power within a constitutional state. Here, I would like to share with you about a new phenomenon of power relation in the world with the emerging hegemonic role of electronic and social media. In the middle of the 20th century, the Montesquieu' doctrine of trias politica of 1689 has been added with the new branch of power, i.e. the media as the fourth estate of democracy. The executive, legislative, judicial power, and the media became four branches of power or four estates of democracy. It changed from a trias politica to a quadru politica.

But today, the actual form of quadru-politics consists of the new domain of powers, i.e. (i) the state, (ii) civil society, (iii) business corporation, and (iv) the media. In Indonesian experience today, there is a new phenomenon of businessmen develop television broadcast company, then establish a political party or manage several political parties by financial support, and finally try to run for the presidency. If there is no regulation that limits the potential conflicts of interest and separates the four branches of power, democracy will be hijacked.

Therefore, the new four estates of democracy should be separated from each other to avoid any potential conflict of interests. The new dominant factors in democratic politics, i.e. (a) the corporate political and economic capitalism and (ii) the electronic and social media hegemony. How to control them for the benefit of the people as a whole. Without separation of power between the new four estates, there will be no more democracy based on the future standard of quality. Democracy and the democratic system will be hijacked or ploughed by corporate capitalism. Therefore, we have to separate the new four estates and to promote the principles of government of the people, by the people, for the people and continue to be with the people (from the people, by the people, for the people, and with the people).

2.6 *The impact of globalization on the new relation between international public law and domestic constitutional law*

In the globalizing world, constitutional law as a subject of study has been changing significantly from a positive and domestic-oriented to a general science of law applicable everywhere in the world. Look, for instance, the phenomenon of EU (European Union) and its constitutional treaty has made the difference and distinction between domestic constitutional and International public law in Europe blurred. We are now on the move from exclusive and domestic-oriented constitutionalism to inclusive regional and international constitutionalism.

Within this new world, the study of comparative constitutional law becomes more and more important. At present, the judicial interpretation of the constitution through comparative reference to other countries' constitution become a common cause. Even in German's Constitutional Court, the EU law is now treated as the same level of the German Constitution that the German Law may not be contrary to the German Constitution as well as the EU Law. The structure of the EU is also new to the conventional political history of our world.

The organizational structure of the EU is just like a state consisting of executive, legislative, and judicial branches of power. EU has also its monetary system, just like an independent state. Therefore, in Europe, there is no more clear distinction between International Public Law and Domestic Constitutional Law. A professor of constitutional law in Europe today is just like an International Law professor.

Because of the above phenomenon, today, we are moving towards universalization of constitutional values around the world and its dynamic relation with the constitutional culture of our local history. All modern constitutions today share common constitutional values which some are borrowed or transplanted from other countries or other international best practices making them all look like similar in substances. However, we have to look into details both the problems of interpretation and implementation of the constitutions with universally contained substances in the cultural context of the respective countries concerned. Universalization of constitutional values is not identical with internationalization nor globalization of values. Universal values may come from outside as well as from our respective cultural history. Therefore, the issues of constitutional culture should become pivotal in the studies of the constitution today. It is now the time to pay more attention to the issues of "cultural constitution and the constitutional culture" of each country so that the institutionalization of constitutional rules would not be conflicting with the living cultural traditions in the respective history.

Indonesia has its own long historical experiences with so pluralistic cultural traditions throughout the country. But most of the ideas adopted into the constitution are transplanted or borrowed from other countries. So it is the task of the scholars to build an intellectual bridge between modern constitutional state institutions to the constitutional culture and cultural living tradition of the people. We have to avoid cultural divide or discrepancy between the political institution and the living tradition of the people. The state and its power is nothing less than the power of the people, for the people, and by the people themselves. Even today, I always add these quotes from Abraham Lincoln that not only the government of the people, by the people, and for the people, but also with the people. The government is always of the people, by the people, for the people, and with the people.

2.7   *From universalism to multiversalism, toward a cosmopolitan legal pluralism*

Another issue following the above trend of universalization of legal and constitutional values and the need for a cultural reading of the law and the constitution is the issue of global pluralism. We live now in a globalizing and borderless world with the complexities of the law, where a single act or actor is potentially regulated by different multiple legal or quasi-legal regimes. We have to live in the complex relationships among international, regional, national, and subnational legal system, where non-state actors such as industry-setting bodies, non-governmental organizations, religious institutions, ethnic groups, and others exert a significant normative pull. We cannot depend anymore upon the old perspectives of sovereigntist territorialism, nor substantive universalist approach that requires people to be conceptualized as fundamentally identical to be brought within the same normative system. We are now moving from universalism toward multiversalism that Paul S. Berman calls it "cosmopolitan legal pluralism" (Paul: 2012) as a useful approach to the design of the procedural mechanisms, institutions, and discursive practices.

A cosmopolitan pluralist approach manages multiplicity without an attempt to erase the reality of that multiplicity. The key solution for the legal scholars on the problem of legal pluralism is not the pluralism itself, but the need for a global comparative study of law and constitution. Legal scholars in every country have to pay more attention to the need for a comparative study. Scholars of the countries with common law tradition have to know civil law legal tradition. Lawyers of one country should know and understand well about the legal system of other countries with intensive interactions between people to people or business to business contacts. Such as Indonesia that has very close relations with the western countries as well as with China, India, Japan, Korea, and the Middle-eastern countries, has to pay more attention and to promote comparative studies with the legal systems of those countries, including the comparative studies on the legal system of every ASEAN member countries, such as Malaysia, Singapore, Thailand, Vietnam, the Philippines, etc.

3 CONCLUSION

Ladies and gentlemen, to conclude, today we have to develop new perspectives in the study of the constitution, that besides continuing to develop (i) the positive constitutional law approach, we have to promote also in a more active way (ii) the studies of comparative constitutions, (iii) the study of science of constitutional law, (iv) constitutional ethics, (v) the constitution of culture and constitutional culture, (vi) economic constitution and the economy of the constitution, and (vii) the study of green and blue constitution. Besides, there are so many aspects of the inner structures of the constitutional law which are also changing rapidly, such as (a) the structures and domains of powers that the principles of separation of power among the executive, legislature and the judiciary are no longer rigid in its implementation. The emergence of press media is growing rapidly and changing the modes of communications and the information system that it becomes the real fourth estate of today's democracy. But the four estates are no longer consisting of the executive branch, the legislature, and the judiciary, but the state, civil society, corporate business, and the media, because of the emergence pivotal roles of the rapid growing of market economy and civil societies globally. To guarantee the quality of today's democracy, we need to separate the new four branches of powers, i.e. the state, civil society, market, and the media so that they will not be in the position of conflicts of interests one to the other.

What would a labour migration and women's rights "champion" do? Examining rhetorics, laws and policies of the Philippines and their impacts on the lives of migrant women teachers in Thailand

Joel Mark Barredo

Institute of Human Rights and Peace Studies, Mahidol University, Salaya, Thailand

ABSTRACT: Due to the size of its population relative to limited resources and opportunities, the Philippines is no stranger to the phenomenon of Filipinos leaving their motherland to search for "greener pastures." Government law and policies over decades on labour migration were meant to be understood as temporary measures to alleviate political, economic, and social ills that the government cannot yet fully address. This study, which focusses on female teacher migrants in Thailand, reveals that despite the existence of laws and agencies, the Philippines has not fully implemented measures to fully document and address the concerns of many OFWs, specifically amongst women. This is evidenced by, *first,* the complete lack of any provision in the vision and objectives of the Embassy of the Republic of the Philippines in Thailand to address issues concerning Filipino migrant workers, more so Filipina migrant workers. *Second*, there are no bilateral agreements concerning the promotion and protection on the rights of migrant workers between the two governments. *Third*, Thailand is, still, not a party to the Migrant Worker's Convention. Thus, the deployment of Filipina migrant workers in a country where their rights are not necessarily secure runs counter to the prescribed policy of the State as enshrined in its laws on migrant workers. Hence, the Philippine state has been remiss in its duties as *parens patriae* to protecting and promoting the rights of the thousands of Filipinas deployed to Thailand.

1   INTRODUCTION

The study then inquires some teacher-respondents on the impacts of relevant laws and programmes to their lives and employment in Thailand. It is revealed that the State yet much to deliver in order for them to fully enjoy their rights as migrant workers. Featured narratives strengthen the argument for the Philippine government to exert better efforts to ensure that no one is left behind in claiming her rights while seeking a better life abroad.

2   DISCUSSION

2.1   *The Philippine state's role as "*Parens Patriae*" in relation to Filipina migrant workers*

The role of the Philippine government involves ensuring that migrant workers' rights in every process of migration are promoted and protected (CIIR, 1987). This is evident in its outright commitment towards international standards related to the rights of migrant workers and women, and national laws such as the Migrant Workers' Act and the Magna Carta of Women. Interestingly, while recognizing the significant contribution of OFWs to the national economy, the State does not promote overseas employment as a means to sustain economic growth and achieve national development (DAWN, 2000). In a nutshell, the Philippine government does not have a structured policy of labour migration; more so, a specific gendered policy on this matter—despite a rather 'empowering' national Migrant Worker's rights law.

This ultimately calls into question the Philippines' capacity to fulfil its responsibility as *parens patriae*, or "parent of the nation," to ensure that all its citizens are not denied their fundamental and inalienable rights as human beings. Specifically, the Philippines, despite being a so-called "champion" of migrant workers' rights, has not initiated moves to fully document the experiences of Filipina migrant workers in Bangkok and its neighbouring provinces. The State, in its role as *parens patriae*, should do more in the discharge of its responsibility to assure the well-being of Filipino migrant workers.

2.2 *The growing migrant labour economy and the rhetorics of the executive branch*

The Philippine government's failure to promote and enforce both a structured and gendered policy of labour migration resonates back to the policy of the Marcos government on exporting labour from the Philippines which was more of a temporary measure to curb unemployment and under-employment. This kind of government action to promote labour migration from the Philippines is linked with the rising oil prices which caused a boom in contract-based migrant labour in the Middle East (O'Neil, 2004). The Marcos era saw labour migration as a solution, albeit a "temporary" one, to "raise production efficiency in all sectors of the economy" (CIIR, 1987). Hence, the outflow of skilled labourers from the Philippines was not then viewed as a permanent cure-all to the economic, social, and political ills ravaging the country. This outflow was seen as a palliative, at best, by the Marcos government.

Yet, this was all just lip service up until today. Quoting Liban, this "temporary measure has, in fact, become a well-established part of economic policy" (Liban, 2003). In fact, presidential rhetoric on economic policy in connection with OFWs has influenced how the other branches of the Philippine government, *i.e.*, the legislative and judicial branches, have responded to the phenomenon of labour migration. In 2009, no less than the Philippine Supreme Court noted how public policy can impact, if not "magnify" the effect of development as brought about by international migration impacts.

> For decades, the toil of solitary migrants has helped lift entire families and communities out of poverty. Their earnings have built houses, provided health care, equipped schools and planted the seeds of businesses. They have woven together with the world by transmitting ideas and knowledge from country to country. They have provided the dynamic human link between cultures, societies and economies. Yet, only recently have we begun to understand not only how much international migration impacts development, but how smart public policies can magnify this effect (Serrano v. Gallant Maritime Services, 2009).

Hence, the initially "temporary" solution to the Marcos government's to the crises it encountered during the 1970s and 1980s has become the "cornerstone policy" in later years, particularly to that of the terms of former Presidents Joseph E. Estrada and Gloria Macapagal-Arroyo (Hau, 2005). This policy has likewise been reiterated by President Benigno S. Aquino, whose administration currently aims to advance the three pillars of national development and good governance objectives, namely: (1) promoting national security; 2) enhancing economic diplomacy, and 3) protecting the rights and welfare of Filipinos overseas.

In former President Estrada's third and final State of the Nation Address (SONA) in 2000, he noted that Philippine "gross international reserves reached an all-time high of $15.44 billion in May 2000, and [was] expected to rise to $17.1 billion by the end of the year" (Estrada, 2000), in which the remittances of OFWs have played a part. Prior to this, and less than two years after the onset of the Asian financial crisis in 1997, President Estrada exhorted OFWs to "help prop up the heavily battered" Philippine economy (Hau, 2005 as cited in De Guzman, 2003, p. 57).

In 2001, former President Macapagal-Arroyo went as far as stating that the Philippines would be "heavily dependent" on OFW remittances "for the foreseeable future (Hau, 2005)" further cementing an economic policy reliant on labor migration and the financial benefits it necessarily brings about (Hau, 2005, as cited in De Guzman, 2003, p. 57). Former President Macapagal-Arroyo was subsequently presented with the "Milkmaid Award" for "milking the maids" by Unifil, a Hong Kong grassroots domestic worker organization, for relying on migrant domestic helpers to keep the country economically afloat (Constable, 2010).

In 2013, President Benigno S. Aquino III offered a toast during the traditional *vin d'honneur* for the celebration of the 115th Anniversary of Philippine Independence to OFWs, among others, for their role in what is essentially successful nation-building. specifically:

> "To our Overseas Filipino Workers, for their dedication and sacrifice—may their hard work redound and further contribute to the Philippines that they can come home to with pride and confidence;

> And to the future and continued success of the Filipino people—may the examples of our heroes be our guide as we tread the straight path to equitable progress" (Aquino, 2013).

Indeed, the Philippines has experienced a long tradition of national chief executives extolling the virtues of a strong and dynamic migrant labour force, despite having virtually no legal framework to support the promotion of a labour-export economy. The gaping hole left by the absence of laws, rules, regulations and policies does not jive with the fact that, as of date, the export-service sector is deemed as the biggest contributor to national economic development. This underscores the urgent need for an integrated and comprehensive upgrading of migrant services of Philippine missions overseas (Ofreneo and Samonte, 2005).

2.3   *Setting priorities: economic gain versus human rights*

It is generally recognized that remittances from Filipino OFWs generate foreign exchange earning ease the country's employment problems and are partly instrumental in the growth of the country's GNP. According to the World Migration Report of 2005, The Philippines received US$ 7.363 billion, or 9.45 per cent of its GDP from migrant workers' remittances (IOM, 2005).

The economic focus shifted to human rights came as a reaction to the sensational execution of a Filipina domestic worker in Singapore. Essentially an offshoot from the Flor Contemplacion case, the Migrant Workers Act of 1995 (Republic Act No. 8042), as amended, brought the promise of better protection to migrant workers in general.

On June 7, 1995 Congress enacted R.A. 8042, which sets that, for among other purposes, sets the Government's policies on overseas employment and establishes a higher standard of protection and promotion of the welfare of migrant workers, their families, and overseas Filipinos in distress.

2.3.1   *'Commitment' to ICMR*

The enactment of the Migrant Workers' Act immediately enabled the Philippine government to be state-party to the United Nations International Convention on the Protection of the Rights of All Migrant Workers and Their Families (ICMR) in July 1995. The law itself strongly resembles principles and operative provisions with the international convention. Prior to the Migrant Worker's Act of 1995, there was no concrete law to protect and promote the rights and welfare of OFWs except in some general provisions in the Philippine Constitution, Philippine Labor code and Rules and Regulations of the Philippines Overseas Employment Administration (POEA) and the establishment of the Overseas Workers Welfare Agency (OWWA) (DAWN, 2000). The law was believed to set in stone a significant shift from previous laws since it de-emphasizes the economic aspects of the Diaspora and creates a higher standard of protection and welfare promotion for overseas workers (Gonzalez, 1998). The existence of the overseas employment program rests solely on the assurance that the dignity and fundamental human rights and freedoms of the Filipino citizen shall not, at any time, be compromised or violated. The state, therefore, shall continuously create local employment opportunities and promote the equitable distribution of wealth and the benefits of development (DAWN, 2000).

It was only in 2009, 14 years after its ratification, that the Government was able to submit its initial report to the Committee on the Protection of Migrant Workers and Members of their Families. Moreover, despite streaks ratifications of international conventions and establishment of relevant laws and policies, recommendations for enhanced protection and promotion of migrant workers had swarmed the Concluding Observations of the committee. Paragraph 16 rather reveals a quintessential reality of labour migration governance in the Philippines: "review its labour migration policy in order to give primary importance to human rights of migrant workers, in line with the State party's own professed goal as set out in RA 8042" (Republic of the Philippines, 2009, Para. 16). Such international observation partly unearths the government's national management of labour migration is softly implemented and is approached from an economic development orientation rather than a rights-based one.

Pertaining to women migrant workers, the Philippine government had listed down existing laws that protect women specifically from the harms of trafficking and sex work. Paragraph 69 of the report declares that the government reinforced "reinforce the constitutional provision of equality between the sexes and pay special attention to women's special needs. Laws on women's health, economic and political participation, those that seek to protect them from violence and prostitution, safeguard their marital and material welfare and laws that seek to protect the girl-child have been passed in the past few years" (Republic of the Philippines, 2009, Para. 69). It has been observed that Filipina migrant workers are mainly perceived to be vulnerable to violence and discrimination in the midst of their contribution to development. Filipina migrant workers are therefore lumped into a category of a marginalized group of OFWs.

The Committee in its conclusion embraces the position of the government towards the vulnerability of migrant workers. It "urges the State party to continue its efforts to promote the enhancement and empowerment of migrant women facing situations of vulnerability. It recommended the government to monitor and identify the situation of Filipina migrant workers, provide local employment opportunities and continue to develop and strengthen gender-sensitive programmes for them (Republic of the Philippines, 2009, Para. 28).

In March 2014, after nearly five years, the Philippine government submitted its second report to the Committee on Migrant Workers' Rights. The report again revealed accomplishments pertaining to the promotion and protection of the rights of Filipina migrant workers. It responded to the concluding recommendations provided in 2009. The government disclosed that it mainly "conducted thorough assessment of the situation of Filipino migrant women, including their income in the informal sector, and taking concrete measures to address the feminization of migration comprehensively in its labour migration policies, and ensuring minimal social protection for Filipino migrant women" (Republic of the Philippines, 2014, Para. 28 (a)). This is manifested through existing laws such as the Magna Carta for Women and the Women's Development and Gender Equality Plan 2013-2016. Both of these policies centre on the strengthening of governance mechanisms that are mandated to promote and protect the rights of Filipina OFWs.

The Philippine government still takes a macro-approach to the diverse realities and roles confronting Filipina migrant workers. While it had the chance to conduct the assessment, it still, as in 2009, perceives Filipina migrant workers as a homogenous sector. The Shadow report submitted by the Center for Migrant Advocacy (CMA) revealed that "women migrants, remain concentrated in gender-stereotypical women's job in the households as domestic workers. They comprise the biggest proportion of workers under one single job category in the annual deployment statistics (CMA, 2014, p.14)." Women are still perceived as agents of development and at the same time boxed into gender typecasts. This prohibits mechanisms to perform more targeted and meaningful solutions to various degrees of challenges that Filipina migrant workers confront in all processes of labour migration.

The Philippine government, up to this date, has been very consistent about its achievements to make migrant workers' rights policies and agencies to be more gender sensitive/embracing.

But at the same time, it sustains the root factors such as gender-bias in deciding on employment, gender discrimination that enable discrimination and disempowerment of women migrant workers by only perceiving women migrant workers merely as marginalized or victims in this labour phenomenon.

2.3.2 'Commitment' to CEDAW and the UPR

Exploring further into the government's global commitment to Filipina migrant workers' rights, in allegiance to the landmark Convention on the Elimination of all Forms of Discrimination Against Women (CEDAW), the Philippine government engendered its perception of "migrant worker."[1] The Philippine State has ratified the CEDAW, in keeping with Section 14, Article II of our 1987 Constitution mandates the State to recognize the role of women in nation building and to ensure the fundamental equality before the law of women and men. As a State Party to the CEDAW, the Philippines bound itself to take all appropriate measures "to modify the social and cultural patterns of conduct of men and women, with a view to achieving the elimination of prejudices and customary and all other practices which are based on the idea of the inferiority or the superiority of either of the sexes or on stereotyped roles for men and women (CEDAW, 1979; Garcia v. Drilon, 2013)."

In its rather outdated yet most recent State report to the Committee on the Elimination of Discrimination against Women (CEDAW) (combined fifth and sixth reports), the Philippine government responded to the previous committee recommendation to "establish a special national focal point to provide information and support services to women before departure to overseas work, as well as in the receiving countries in cases of need. (Paragraph 25)" (State Report of the Philippine State Party). It asserted that it has been implementing the provisions of the Migrant Worker's Act of 1995 by (1) putting in place additional welfare services for OFWs beginning 2002 under the program "International Social Welfare Services for Filipinos" of the Department of Social Welfare and Services and (2) render direct services (transportation, shelter, and food assistance); case management, including referrals to other social services, counseling, group therapy, and filing of cases; jail and hospital visits; values enhancement and skills training; and advocacy and social mobilization (Paragraph 15, Fifth and Sixth Philippine reports to the CEDAW).

The report was submitted before the establishment of General Recommendation 26 on Women Migrant Workers. Yet it had already set a welfare-like stance in addressing women migrant workers' issues. Moreover, it also had placed migrant women in a more vulnerable position. In paragraph 16(b) of the country report, women migrant workers have to be informed more about abuses (trafficking, illegal recruitment) rather than reinforcing rights and their freedoms while in the process of migration. Certain professions are also highlighted to be more vulnerable to abuse such as entertainers, domestic workers and other low skilled jobs (which in reality are embraced by the majority of Filipina migrant workers. At this vantage point, the Philippine government seemed to focus more on sensational cases of female labour migration—leaving out women from 'safer and more secure professions' such as teachers.

The second (and policy/law heavy) Universal Periodic Review National Report of the Philippine Government (2012), which is the most recent international human rights State Report from the Philippines, mentioned very little about migrant worker's rights, less so with Filipina OFWs. It only declared certain amendments to the Migrant Worker's Act which is not well articulated in the document.

At the international level, the Philippine state perception embraces that of the mainstream State-view on female migrant workers—one that is reactionary, focused on macro issues and approached towards vulnerability and disempowerment. It is consistent with the *parens*

---

1. The concept of migrant worker was defined in the Migrant Worker's Act of 1995. It "refers to a person who is to be engaged, is engaged or has been engaged in a remunerated activity in a state of which he or she is not a legal resident to be used interchangeably with Overseas Filipino worker (OFW)" (Migrant Workers Act of 1995, Sec. 3).

patriae role that it assumes. In this spirit, Filipina migrant workers have to endure limited (dis-empowered) identities and abuse before they can fully access and enjoy their rights. Spaces for claiming/bargaining for their rights have already been washed down at this point.

2.4 *The slow yet steady recognition of women as essential to Filipino "Nation-Building"*

Despite an international stance, it is still imperative to explore and assess the fulfilment of women migrant worker's rights at the national/local levels. At the onset, laws and policies related to women and labour migration set by the Philippine government are striking and powerful. Corrective labour and social laws on gender inequality have emerged with more frequency due to our country's commitment as a signatory to the CEDAW (Philippine Telephone and Telegraph Company v. National Labor Relations Commission, 1997). These laws include Republic Act No. 6727 which explicitly prohibits discrimination against women with respect to terms and conditions of employment, promotion, and training opportunities; Republic Act No. 6955 which bans the "mail-order-bride" practice for a fee and the export of female labor to countries that cannot guarantee protection to the rights of women workers; Republic Act No. 7192 also known as the "Women in Development and Nation Building Act," which affords women equal opportunities with men to act and to enter into contracts, and for appointment, admission, training, graduation, and commissioning in all military or similar schools of the Armed Forces of the Philippines and the Philippine National Police; Republic Act No. 7322 increasing the maternity benefits granted to women in the private sector; Republic Act No. 7877 which outlaws and punishes sexual harassment in the workplace and in the education and training environment; and Magna Carta of Migrant Workers, as previously discussed" which prescribes as a matter of policy, *inter alia*, the deployment of migrant workers, with emphasis on women, only in countries where their rights are secure. Likewise, it would not be amiss to point out that in the Family Code, women's rights in the field of civil law have been greatly enhanced and expanded.

Most recently, the concept of the Filipina migrant worker is reinforced through the ratification of the Magna Carta for Women of 2009. It states that "Filipino women who are to be engaged, are engaged in a remunerated activity in a state which they are not legal residents, whether documented or undocumented, pursuant of Section 3(a) of the Migrant Workers and Overseas Filipinos Act of 1995". Moreover, they are lumped together with abused women, women from minority groups and indigenous groups as part of the marginalized sector. Inclusion to this means that the "State shall, at all times provide for the protection, participation and empowerment of these women."

In addressing violations and abuses, past administrations took steps to impose labour migration bans on certain countries. This is most significant at the height of Flor Contemplacion (Singapore) and Sarah Balabagan (United Arab Emirates) cases. In principle, as mandated by the Migrant Worker's Act (2005), "The State shall deploy overseas Filipino workers only in countries where the rights of Filipino Migrant Workers are protected (Section 4)". Oishi argued though that these bans were not actually to protect the rights of Filipina OFWs but to reduce strains in terms of diplomatic relations with migrant-receiving countries (Oishi, 2005).

2.5 *A different understanding and application of the human rights of Filipina migrant workers*

Amidst the advancement of programmes on migrant workers, human rights are rather differently understood and applied by the Philippine government's mandated agencies. Ms. Louella Callanza, Overseas Welfare Workers' Agency (OWWA) Officer III, summed up the main objective of her agency which is "to promote and protect the interest and well-being of OFWs." While the Migrant Worker's Act, a rights-based law, guides the agency, Ms. Callaza (personal interview, 2013) confirmed that "rights and welfare are the same, the Migrant Worker's Act is a guideline." Thus human rights are equally perceived as welfare—which is the provision of social benefits and programmes. So-called "benefits" from OWWA can only be enjoyed by registered active members who have paid their yearly dues to the Philippine government. Thus, an accommodating space to fully claim rights is limited. To make matters worse, this welfare programme totally excludes migrant workers who are undocumented. Furthermore, the burden to claim these benefits is left on the OFW, "you can always ask for help". In terms of gender-specific programmes, OWWA offers none (Callaza, 2013).

In Thailand, which is relatively a low-level migrant-receiver country, the Philippine embassy/consulate, based in Bangkok, had recently relieved its labour attaché or the Philippine Overseas Labour Office (POLO) due to the recent floods which struck Central Thailand. Since then, they never reinstated a focal person on migrant worker affairs. Ms. Callanza mentioned though that there is not much relevance to sustaining POLO in Thailand. She shared that the Philippine government is keener on high-frequency receiving countries and regions such as Hong Kong, Singapore and the Middle East (Callaza, 2013). Neither is there an officer who specifically attends to women's issues.

2.6 *Acknowledging the strain between what the law "Is" and what the law "Should be"*

The Philippine embassy and consulate furnished a Citizen's Charter, which envisions the body to be:

> "[T]he partner of Filipinos in Thailand in the pursuit of national interest and in the promotion and protection of their rights and well-being.(a) The Consular Section, as the frontline unit of the Embassy, aims to be customer sensitive, highly responsive, and innovative. (b) We are conscientious of the needs of our Filipino citizens and foreign clients whom we serve. (c) We are committed to maintaining a highly responsive system which ensures the effective delivery of services. (d) We are diligently looking for innovative and creative ways to further improve our services (Embassy of the Republic of the Philippines in Bangkok, 2013)."

The objectives of the Embassy/Consulate of the Republic of the Philippines in the Kingdom of Thailand are the following: Consular functions, including the issuance of passports, travel documents and visas, legalization of documents, and reports of civil registry records, are a critical component of the Embassy's services to the public. Cognizant of this fact, the Consular Section shall endeavour to perform such functions in the most efficient and timely manner. The focus shall be customer satisfaction, which can be guaranteed by reducing the supporting documents to the barest minimum; creating an e-passport room that is clean and well-maintained; and assigning a staff complement that is friendly and focused on delivering exceptional customer service (Embassy of the Republic of the Philippines, 2013.)

An examination of this document reveals that there is no mention of any duties to address issues concerning Filipino migrant workers in general. Delving deeper, it is crucial to note that the governments of the Philippines and Thailand do not have bilateral agreements concerning the promotion and protection on the rights of migrant workers. Thailand is neither a State-party to the Migrant Worker's Convention. Based on the foregoing, there seems to be no strong indication that the Filipino government would negotiate for stronger ties with Thailand leading to the fulfilment of rights of Filipina OFWs. Thus, the prescribed policy of the Philippine state which involves the deployment of migrant workers, with emphasis on women, only in countries where their rights are secure, has not been applicable to Filipina migrant workers based in Bangkok and its neighbouring provinces. Hence, the Philippine state has been remiss in its duties as *parens patriae* to protecting and promoting the rights of the thousands of Filipinas deployed to Thailand, and are currently working there, whether documented or not.

Limited understanding and weak enforcement of mandates that protect and promote the rights of Filipina migrant workers specifically in Thailand strengthens the argument that the Philippine government strongly underscores economic benefits brought by its OFWs. Filipinas workers in Thailand, therefore, as far as the government is concerned, have to settle with their "role in nation building" as set by the 1987 Philippine Constitution.

In order to expand the analysis of the impact of laws and policies by the Filipino government on labour migration, it is key to capture insights from recipients of these rights, women OFWs. This study selected nine respondents who worked as teachers in Bangkok and its neighbouring provinces at the time of data collection. Their narratives would reveal the level of impact of existing laws, policies and mechanisms in place in both their personal and professional lives.

Gonzalez (1998) provided an explanation of the construction and implementation of public policies affecting the rights of Filipino Migrant Workers. In the Philippines, public policy-makers have formulated many measures from the more prudent/conservative approach as applied in the local context, addressing the symptoms of countless policy issues rather than first seeking to define strategic long-term measures. It is sad to say but in the Philippines, public policy has seldom been rational or comprehensive. The situation becomes even worse when it reaches the implementation since Filipinos are fond of saying "pwede na ito, bahala na" (this will do, I'll leave it to God) (Gonzalez, 1998).

This attitude and world view are very apparent in the Philippine state agency presence in Thailand. When asked during a rather closed interview about how the Philippine embassy/consulate can address issues related to Filipina OFWs, the respondent officer kept on repeating the statement, "they are adults." One has to read between the words in order to fully digest what these words really entail. On the one hand, Filipina OFWs are left on their own to adapt ways in dealing with their situation. On the other, while equipped with rather weak mechanisms for migrant protection rights, the Philippine government seems to wash off its hands from dealing with matters concerning its citizens.

At this point, dealing with a weak state organ and revelations about respondent teachers low knowledge on their rights, it is crucial to look at how migrant teachers perceive the role of the embassy in delivering the mandate of promotion and protection of their rights.

It is interesting to note that for most teacher-respondents, the Philippine state body mainly plays the role of a consular body—one that generally processes and regulates national documents. Ma'am Luz shared that "I went there to process my passport and that's it (Ma'am Luz, Personal Interview, 2013)." Ma'am Diana felt the same way, "I don't feel the embassy/consulate's presence so much. I only visit the embassy when renewing the passport (Ma'am Diana, Personal Interview, 2013)." Such views would not be surprising. The Citizen's Charter mainly mandates the Philippine state body to perform this duty as a consular agency. Ma'am Vivian, who carries the same view, confirmed that the embassy/consulate has been effective with it's standard 'responsibility', "They are able to deliver documents whenever I need them (Ma'am Vivian, Personal Interview, 2013)."

Apparently, this dissatisfies or dislocates one from fully claiming her rights. In the interviews conducted, amidst low awareness and education of rights, teacher-respondents seem to expect more from the Philippine embassy/consulate's presence and responsibility for its nationals working in the Thai Kingdom.

Ma'am Aida has been very sharp about the uneven dynamics that play between the OFW and the embassy. She mentioned that "I have gotten no help from the Philippine embassy, they only react or address issues when one asks help from them (Ma'am Aida, Persona Interview, 2013)." This resonates that the Philippine organ plays a reactionary role in challenges or problems referred to it. Ma'am Luz deems that it is more of a burden (and costly) for her to approach the embassy whenever she has problems, "Was they helpful?, Yes, but I had to pay a lot of money (Ma'am Luz, 2013)."

For some, such negative orientation had led to a rejection of any support or recognition that the Philippine Embassy can address their issues and eventually protect and promote their rights. Ma'am Vivian shared, "The Philippine embassy/consulate did not do anything to help me. I am neither asking any help from them. What they only want is my contributions as an OFW (Ma'am Vivian, Personal Interview, 2013)." Moreover, there has been a sentiment that such interventions from the government body even add on to the injuries afflicted by their work. Ma'am Minda believes that it is not an option in solving any of her personal and professional problems, "It is much easier and better to solve my problem on my own, I just feel irritated and disappointed whenever I come up and find no help from them (Ma'am Minda, Personal Interview, 2013)."

Ma'am Sally, who was the most optimistic of all the respondent-teachers, already understood the limitations of the embassy. She felt that she possesses the duty of a citizen to provide support to the Philippine organ based in Bangkok, "I volunteer for the embassy for gatherings, such as emcee for events. It is my way of helping the Philippines (Ma'am Sally, Personal Interview, 2013)."

It is clear, at this stage, that the Philippine state organ in Bangkok, Thailand has to step up in promoting and protection of the rights of migrant workers, specifically Filipinas. Furthermore, the "leave it to God" syndrome is slowly seeping into the minds of teacher-respondents. Yet, there still exists a gleam of hope that the embassy has to do something about the gaps that Filipina migrant teachers are currently facing.

The *parens patriae* effect, both in-country and abroad, has been encouraging public dependency towards the Filipino government. It has been a general practice amongst citizens that the State, as the father of the land, would be able to solve problems of varied scales. This remains true among teacher-respondents based in Bangkok and its neighbouring provinces. While the parlance of rights is missing in aspirations for a rights-oriented duty bearer, the select respondents have provided specific and targeted recommendations to solve their personal and professional challenges.

Ma'am Luz who is disappointed with her financial status in Thailand because "the salary I get is not enough" is still hopeful that the Philippine state can provide her assistance. She said, "How I wish that the embassy is able to help us get benefits from the schools. We should have a standardized increase in salary. In terms of benefits, we do not have health insurance and social security (Ma'am Luz, 2013)."

Many responses revealed that rights are reduced to assistance for job security. There has been no mention about other matters involving better access to health services, right to self-determination or freedom from discrimination. It must be noted that these are the effects of a limited understanding of the range of rights that a migrant worker and woman should be enjoying.

Ma'am Anne, whose financial situation back home had pushed her to work in Bangkok and is having a hard time dealing with the language barrier in her school and other social circles, shared that "the embassy must provide help for those who are working like assisting in jobs and ensure security (Ma'am Anne, Personal Interview, 2013)."

Such perception is also carried by Ma'am Lorna. While she finds it hard to deal with the language barrier, attitude of the students towards their education and work ethics of Thai and western co-workers, she still sees job security as an utmost priority that the Philippine Embassy must push for. She shared, "they can take the initiative to negotiate with agencies and schools for teacher benefits. They also have to conduct job fairs regularly." She also yearns for assistance to ensure her personal security; "they should make sure that teachers are able to avail accident and health insurance. (Ma'am Lorna, Personal Interview, 2013)."

Some teacher-respondents, like Ma'am Diana, albeit her limited knowledge in human rights affairs, pointed out the need to partner and negotiate with the Thai government in pursuit of better assistance and protection of Filipina OFW rights. She asserted, "They must work with the Thai government about the salary scale of Filipino workers (Ma'am Diana, 2013)." Although, she doubts that this may happen in their favour. "I don't think that this is the government's priority. I think as long as one is holding the Philippine passport, discrimination will continue (Ma'am Diana, 2013)." Ma'am Anne is also aware that the Philippine government must take the initiative to partner with the Thai government with respect to addressing their concerns; "The Thai government is not aware of the needs of foreign workers. Visa problems are consistent amongst Filipinos and other nationalities (Ma'am Anne, 2013)."

At the end of the day, Filipina teacher-respondents need and want more in life—either to explore better opportunities abroad or to move back to the Philippines. The cycle of dislocation-dissatisfaction-desire remains strong as they experience and perform their duties. Perhaps due to personal and professional insecurities, it has been apparent that working and living in Thailand would not be a permanent host country.

Ma'am Lorna, who has worked for three different schools in a matter of five years, thinks that it is time to move on soon; "I will just finish this academic year and will have to find a better job in another country. If not, I will go back to the Philippines and work there (Ma'am Lorna, 2013)." Ma'am Diana, who gave birth to and is raising her son in Bangkok, felt that working in Thailand may impede her dreams for her child; "I do not think that my child will benefit from my work and the social environment if we stay longer in Thailand (Ma'am Diana, 2013)." Ma'am Vivian will only be staying for two more years. She is currently applying for a visa to the United States of America to be with her fiancé. Ma'am Anne plans to also take the same route, yet still for the job and economic reasons; "I will look for greener pastures soon in other western countries (Ma'am Vivian, 2013)."

After all what was said and learned, despite struggles in asserting the human rights parlance and practice of both Filipina migrant teachers and the Philippine state organ in Thailand, everything still reverts back to the aspiration of attaining 'greener pastures'. While there are difficulties and matters that may lead to abuse and violations, one aspect that is observed is that these women remain to be firm and strong to their purpose/s for migrating. The awareness of dislocations is present, but these women have been trying to manage and address factors that may impede them from reaching their goals—which, at the end of the day, is still their strongest source of strength.

3 CONCLUSION

3.1 Are legal and rhetorical "championing" migrant workers' rights enough?

The Philippine government, based on policies and laws being implemented, can well be distinguished as human rights or women's or women migrant workers' rights champion. It's Migrant Worker's Act of 1995 and the Magna Carta of Women could strongly serve as global benchmarks in terms of content and strong provisions on the fulfilment and protection of rights amongst Filipina migrant workers. These laws do address the urgent need to lift women migrant workers from disempowering, restricting and abusive circumstances.

The *parens patrae* syndrome moulded the Philippine government into a 'messianic' state agency. This had led the government to mainly perform two main duties. On the one hand, relevant agencies such as OWWA tend to focus more on welfare services. Migrant workers have to undergo distress in order to access support from the agency; this is on top of the fact that they have to be active registered members and be able to satisfy requirements. Claiming of "rights" is limited to their contribution and legal status. Moreover, it does not have gendered programmes catering to Filipina migrant workers. On the other hand, the government has been addressing issues related to sensational cases—abuses, criminal circumstances involving OFWs. As such, it employs a reactionary stance. The space for rights-bargaining of Filipina migrant workers, regardless of their socio-economic experience, is sidelined. Flor Contemplacion, unfortunately, had become the poster girl of how a Filipina migrant worker should be and undergo in order to be rescued by the government. One has to first be abused, bruised and marginalized before the government answers to her call. It thus ignores other facets and identities of Filipina labour migration. This is most evident in the Philippine embassy/consulate in Bangkok, Thailand. Moreover, the absence of a bilateral agreement with the Thai government is a breach of its mandate to protect and fulfil the rights of migrant workers. It is ignorance of harms that may potentially cause systemic job insecurity and discrimination of Filipina teachers in their respective locales.

Moreover, amongst these select respondents, there seems to be a low understanding of their rights and freedoms. Moreover, negotiating and claiming their rights with the Philippine government is perceived as more of a burden for them. Some teacher-respondents in Bangkok and its neighbouring provinces refuse to bargain for their rights—as it might further compromise the economic benefits that they are currently enjoying. This ignorance could potentially turn out into maltreatment of future Filipina teachers aspiring for better opportunities in Thailand. They might be caught in a situation where labour contracts solely define them—ignoring their needs, wants and potentials as teachers or more so, even as human beings.

REFERENCES

Aquino, Benigno III (2010) *Statement of President Aquino during the ASEAN Leader's Retreat in Hanoi, Vietnam* <available: http://www.gov.ph/2010/10/28/statement-of-president-aquino-during-the-asean-leaders-retreat-hanoi-vietnam/> (accessed: 20 February 2011).

Aquino, Benigno III (2013) Toast of President Benigno S. Aquino III during the Vin d'Honneur for the celebration of Philippine Independence <available: http://www.pcoo.gov.ph/speeches2013/speech2013_june12a.htm> (accessed: December 2013).

Bautista, Liberato and Elizabeth Rifareal (Ed.) (1990) *And She Said No!: Human Rights, Women's Identities and Struggles*, National Council of Churches in the Philippines: Quezon City.

Beltran, Ruby and Gloria Rodriguez (2006) *Filipino Women Migrant Workers: At the Cross Road and Beyond Beijing*, Giraffe Books: Quezon City.

Cabilao, Minda (1995) *Labor Migration: Issues fro DFA Personnel in Servicing Filipino Migrant Workers*, CIRSS: Manila.

Cadias, Abel (2011) "The Filipino Teaching Community in Thailand: Towards a Credible Institute" <available: http://abelcadias.blog.co.uk/2011/02/07/the-filipino-teaching-community-in-thailand-towards-a-credible-institute-10531766/> (accessed: September 2012).

Catholic Institute for International Relations (1987) *The Labour Trade: Filipino Migrant Workers Around the World*. Kaibigan: Mandaluyong.

CEDAW (2008) General Recommendation No. 26 on Women Migrant Workers <available: http://www2.ohchr.org/english/bodies/cedaw/docs/GR_26_on_women_migrant_workers_en.pdf> (accessed March 2014).

Center for Migrant Advocacy (2012) *Submission to the UN Committee on Migrant Workers: For the List of Issues Prior to Reporting (LOIPR) of the Philippine Government in Preparation for the Committee on Migrant Workers' 16th Session (16-27 April 2012)*, CMA: Manila.

Center for Migrant Advocacy (2014) *Philippine NGOs- Trade Unions Submission of Information: For the Second Periodic Report of the Philippines on the Implementation of the International Convention on the Rights of All Migrant Workers and Members of their Families*, CMA: Quezon City <available: http://tbinternet.ohchr.org/Treaties/CMW/Shared%20Documents/PHL/INT_CMW_NGO_PHL_16707_E.pdf> (accessed: April 20, 2014).

DAWN (2000) *A Critical Assessment of the Migrant Workers and Overseas Filipinos Act of 1995 (RA 8042)*, DAWN: Makati.

De Guzman, Odine, (2013) "Overseas Filipino Workers, Labor Circulation in Southeast Asia, and the (Mis)management of Overseas Migration Programs", In Kyoto Review of Southeast Asia, <available: http://kyotoreview.cseas.kyoto-u.ac.jp/issue/issue3/article_281.html> (accessed October 2003).

Dizon-Anonuevo, Estrella and Augustus T. Anonuevo (ed.) (2002), *Coming Home: Women, Migration and Reintegration. Philippines*, BalikBayani Foundation Inc and ATHIKA: Quezon City.

Ejercito-Estrada, Joseph (2000) Third State of the Nation Address of His Excellency Joseph Ejercito-Estrada, <available: http://www.gov.ph/2000/07/24/joseph-ejercito-estrada-third-state-of-the-nation-address-july-24-2000/> (accessed December 2013).

Embassy of the Republic of the Philippines in Bangkok, Thailand, Citizen's Charter for Good Governance <available: www.bangkokpe.com> (accessed October 2013).

Garcia v. Drilon G.R. No. 179267. <available: http://www.lawphil.net/judjuris/juri2013/jun2013/gr_179267_2013.html#rnt84> (accessed December 2013).

Gonzalez III, Joaquin L. (1998) *Philippine Labour Migration: Critical Dimensions of Public Policy*, ISEAS: Singapore.

Hau, Caroline (2005) "Rethinking History and 'Nation-Building' in the Philippines," In *Nation Building: Five Southeast Asian Histories*, edited by Wang Gungwu, Institute of Southeast Asian Studies: Singapore.

International Organization for Migration (2005) "International Migration Trends: Facts and Figures" In *World Migration Report 2005*, available <http://images.gmanews.tv/html/research/2007/12/world_migration_report_2005.htm> (accessed April 2013).

Jayme-Lao, Melissa (1999) "Foreign Policy and the Philippine Experience", in *Politics and Governance: Theory and Practice in the Philippine Context*, Ateneo de Manila: Quezon City.

Macapagal-Arroyo, Gloria (2008) "Presidential Speech speech during the groundbreaking ceremony of the Pamayanang Maliksi-Cavite Mass Housing Project and the Women's Month provincial kick-off ceremony "CEDAW ng Bayan: Kayamanan ng Kababaihan", Office of the President Website, available <http://www.op.gov.ph/index.php?option=com_content&task=view&id=4649&Itemid=27> (accessed February 8, 2010).

Macapagal-Arroyo, Gloria (2009) *Presidential Speech during the Conferment of the 2009 Galing Pook and Bagong Bayani Awards*, available <http://www.op.gov.ph/index.php?option=com_content&task=vie w&id=27073&Itemid=27>, (accessed April 2010).

O'Neil, K. (2004, 1 January). Labor Export as Government Policy: The Case of the Philippines, <available: http://www.migrationpolicy.org/article/labor-export-government-policy-case-philippines/> (accessed September 2016).

Ofreneo, Rene and Isabelo Samonte (2005) *Empowering Filipino migrant workers: policy issues and challenges*, ILO: Geneva, <available: http://www.ilo.org/wcmsp5/groups/public/—asia/—ro-bangkok/docu ments/publication/wcms_160550.pdf> (accessed October 2013).

Overseas Development Institute (2001) "ODI Briefing Paper: Economic Theory, Freedom and Human Rights: The Work of Amartya Sen". ODI. 7 October 2013. <available: http://www.odi.org.uk/sites/ odi.org.uk/files/odi-assets/publications-opinion-files/2321.pdf> (accessed October 7 2013).

Parrenas, Rhacel (2003) *Servants of Globalization: Women, Migration and Domestic Work*. Ateneo de Manila University Press: Quezon City.

Philippine Telephone and Telegraph Company v. National Labor Relations Commission, G.R. No. 118978 (May 23, 1997).

Piper, Nicola (2005) *Gender and Migration: A Paper prepared for the Policy Analysis and Research Programme of the Global Commission on International Migration*, GCIM: Geneva.

Republic of the Philippines (1987) *Philippine Constitution*.

Republic of the Philippines (1995) *Republic Act No. 8042: Migrant Workers and Overseas Filipinos Act of 1995*.

Republic of the Philippines (2004) "Consideration of reports submitted by States parties under article 18 of the Convention on the Elimination of All Forms of Discrimination against Women Combined fifth and sixth periodic reports of States parties: Philippines", UN Committee on the Elimination of Discrimination Against Women, <available: http://daccess-dds-ny.un.org/doc/UNDOC/GEN/N04/459/ 70/PDF/N0445970.pdf?OpenElement> (accessed: March 2014).

Republic of the Philippines (2008a) "Consideration of Reports Submitted by State Parties Under Article 73 of the Convention: Intial reports of State Parties due in 2004: Philippines", UN Committee on the Protection of the Rights of all Migrant Workers and their Families, UN: Geneva.

Republic of the Philippines (2008b) *"National Report Submitted in Accordance with Paragraph 12 (a) of the Annex to Human Rights Council Resolution 5/1* (UPR Report)*, UN: Geneva.

Republic of the Philippines (2009) *Magna Carta of Women*.

Republic of the Philippines (2012) "National report submitted in accordance with paragraph 5 of the annex to Human Rights Council resolution 16/21: Philippines", UN Human Rights Council, <available: http://daccess-dds-ny.un.org/doc/UNDOC/GEN/G12/123/16/PDF/G1212316.pdf?OpenEle ment> (accessed March 2014).

Republic of the Philippines National Statitics Office (2013) "Total Number of OFWs is Estimated at 2.2 Million (Results from the 2012 Survey on Overseas Filipinos" <available: www.census. gov.ph/content/total-number-ofws-estimated-22-million-results -2012-survey-overseas-filipinos> (accessed November 18, 2013).

Republic of the Philippines (2014) "Consideration of Reports Submitted by State Parties Under Article 73 of the Convention pursuant to the simplified reporting procedure: Second reports of State Parties due in 2011: Philippines", UN Committee on the Protection of the Rights of all Migrant Workers and their Families, UN: Geneva.

UP Center for Women Studies (2006) *A Gendered Review of Selected Economic Laws in the Philippines*, UP: Quezon City.

Vittin-Ballima, Cecile (2002) "Migrant workers: The ILO standards" In *Migrant Workers (Labor Education 2002/4 No. 12)*, ILO <available: http://www.ilo.org/wcmsp5/groups/public/@ed_dialogue/ @actrav/documents/publication/wcms_111462.pdf> (accessed October 2013).

Women's Legal Bureau (2006) *Philippine NGOs Shadow Report to the 36th Session of the committee on the Elimination of Discrimination Against Women*, WLB: Quezon City.

INTERVIEWS

Callanza, Lourdes (2013) Personal Interview.
"Ma'am Anne" (2013) Personal Interview.
"Ma'am Diana" (2013) Personal Interview.
"Ma'am Lorna" (2013) Personal Interview.
"Ma'am Luz." (2013) Personal Interview.
"Ma'am Sally" (2013) Personal Interview.
"Ma'am Vivian" (2013) Personal Interview.

Advancing Rule of Law in a Global Context – Susetyo, Rinwigati Waagstein & Budi Cahyono (eds)
© 2020 Taylor & Francis Group, London, ISBN 978-1-138-32782-5

Digital identity and personal data protection: Analysis of rights to erasure and data portability in Indonesia

Edmon Makarim & S. Kom
Faculty of Law, Universitas Indonesia, Indonesia

ABSTRACT: With the globalization of e-commerce, the use of online identity together with personal data and an e-authentication system is certainly a necessity. Regionally and globally, cybersecurity and personal data protection have become hot topics in the international fora. Interesting dynamics have appeared in the European Union, which has changed some directives into regulations, such as Regulation 910/2014 on e-Identification and Trust Services, and Regulation 679/2016 on General Data Protection Regulation. The right to erasure and the right to portability are two important issues for e-system providers. Similarly, Indonesia has also revised Law No. 11 Year 2008 on Information and Electronic Transactions into Law No. 19 Year 2016, which contains the right to be forgotten but unfortunately does not yet have provisions on the right to portability. Indonesia needs to review its legal provisions on the interoperability of e-identity (e-ID) and the e-authentication system, and the government's duties and responsibilities towards the protection of such personal data, as one of the big users of data. Therefore, legal research is required to clarify the legal framework associated with e-identification and authentication systems (e-IDAS) that use online digital identities and the personal data of Indonesian citizens. This study found that the provisions of the national law on online identity and personal data are still relatively incomplete and spread out across various laws and regulations. In the future, reform must take place in order to protect the national interest in the rise of cross-country electronic transactions, especially in the midst of the ASEAN Economic Community.

Keywords: Digital Identity, Electronic Authentication, Electronic Identity Management, Personal Data Protection, Privacy, Trust Services

1 INTRODUCTION

Recently, the Indonesian government has promulgated Presidential Regulation No. 74 of 2017 on the Electronic Commerce Roadmap that focuses on several aspects to encourage the growth of e-commerce – namely, (i) funding, (ii) taxation, (iii) consumer protection, (iv) education and human resources, (v) communications infrastructure, (vi) logistics, and (vii) cybersecurity. Despite the mention of consumer protection as well as cybersecurity in e-commerce, it seems that the government still has not paid close attention to basic security needs for facilitating e-transactions. For the protection of consumers and business actors, the validity of e-transactions depends heavily on the clarity of the law on information security and its communication. Both require the implementation of electronic identification by using the digital identity of the parties and the electronic authentication system (e-IDAS), which depend on personal data protection. In subsequent developments, it requires trustworthiness in the e-system to ensure such electronic transactions will be valid.

In the past decades, both regionally and globally, the issue of cybersecurity in e-commerce cannot be separated from the legal issues of privacy and personal data protection in an electronic transaction, whether for commercial activities or for accessing public services. Many countries have come to realize the urgency of protecting the personal data of their people as the most invaluable national asset. This means making the public more aware of not being exploited or not being the object of profiling by others. For the sake of global e-commerce, e-transactions need the interoperability of the various e-identification systems in the market and the legal guarantee of personal data protection as well.

Ironically, the strong correlation in the authenticity linkage between the e-IDAS and the personal data protection were seemingly not clear yet in the national legal system. Therefore, a legal study is needed to show an important correlation in order to explain the trusted framework for the implementation of e-IDAS in Indonesia. This review will address some of the following issues: (i) how the global dynamic trends to regulate the legal system of e-identification and its e-authentication, (ii) how the personal data protection law relates to the implementation of e-IDAS, in particular, the right to data portability and the right to erasure of personal data, and (iii) how the clarity of the national legal system in accommodating the Right to Erasure and the Right to the Data Portability.

2 DISCUSSION

2.1 *The right of identity and personal data*

In the context of legal relationships, a person's identity is, in fact, a representation of the inherent uniqueness of that person. Accordingly, personal identity is an individual right which could be classified as intangible property. Identity is an attribution of a person's personalization which is also a valid proof of the legal capacity required to make a contract. If identity is seen in its object, which constitutes a set of personal data, then a person has the right to such personal data attached to him/her of ownership, possession, and legitimate legal interest in the security and comfort of his/her private life, called privacy. Thus, the right to disclose or to restrict a set of personal data to the public is the personal right inherent of the data subject. However, the individual right is certainly not absolute; it can be disregarded if there is a legal necessity for protecting the public interest.

As property – for example, the status of a portrait as an image of a person is protected under the intellectual property regime – a set of personal data could be enjoyed as a privilege to the Subject Data because it is naturally inherited to that person. Digital identity could be classified as a digital property whose legal perception is as an immovable object or the right of an immaterial object. The conception of property rights is attached to the Data Owner (Subject Data), not to those who possess and control the personal data. Moving or copying the data does not represent the transfer of its ownership. Therefore, anyone who obtains Identity actually acts as a custodian or trustee as the controller of the personal data – known as the Controller or Processor – and is a responsible party to ensure the procuring and/or processing of such data is relevant to the particular purpose based on the approval of the Data Subject. Implemented in electronic communications, any person or any Provider of the System that obtains the identity of any person's personal data shall keep the material order that he/she will not act unlawfully. Any breach of such matter has the potential to disrupt the privacy of the individual concerned and is also a violation of the material right itself. In other words, the mastery of the personal data of others does not mean that the controlling party becomes the owner and takes economic value from him/her, but he/she must honour the mandate or legal obligation to always keep the ownership of the subject of the data. All unlawful acts in data acquisition and use may be perceived as theft or fraud.

Similarly, in the context of public law, the transfer of personal data to the government, and its use by the government, are, in fact, a form of the goodwill of every citizen to his/her country in fulfilling its legal obligations. On the other hand, the government is responsible as per the constitutional mandate to protect the citizens and their nation. State governments should ensure that any acquisition and use of personal data of citizens or residents should be carried out responsibly according to their legal interests or within the context of governmental affairs itself. Use beyond the relevance of the governmental affairs in question is an abuse of power.

From a macro perspective, national security is in fact directly proportional to the security of every person. If every resident or citizen can be profiled, including all the behaviours of his/ her daily life (profile), by other nations, that poses a serious threat to the country itself, and is an important part of the scope of cyberspace utilization related to national cybersecurity. In essence, the organization of Identity and its personal data is, in fact, a protection of the national asset strategies connected to the Internet, including the personal data of each resident, who should not easily become the profiling object of the nation or other countries.

Instead of e-commerce, access to and interoperability of e-IDAS are also essential for access to public services and government administration. In the practice of state administration, e-IDAS becomes the determinant in the process of the legalization or authentication of public documents, because the personal data of officials will also be shown to the extent of authorization of authority at the time of signing with their identity. Therefore, cybersecurity cannot be separated from the discussion of national digital identity management itself, because the assets of a nation include not only its tangible physical natural resources, but also its intangible population data. The mandate of safeguarding population and personal data is, in fact, a major constitutional obligation for the government. Not surprisingly, along with the development of Information Technology, several countries in the European Union have legal provisions for the protection of personal data in their constitutions.

The Association of Southeast Asian Nations (ASEAN) also incorporates the protection of personal data in its human rights declaration. The ASEAN ICT Master Plan for 2015–2020 has an ambition per its evolution: the development of technology about identity and personality, as well as the provision of a trusted framework for building a good identity ecosystem, so that transactions and information exchanges will be safe, secure, and trustworthy. The Vision AIM 2020 aims for the following characteristics in such a framework: (i) digital: digitally enabled programmes for continual education and upgrading to equip ASEAN citizens with the latest infrastructure, technology, digital skill sets, information, applications, and services; (ii) secure: a safe and trusted ICT environment in ASEAN, providing reassurance in the online environment by building trust in online transactions via a robust infrastructure; (iii) sustainable: responsible and environmentally friendly use of ICT; (iv) transformative: a progressive environment for the disruptive use of technology for ASEAN's social and economic benefit; (v) innovative: a supportive entrepreneurial environment that encourages innovative and novel uses of ICT; and (vi) inclusive and integrated: empowered and connected citizens and stakeholders.

In fact, several Identity providers are already in the market – namely (i) the identity of a person as a legal subject provided by the government in the form of a national ID card; (ii) the identity of a person as conveyed by a company or community, such as credit or debit cards or golf community membership or other community identity cards; and (iii) online digital identities as provided by the providers of electronic systems on the Internet, e.g., Yahoo!, Google, Facebook, and so on. They are called soft identity providers. Therefore, nullifying all existing identity for the growth of e-commerce is obviously not a good policy, since the use of identity for commercial purposes is the right of the person concerned, not an obligation as is the use of identity for access to public services. Through technical dynamics and the media, it is evident that the identity system has evolved into an electronic form that is not only off-line (chip), but also online via the Internet. This evolved identity system will look to the Registration and Authentication process.

Along with globalization comes the issue of how different identity systems can be recognized and applied in other countries (across borders). This requires clear interoperability standards for identification and authentication systems to be recognized and accepted by other systems. Therefore, a policy that can integrate system diversity, or Federated Identity Management (FIM), is certainly based on an open identity system environment (Open Identity). It requires a Trust Services system to validate identification. With the interoperability of the identity system, fraud via cyberspace can be eliminated, and electronic transactions are safer.

In comparison, the United States of America prefers to direct the state to the task of providing a legal framework for interoperability under the policy of the National Strategies for

Trusted Identity in Cyberspace (NSTIC), by providing some conditions for it. The realization of this vision is the user-centric 'Identity Ecosystem' described in this Strategy. It is an online environment where individuals and organizations can trust their digital identities and the digital identities of devices. The Identity Ecosystem is designed to securely support transactions that range from anonymous to fully authenticated and from low value to high value. The Identity Ecosystem, as envisioned here, will increase the following: (i) convenience for individuals, who may choose to manage fewer passwords or accounts than they do today; (ii) efficiency for organizations, which will benefit from a reduction in paper-based and account management processes; (iii) ease of use, by automating identity solutions whenever possible and basing them on technology that is simple to operate; (iv) security, by making it more difficult for criminals to compromise online transactions; (v) confidence that digital identities are adequately protected, thereby promoting the use of online services; (vi) innovation, by lowering the risk associated with these services and developing their online presence; (vii) choice, as service providers offer individuals different yet interoperable identity credentials and media.

Similar efforts are also happening in the European Union; with its regulation 910/2014 on e-IDAS, each country can run mutual recognition while maintaining the Quality Level Assurance of the exchanged system identity. However, beforehand, they have conducted data collection on QAA standards from several identity systems that exist in each country under the STORK Project. Based on the project, several mapped levels are inputted in the regulation e-IDAS; the levels range from Low-Level and Substantial Assurance to High-Level Assurance. For interoperability, only the Substantial-Level and High-Level Assurances are acceptable. Each European country has different policies on national ID cards, but the e-IDAS regulation is not intended to synchronize them. The regulation is intended to answer the interoperability needs of member countries to facilitate the access of their citizens, so that each country's citizens may use its existing Identity System to access services in other countries. This does not necessarily mean that a national ID card is mandatory, but it could also be used to access another identity system (e.g., health card), as long as its Assurance level meets the requirements. Connectix, a Dutch company, has appeared to emerge as the natural gateway to connect all these identities. The company seems to be meditating to provide the necessary identity services for the member states of the European Union.

The European Union's member countries must also comply with European Regulation No. 679/2016 on General Data Protection Regulation (GDPR). This regulation implements extraterritorial jurisdiction. European personal data protection laws govern how each use of personal data (processing) must conform to the GDPR, which explicitly clarifies the rights of everyone (subject data) concerning any use of his/her personal data: (i) the right to be informed; (ii) the right to access; (iii) the right to verify; (iv) the right to data portability; (vi) the right to object; and (vii) the rights each individual has in relation to automated decision-making and profiling (profiling means any form of automated processing of personal data relating to a natural person, in particular, to analyse or predict aspects of natural personality at work, economic situation, health, personal preferences, interests, reliability, behaviour, location, or movements).

Furthermore, the Guidelines on the Right to Data Portability made by the Working Party of Article 29 on Data Protection are concerned with several important elements – namely, (i) the right to obtain data; (ii) the right to transmit personal data; and (iii) the right to access means of data portability and control. The Guidelines describe the terms of applying the right to portability: (i) such personal data must be processed automatically; (ii) personal data shall be subject to the data subject and on the basis of consent; and (iii) no prejudice is tolerated against the rights and freedoms of third parties or the relying party. In the case of obtaining data, any subject of data which has submitted data to the providers of electronic systems shall have the right to obtain electronic data that is commonly used to access the systems of others (commonly used) in a machine-readable format. Technically, the commonly used is in the form of an Application Programming Interface (API) as a set of subroutine definitions, protocols, and tools for building software and applications. It refers to the interfaces of applications or web services made available by data controllers so that other systems or applications can link and work with their systems. Furthermore, after the data is obtained, the subject of data then can use it by transmitting it directly to other parties or directly from one system to another 'without hindrance'. It also provides empowerment to the subject of data as a consumer to be unlocked to the company. The policy also encourages the growth of service industries to facilitate mediation for it.

On the technical level, the interoperability of the data will meet several conditions:

(i) First condition: <u>personal data concerning the data subject</u>. Only personal data is in the scope of a data portability request. Therefore, any data which is anonymous or which does not concern the data subject will not be in scope of a data portability request.

(ii) Second condition: <u>data provided by the data subject</u>. Data which is actively and knowingly provided by the data subject is included in the scope of the right to data portability (for example, mailing address, user name, age, etc.). Observed data are 'provided' by the data subject by virtue of the use of the service or the device. Such data may include a person's search history, data traffic, and location data. It may also include other data such as heartbeats tracked by fitness or health trackers. In contrast, inferred data and derived data are provided by the data subject. These personal data do not fall within the scope of the right to data portability. In general, given the policy objectives of the right to data portability, the phrase 'provided by the data subject' must be interpreted broadly, and only to exclude 'inferred data' and 'derived data' received from a service provider (for example, algorithmic results).

(iii) Third condition: The third condition intends to <u>avoid retrieval and transmission</u> of data containing the personal data of other (non-consenting) data subjects to a new data control (Article 20 (4) of the GDPR). Such an adverse effect would occur, for example, if the transmission of data from one data controller to another under the right to data portability would prevent third parties from exercising their rights as data subjects under the GDPR (such as the rights to information, access, etc.). The data subject initiating the transmission of his/her data to another data controller either consents to the new data controller for processing or enters into a contract with them. Where the data is set, another ground for the lawfulness of processing must be identified. For example, a legitimate interest under Article 6 (1) (f) may be pursued by the data controller to whom the data is transmitted, in particular when the subject of the data controller is to provide a service to the data subject that allows the latter to process personal data for a purely personal or household activity.

Data portability and deletion rights make it an obligation for providers to inform prior (informed) owners of the data on how they perform the obligation. Basically, they must act quickly (without undue delay) in response to a Data Subject's request for data portability. In general, on the other side, there are also views against the right to be forgotten and the right of data portability. Opponents of these rights feel that the right to be forgotten contradicts the freedom of the press because it is counterproductive to the right to convey news about a person. Similarly, some business actors in the United States feel the right of data portability will conflict with a conducive competition climate, especially related to the commercial value of customer data protected with intellectual property rights as company secrets.

In 2016–2017, the United Nations Commission on International Trade Law (UNCITRAL) established a Working Group to discuss the legal issues of digital identity and trust services, which, as reported in March 2017, stated that in general the legal principles in ID management and trust services are: (i) party autonomy; (ii) technological neutrality; (iii) functional equivalence; and (iv) non-discrimination. It further agreed that the identification process requires several steps: (a) verification of the identity document; (b) verification of whether the person presenting the document was the person identified in that document; and (c) correctness of the steps undertaken and the judgement used in identifying the person. The use of ID management does not by itself answer the issues of forgery, hacking, and a good faith guarantee of the relying party. The Working Group recognized that the determinants are level of assurance and risk management and transparency mechanisms; in the use of personal data, such identity must be in accordance with its objectives and become a determinant of interoperability. The Working Group had not succeeded in furthering the research as of 2017, and agreed to explore further in the next year's session.

It is important to note that in the interest of cross-border interoperability, the Working Group is not authorized to determine technical standards but rather to focus on policy aspects only. Given that some countries also still rely on paper-based identities, then digital identification should meet the functional equivalence requirement so that it can be equated with written evidence. The Working Group also made references to the Intra-ASEAN Secure Transaction Framework, applicable to the public and private sectors based on ISO 29115 standards. It explained that the objective of non-regulatory schemes in ASEAN is to promote legal recognition of identification and authentication across ASEAN countries. But many challenges are faced in this regard, and UNCITRAL is a good effort to deal with these problems by developing harmonization of provisions for global consensus.

The UNCITRAL expert team on ID management and trust services does not impose specific solutions on the commercial side, but rather provides a set of options to meet the needs of risk management. Commercial parties should be free to link various securities to different levels of assurance. However, the value of ensuring a common understanding of identity affirmation in the IBM scheme against a set of trust levels should not be questioned. Thus the availability of a common frame of reference in which the ID management system can be mapped is considered essential for international trade.

2.2 *Identity management and personal data protection in Indonesia*

Pursuant to the global discussions, some lessons can be learned from Indonesia, which is in the process of reforming national e-ID management and protections for personal data. In the 1945 Constitution of the Republic of Indonesia, there is no explicitly stated terminology such as 'privacy and personal data protection'. It is actually part of the protection of private life guaranteed by Articles 28G and 28H. Furthermore, data protection is also regulated in Articles 21 and 31 of Law No. 39 of 1999. In addition to the 1945 Constitution and Law No. 39, regulation on identity, privacy, and personal data in Indonesia cannot be separated from the enforcement of Law No. 8 of 1999 on Consumer Protection (Consumer Law), Law No. 24 of 2013 on the revision of Law No. 25 of 2016 on Population Administration (Adminduk Law), and Law No. 19 of 2016 on the revision of Law No. 11 of 2008 on Information and Electronic Transactions (UU-ITE).

2.2.1 *Identity and personal data protections as consumer rights*
Subject to Article 4 of the Consumer Law, any subject of data that becomes the user of the electronic system shall be protected by its right as a customer in its electronic and non-electronic transactions. Specifically, this right includes: (i) the right to comfort and safety in consuming goods and/or services; (ii) the right to choose goods and/or services and to obtain the goods and/or services in accordance with the exchange rates and the conditions and promised warranties; and (iii) the right to true, clear, and truthful information about the condition and guarantee of goods and/or services. In fact, consumers using electronic systems and electronic transactions not only have paid the price of the goods and/or services used, but also have provided identity and personal data that is not priceless to the electronic service provider, the value of which is not yet counted in the exchange rate. Consumers should, therefore, obtain correct, clear, and honest information about the conditions and guarantees as given in the privacy statement presented by the applicant of the electronic system operator in question. Business actors, pursuant to Article 7 of the Consumer Law, are obliged to: (i) provide true, clear, and honest information regarding the conditions and warranties of goods and/or services and provide explanations of use, repair, and maintenance; (ii) guarantee the quality of goods and/or services produced and/or traded under the provisions of the applicable quality standards of goods and/or services; and (iii) compensate, indemnify, and/or reimburse for losses arising from the use of traded goods and/or services. In addition, pursuant to Article 15 of the Consumer Law, business parties are also prohibited from offering goods and/or services by coercion or other means that may cause physical or psychological disturbance to consumers. Thus, the business actor has an obligation not to compel or manipulate the consumer. The consumer, as the owner of personal data, may request the business actor, as the data

controller, for removal or for the interoperability rights to his/her personal data so that the security of the personal data is guaranteed.

2.2.2 *Identity and personal data in the administration of population*

In the context of the implementation of identity, explicitly the terms 'Identity' and 'Personal Data' are mentioned in the Adminduk Law, which is intended to build a population database in the Information System of Population Administration (SIAK), but they are implicitly also regulated in the ITE Act essentially as electronic signatures and certificates. Article 1 paragraph (22) of the Adminduk Law defines 'Personal Data' as certain personal data which is stored and maintained, and its confidentiality protected. Article 84 paragraph (1) states that the Personal Data to be protected contains: (a) description of physical and/or mental disabilities; (b) fingerprint; (c) iris scans; (d) signature; and (e) other data elements that could be disgraceful or harmful to the person. Further, Article 2 of the Adminduk Law states in its letter (c) that every resident has the right to obtain the protection of Personal Data; in letter (d) that every resident has the right to legal certainty over ownership of documents; in letter (e) that information will be protected concerning data on the results of the Population Registry and the Civil Registry of him/her and/or his/her family; and in letter (f) that every resident has the right to redress and restitution as a result of errors in the Registration of Population and Civil Registration or the misuse of Personal Data by the Implementing Agencies.

However, pursuant to Article 87, the Personal Data User of the Population (both public and private agencies) may obtain and use Personal Data from the Operator and the Implementor, who has access rights. This shows that the government acts alone in the delivery of personal data to private parties, without the approval of the Data Subject. In the context of identity, the government applies an electronic ID card (KTP-el), which is a citizenship card with a chip that is the official proof of identification of the population. Each resident shall have a permanent and lifelong Population Identity Number (NIK) granted by the government and issued by the implementing authority to each resident after the registration of the data (Article 13).

The NIK should be included in each Population Document and used as the basis of issuance of public documents such as a passport, driver's license, taxpayer's principal number, insurance policy, land title certificate, and other identity documents. Every citizen of Indonesia and all foreigners who have a permanent residence or who have been married seventeen years to an Indonesian citizen shall have an ID card (Article 63 of the Adminduk Law). In the KTP-el are stored chips that contain electronic records of individual data (Article 64 paragraph (6)). As stated in the general explanation of the revision of the Adminduk Law, with the application of ID cards, residents can no longer have more than one ID card, since in the KTP-el are loaded the security code and the electronic data of the residents, among others, in the form of iris scans and fingerprints.

After the revision of the Adminduk Law, the period of validity of the applicable five-year ID cards was changed to a lifetime, insofar as there is no change in the elements of the population data or the resident's domicile (Article 64 paragraph (4)). This needs to be done in order to obtain ease and smoothness in public service either by the government or by the private sector, and also to get the financial savings from not renewing all cards every five years. In line with the establishment of the demographic database, the Adminduk Law is expected to clarify the regulation of access rights for the utilization of population data for officers serving as the operator, the implementing agency, or the user. 'Users' includes state institutions, ministries/non-ministerial government institutions, and/or Indonesian legal entities (explanation section 79 paragraph (2)).

According to Article 64 of the Adminduk Law, Indonesia embraces a centralized model in which all other identity documents must refer to the NIK as a single identity number valid as an official identity for access to all public services. To carry out all public services, the government shall integrate existing identity numbers which can be used for public services no later than five years after the law is ratified. The KTP-el function is stepped up gradually into a multipurpose KTP where individual data loaded in the chip adapt to current needs (explanation of Paragraph (6)). In practice, however, the multipurpose KTP could not be used because it is not equipped with digital signatures or certificates as expected due to the lack of space in the chip.

In accordance with the ITE Act, digital identity is known as the existence of an electronic signature (TTE), a signature consisting of electronic information embedded into, associated with, or related to other electronic information used as verification and authentication tools. It is further stipulated in Government Regulation No. 82 of 2012 on Electronic Systems and Transactions (PP-PSTE), which in its enforcement may enable the support of notary services to strengthen electronic signatures' legal authenticity. Technically, an actual electronic signature is to function as an authentication and verification tool on (a) signatory identity and (b) the integrity and authenticity of electronic information (Article 52 PP-PSTE). The electronic signature functions as a manual signature in terms of representing the signatory's identity. The manual signature may be authenticated by verification or examination of the electronic signature specimen of the hand signature. In electronic signatures, electronic signature data serves as an electronic signature specimen of the hand signature. The electronic signature shall be submitted to competent experts to examine and verify that electronic information signed with an electronic signature does not change after it is signed. In its operation, TTE is known to have two types: (i) those types which are supported by an electronic certificate, and (ii) those types which are not supported by electronic certificates. Furthermore, an electronic certificate is then divided into three categories: (i) registered electronic certificates, (ii) certified electronic certificates, and (iii) electronic certificates that are supplied to the Government Electronic Certificate (Root CA). The requirements and mechanisms for the operation of electronic signatures and certificates are stipulated in the PP-PSTE.

In order to guarantee the implementation of good IT governance, the UU-ITE mandates every electronic system provider to always assume 'presumed liability', legal responsibility that ensures that the electronic system it operates is reliable, secure, and responsible (Article 15 ITE Law). The operator of an electronic system shall protect its users and the public at large from the harm caused by its electronic system (Article 27 PP-PSTE). Then, to ensure legal liability in order to protect the public interest, all operators of electronics systems used for public services must: (i) register their electronic systems; (ii) use electronic signatures and certificates in their electronic transactions; and (iii) guarantee electronic systems before operating public services (Articles 5, 41, and 30 PP-PSTE). Furthermore, each electronic system provider must also follow the technical arrangements of the Ministry of Communication and Information Technology, the Ministerial Regulation Concerning the Electronic System Security Standard ISO/SNI 27001 (Ministerial Regulation of Communication and Information Technology 4/2016), and the Ministerial Regulation on the Protection of Personal Data 20/2006, among others.

From the perspective of good e-government planning, the SIAK and KTP-el programmes are merely based on government-centric/non-user-centric paradigms because the government does not survey in advance or attempt to dig up what the community needs, or to create a system that is closer to the community. The prominent urgency at the time was to facilitate the 2014 general election, and the KTP-el seemed politically imposed a bit because in the initial draft, it should have not only registered citizens, but also included the availability of card reader devices for authentication. However, the KTP-el was implemented, although the availability of card readers was not assured. As a result, although the registration process has been conducted (more than 130 million population), it is still not guaranteed that the ID card has been received directly and held by the correct residents. This has resulted in the system of authenticity being of a low level because the first authentication was not done well.

The purity and validity of the personal data contained in the SIAK has confronted similar issues. Cleaning up the duplication of personal data for the achievement of a single identity number apparently tends to be conducted unilaterally without the approval of the subject data. Moreover, the notary has not yet been facilitated as an urgent public official to check the identity of the parties, whereas the notary is actually a strategic partner for the Ministry of Home Affairs to clean up the duplication of the data. Implementation of the electronic ID card (KTP-el) is still constrained by the cost of procurement and card printing, so many residents have not obtained ID cards.

The revision of UU Adminduk 2013 states that the electronic ID card that has been printed is valid for life, which is contrary to technological limitations for the durability of the card media itself in storing information, especially if biometric personal data is stored on it. For example, a face in a photo as real biometric data will always change according to age. The government is required to maintain such updated data well and/or to the extent possible provide access to residents to report updates on personal data and also to obtain reports of the use of personal data online. Moreover, the government acts like it is the owner of the personal data of the inhabitants, so that the government seems entitled unilaterally to get the economic value of every instance the data is accessed by private parties. It is not possible to give direct access to Data Subjects to update their personal data, and the absence of a report of the use of personal data indicates the personal data is vulnerable to abuse by the government.

Meanwhile, although the provisions of the UU-ITE and the PP-PSTE state that the electronics system for public services must be registered and get a certificate of merit, the SIAK and the KTP-el are still not registered, do not use electronic signatures and certificates, and have not obtained the certification of feasibility from the Ministry of Communication and Information Technology. Based on these records, the trustworthiness aspect (trust) of the KTP-el is not good because it does not fulfil Article 15 UU-ITE or the PP-PSTE. This is reinforced by the news that shows the number of *madalah* in its implementation, especially the problem of corruption, and the high number of residents who have not received an electronic ID card.

Due to the issues with data integration and the interoperability of the SIAK with the tax information system, the directorate general of taxation also took the initiative to create a system of identity with the Kartin1 programme, which is expected to be used as a platform to provide interoperability for personal data in order to get any information related to taxation investigation. Meanwhile, the business community is also developing platforms for online digital identity, one of which is www.privy.id, which seeks to protect privacy in electronic transactions.

Regarding the latest development of information technology that provides authentication of biometric data, e-system providers will not want to also apply the latest technology features, one of which is face character recognition. This is one solution because of the government's limitations in the updating of photos on the KTP-el, so that the platform will ask users to upload their photo and then do their own checking of the photo with the character of the photo on the KTP-el.

Referring to the provisions of Article 26 UU ITE and Article 15 PP-PSTE, and the substance of the Regulation of the Minister of Communication and Information Technology 20/2016, basically there is sufficient regulation of personal data protection, but unfortunately, it is still not comprehensive compared to best or good practices in personal data protection law. Principles that have been raised include consent, the availability of data, integrity, and confidentiality. In the revision of the UU ITE and the Regulation of the Minister of Communication and Information Technology 20/2016, the regulation on the right to erasure and the right to portability of personal data does not apply the principle of 'without undue delay'. Procedural limitations make the implementation of the right difficult. The business players can seek any reason to state that personal data is still relevant to them. Similarly, the existence of an identity-based card system does not follow the principle of personal data protection in general. The Ministry of Home Affairs can argue that it has special legal power to carry out the mandate from the country (*lex specialis*), so it can be excluded from some basic principles of personal data protection.

3   CONCLUSION

Looking closely at the dynamics, it is clear that the global legal trend towards identity management, particularly in the use of identification and authenticity systems, appears to lead to open system policies, but it still gives attention to the clarity of relatively high levels of security. Mutual recognition is based on the principle of registration and examination of the level of security quality. In the interoperability of e-IDAS systems, it is done with compliance with personal data protection laws, especially regarding the existence of the right to data portability and the right to erasure. Both rights are basically prioritized as data subject requests that must be followed up quickly, 'without undue delay'.

The linkage between personal data protection law and the implementation of the e-IDAS is absolute because the right of deletion and the right to portability of personal data will be highly dependent on the guarantee of compliance with general data protection law principles.

In the Indonesian context, the setting of online digital identity and the protection of personal data are still strongly perceived in the government-centric paradigm in which the Ministry of Home Affairs feels it rules as a *lex specialis* against the protection of personal data. Meanwhile, securing the right to delete personal data will be difficult because it depends on the process of appealing to the courts, in which business actors can be protected on the ground that the personal data they retain is still relevant to them. The related right to the portability of data can also be called unfavourable because of the unknown rights in the legislation explicitly. No arrangement obligates operators of electronic system providers to respond to requests from data subjects on the right to the portability of such data.

3.1 *Recommendations*

1. Indonesia must improve its personal data protection law if it does not want to be regarded as having lower personal data protection standards than Europe. This will affect e-commerce relations between Indonesia and Europe.
2. Indonesia, even though it has implemented government-centric and centralized identification efforts, should also show openness to interoperability with other digital identity systems. This can be facilitated by the Minister of Communications and Informatics Regulation concerning the implementation of digital identity and the operation of electronic signatures and certificates with reference to international dynamics in an effort to establish a good identity ecosystem.

European GDPR	Indonesian (EIT Law)
Section 3: Rectification and erasure	**Article 26 of Law 19/2016 on the revision of Law 11/ 2008 on Electronic Information and Transaction Act**
Article 16: Right to rectification	(1) Unless otherwise provided by laws and regulations, the use of any information through electronic media concerning the personal data of a person shall be made with the consent of the person concerned.
The data subject shall have the right to obtain from the controller <u>without undue delay</u> the rectification of inaccurate personal data concerning him or her. Taking into account the purposes of the processing, the data subject shall have the right to have incomplete personal data completed, including by means of providing a supplementary statement.	
	(2) Anyone whose rights are violated as referred to in paragraph (1) may file a lawsuit for damages incurred under this Act.
Article 17: Right to erasure ('right to be forgotten')	(3) Each Electronic System Provider shall <u>remove any irrelevant</u> Electronic Information and/ or Electronic Documents under its control <u>at the request of the Person concerned pursuant to</u> a court decision.
1. The data subject shall have the right to obtain from the controller the <u>erasure of personal data</u> concerning him or her <u>without undue delay</u> and the controller shall have the obligation to erase personal data <u>without undue delay</u> where one of the following grounds applies:	
	(4) Each Electronic System Provider shall provide mechanisms for the elimination of Electronic Information and/or Electronic Documents that are not relevant in accordance with the provisions of laws and regulations.
(a) the personal data are <u>no longer necessary</u> in relation to the purposes for which they were collected or otherwise processed;	
(b) the data subject <u>withdraws consent</u> on which the processing is based according to point (a) of	(5) Provisions on procedures for the elimination of Electronic Information and/or Electronic

(Continued)

European GDPR	Indonesian (EIT Law)
Article 6(1), or point (a) of Article 9(2), and where there is no other legal ground for the processing;	Documents as referred to in paragraph (3) and paragraph (4) shall be regulated in government regulations.

European GDPR

Article 6(1), or point (a) of Article 9(2), and where there is no other legal ground for the processing;

(c) the data subject objects to the processing pursuant to Article 21(1) and there are no overriding legitimate grounds for the processing or the data subject objects to the processing pursuant to Article 21(2);

(d) the personal data have been unlawfully processed;

(e) the personal data have to be erased for compliance with a legal obligation in Union or Member State law to which the controller is subject;

(f) the personal data have been collected in relation to the offer of information society services referred to in Article 8(1).

2. Where the controller has made the personal data public and is obliged pursuant to paragraph 1 to erase the personal data, the controller, taking account of available technology and the cost of implementation, shall take reasonable steps, including technical measures, to inform controllers which are processing the personal data that the data subject has requested the erasure by such controllers of any links to, or copy or replication of, those personal data.

3. Paragraphs 1 and 2 shall not apply to the extent that processing is necessary:

(a) for exercising the right of freedom of expression and information;

(b) for compliance with a legal obligation which requires processing by Union or Member State law to which the controller is subject or for the performance of a task carried out in the public interest or in the exercise of official authority vested in the controller;

(c) for reasons of public interest in the area of public health in accordance with points (h) and (i) of Article 9(2) as well as Article 9(3);

(d) for archiving purposes in the public interest, scientific or historical research purposes or statistical purposes in accordance with Article 89(1) in so far as the right referred to in paragraph 1 is likely to render impossible or seriously impair the achievement of the objectives of that processing; or

(e) for the establishment, exercise or defence of legal claims.

Article 20: Right to data portability

1. The data subject shall have the right to receive the personal data concerning him or her, which he or she has provided to a controller, in a structured, commonly used and machine-readable format and have the right to transmit those data to another

Indonesian (EIT Law)

Documents as referred to in paragraph (3) and paragraph (4) shall be regulated in government regulations.

Article 15 of Government Regulation 82/2012 on electronic system and transaction.

(1) The Operator of the Electronic System shall:

a. keep the secrets, wholeness, and availability of the Personal Data it manages;

b. ensure that the acquisition and use of Personal Data is subject to the consent of the owner of Personal Data unless otherwise provided by law and regulation; and

c. guarantee the use or disclosure of data is made on the basis of the consent of the owner of such Personal Data and in accordance with the purpose communicated to the owner of the Personal Data at the time of data acquisition.

(2) In the event of a failure in the confidential protection of the Personal Data it administers, the Operator of the Electronic System shall notify the owner of such Personal Data in writing.

(3) Further provisions regarding the guidelines for the protection of Personal Data in Electronic Systems as referred to in paragraph (2) shall be regulated in a Ministerial Regulation.

Infocomm Ministerial Regulations 20/2016 Protection of Personal Data in Electronic System

Article 11

(1) Electronic Systems used to collect and collect Personal Data shall:

a. have interoperability and compatibility capability; and

(*Continued*)

(*Continued*)

European GDPR	Indonesian (EIT Law)
controller without hindrance from the controller to which the personal data have been provided, where: (a) the processing is based on consent pursuant to point (a) of Article 6(1) or point (a) of Article 9(2) or on a contract pursuant to point (b) of Article 6(1), and (b) the processing is carried out by automated means. 2. In exercising his or her right to data portability pursuant to paragraph 1, the data subject shall have the right to have the personal data transmitted directly from one controller to another, where technically feasible. 3. The exercise of the right referred to in paragraph 1 of this Article shall be without prejudice to Article 17. That right shall not apply to processing necessary for the performance of a task carried out in the public interest or in the exercise of official authority vested in the controller. 4. The right referred to in paragraph 1 shall not adversely affect the rights and freedoms of others.	b. use legal software. (2) Interoperability and compatibility capability as referred to in paragraph (1) letter a in accordance with the provisions of laws and regulations. (3) Interoperability, as referred to in paragraph (2), represents the ability of different Electronic Systems to work in an integrated manner. (4) Compatibility as referred to in paragraph (2) constitutes the compatibility of Electronic Systems to one with other Electronic Systems

APPENDIX

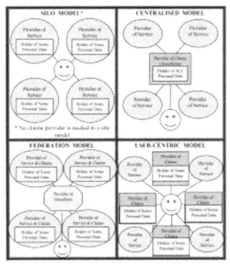

Table A2.1. Features of Technology Models for IdM systems

	Siloed	Centralised	Federated	User-Centric
Method of Authentication	The user authenticates to each account when he wishes to use it.	The user authenticates to one main account.	The user authenticates to an identity provider, with this one authentication serving for the federation.	The user authenticates to identity providers, and service providers have to rely on that authentication.
Location of Identity Information	Identity information is stored in separate service provider accounts	Identity information is stored in the one main account, a super account	Service providers in the federation keep separate accounts in different locations. They may have agreements for sharing information	Identity information is stored by identity providers chosen by the user. The user can help prevent the buildup of profiles that others hold about him
Method of linking accounts/ learning if they belong to the same person	There is no linking between accounts and no information flow between them	Linking between accounts is not applicable. (A user's full profile resides in that single place.)	The identity provider can indicate what identifiers for accounts with federation members correspond to the same person.	Uses of cryptography can prevent linkages between a user's different digital identities, leaving the user in control.

258

	SILOED	CENTRALIZED	FEDERATED	USER CENTRIC
Trust Characteristics (who is dependent on whom, for what)	The user is reliant on the service provider to protect their information, even if limited. The absence of information sharing has privacy advantages.	The user is reliant on the service provider to maintain the privacy and security of all of his or her data.	Users have rights from contracts, but they may be unfamiliar with options. The federation has leverage as it is in possession of the user's information.	Users can keep accounts separate and still allow information to flow, but bear greater responsibility.
Convenience	Siloed accounts are inconvenient for users and service providers due to multiple authentications, redundant entry of information, and lack of data flow.	This arrangement is easy for the user since he or she only has to deal with one credential to call up the account and since he or she has to authenticate just once.	Other members of the federation avoid the burden of credential management. Organizations that provide services to a user can coordinate service delivery.	Users may be ill equipped to manage their own data (also a vulnerability) and may need training and awareness-raising.
Vulnerabilities	Siloed systems offer the advantage of having limited data on hand, thus creating less of an incentive for the attack. They also have a better defined and stronger security boundary to keep attackers out and limit exposure from failures.	The central party controls the person's entire profile; other entities have little to check that profile against, and an insider could impersonate the person or alter data. Currently, there is no way to safeguard data after it has been shared.	Users have little input into the business–partner agreements. Some service providers will set up federation systems to exploit users. Currently, there is no way to safeguard data after it has been shared.	Concentration in the market for identity providers could leave them with much power. Currently, there is no way to safeguard data after it has been shared.

Article 28G	(1) Everyone is entitled to **personal**, family, honour, dignity and property protection under his control, and is entitled to a sense of security and protection from the threat of fear to do or not to do something that is a human right. **)
	(2) Everyone has the right to be free from torture or degrading treatment of human dignity and is entitled to political asylum from another country. **)
Article 28H	(1) Everyone shall have the right to live in physical and spiritual prosperity, to live, and to obtain a healthy and healthy living environment and to be entitled to health services. **)
	(2) Everyone shall be entitled to special facilities and privileges to obtain equal opportunities and benefits in order to achieve equality and fairness. **)
	(3) Everyone is entitled to social security that enables his complete development as a dignified human being. **)
	(4) Everyone has the right to **own** property and such property shall not be arbitrarily taken over by any person. **)

1. Personal data shall be:

(a) processed lawfully, fairly and in a transparent manner in relation to the data subject ('lawfulness, fairness and transparency');

(b) collected for specified, explicit and legitimate purposes and not further processed in a manner that is incompatible with those purposes; further processing for archiving purposes in the public interest, scientific or historical research purposes or statistical purposes shall, in accordance with Article 89(1), not be considered to be incompatible with the initial purposes ('purpose limitation');

(c) adequate, relevant and limited to what is necessary for relation to the purposes for which they are processed ('data minimisation');

(d) accurate and, where necessary, kept up to date; every reasonable step must be taken to ensure that personal data that are inaccurate, having regard to the purposes for which they are processed, are erased or rectified without delay ('accuracy');

(e) kept in a form which permits identification of data subjects for no longer than is necessary for the purposes for which the personal data are processed; personal data may be stored for longer periods insofar as the personal data will be processed solely for archiving purposes in the public interest, scientific or historical research purposes or statistical purposes in accordance with Article 89(1) subject to implementation of the appropriate technical and organisational measures required by this Regulation in order to safeguard the rights and freedoms of the data subject ('storage limitation');

(f) processed in a manner that ensures appropriate security of the personal data, including protection against unauthorised or unlawful processing and against accidental loss, destruction or damage, using appropriate technical or organisational measures ('integrity and confidentiality').

2. The controller shall be responsible for, and be able to demonstrate compliance with, paragraph 1 ('accountability').

Section 2

(1) Protection of Personal Data in Electronic Systems includes protection against the acquisition, collection, processing, analyzing, storage, appearance, announcement, delivery, dissemination and destruction of Personal Data.

(2) In implementing the provisions referred to in paragraph (1) shall be based on the principles of good Personal Data protection, including:

a. respect for Personal Data as privacy;

b. Personal Data is confidential in accordance with the Agreement and/or in accordance with the provisions of the laws and regulations;

c. based on consent/approval;

d. relevance to the purpose of acquisition, collection, processing, analyzing, storage, appearance, announcement, dispatch and dissemination;

e. certified Electronic System that could be used;

f. goodwill to promptly notify the Owner of Personal Data in writing of any failure to protect Personal Data;

g. availability of internal rules of Personal Data protection management;

h. legal responsibility for any Personal Data that is in User's control;

i. ease of access and correction of Personal Data by the Owner of Personal Data; and

j. integrity, accuracy and validity as well as personal data updates.

(3) The privacy referred to in paragraph (2) a is the freedom of the Owner of the Personal Data to declare a secret or not to declare the Secrets of his Personal Data, unless otherwise provided in accordance with the provisions of the law.

(4) The approval as referred to in paragraph (2) letter b shall be granted after the Owner of the Personal Data stipulates confirmation of the truth, confidentiality and personal data management purposes.

(5) The validity, as referred to in paragraph (2) letter j, is legality in the acquisition, collection, processing, analyzing, storage, appearance, announcement, delivery, dissemination and destruction of Personal Data.

Article 21

Every person is entitled to a personal unity, both spiritual and material, and therefore should not be the object of research without his consent.

Elucidation of Article 21

What is meant by 'being the object of research' is the activity of placing a person as a party to be asked for comments, opinions or information concerning his personal life and personal data and recorded images and sound.

Article 31

(1) The residence of any person shall not be disturbed.

(2) Stepping on or entering a yard of residence or entering a house contrary to the will of the person inhabiting it, is permissible only in matters established by law.

Elucidation of Article 31

Paragraph (1) Referred to as 'not to be disturbed' is a right related to private life (privacy) within its residence.

Paragraph (2) Self explanatory

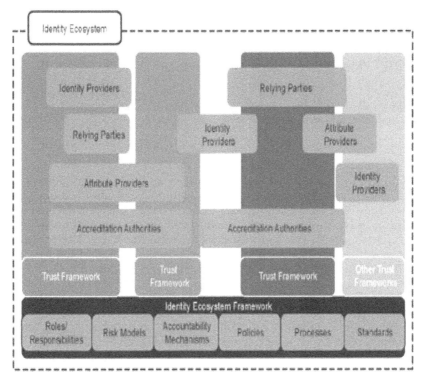

The Identity Ecosystem - Conceptual Model   Image: NIST

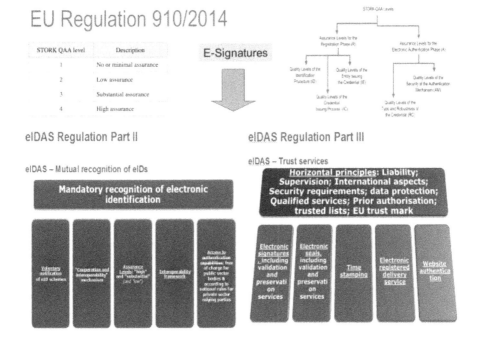

Advancing Rule of Law in a Global Context – Susetyo, Rinwigati Waagstein & Budi Cahyono (eds)
© 2020 Taylor & Francis Group, London, ISBN 978-1-138-32782-5

Human rights certification in Indonesia: Problems and implications

Patricia Rinwigati Waagstein
Faculty of Law, Universitas Indonesia, Indonesia

ABSTRACT: To follow up the UN Guiding Principle on Business and Human Rights, companies have made efforts to integrate human rights into day-to-day business activity. One
such effort is a certification mechanism. This interest appears to be a trend, and it continues
to grow. The question is whether certification can be used effectively to measure businesses'
human rights performance. This paper attempts to answer these questions by analysing two
regulations issued by the Ministry of Maritime Affairs and Fishery: Ministerial Regulation
No. 35/PERMEN-KP/2015 on the human rights system and certification of all fishery business, as well as Ministerial Regulation No. 2/PERMEN-KP/2017. These regulations require
all fishery businesses to acquire a human rights certification. A punishment is imposed for
non-compliance. It is argued in this paper that the certification mechanism has provided
a 'novel' regulatory tool for pushing the fishery industry to respect human rights. Its implementation may lead to various problems, however, if it is not handled with care.

1   INTRODUCTION

One of the greatest challenges in the discourse of business and human rights is how to integrate
human rights into business. A wave of international agreements, proposals, and campaigns has
emerged which is associated with human rights and business concepts. The regulatory or legalis-
tic approach is one of the most popular ones, and the discourse among scholars and stakeholders reveals that this approach is absolutely necessary. Other such soft regulations have also
been introduced and applied in practice to complement state regulation. The question is how to
regulate corporations in order to ensure effective human rights protection.

The issue of certification as a type of regulation is not novel. Certification has been used as
a tool to provide common standards and to ensure the uniformity of norms. The usual certification process is that a certification body or certified professionals will audit an organization's
management system, validate information based on certain common standards or best practices, and issue a certificate confirming that the organization conforms to the standards. Simultaneously, certification can also serve as documented evidence of a professional's
qualifications, competencies, and experience.

In the context of human rights, it is not really clear whether a certification system can be
used to integrate human rights protections into business effectively. The discussion was initiated during the preparatory works of ISO 26000. During that time, the topics of discussion
covered the legitimacy of the institution, the mechanism, and its consequences, in addition to
the development of guidelines. It was finally decided that ISO 26000 is intended to develop
standards and does not carry auditing and certification to its standards, or accreditation of
the certification bodies that operate independently of ISO (Head, 2010). The aim is to provide
a certain learning curve association with ISO 26000 rather than a management system with
rewards as in the certification system. The idea is to develop guidelines and a platform for
learning and communication among corporations, institutions, and organizations, as well as
other stakeholders. In addition to ISO 26000, other initiatives have emerged for assessing
human rights performance. Nevertheless, such initiatives are voluntary in nature and conducted on an ad hoc basis by an internal corporation unit and/or a third party.

The situation is a bit different in Indonesia. The trend continues to impose mandatory norms surrounding the issue of human rights and business. After the passing of Law No. 40/2007 on limited liability corporations requiring all limited liability corporations to implement corporate social and environmental responsibility, the Ministry of Maritime Affairs and Fishery issued two separate regulations requiring all fishery businesses to acquire human rights certification. A sanction was imposed for non-compliance. This initiative was driven by two human trafficking cases in Benjina and Maluku involving illegal fishery, abuse, and even alleged slavery on and near the remote island of Benjina and Maluku (International Organization for Migration, 2016).

Beaten to death according to local media.

This paper does not try to answer the question of whether certification is an appropriate mechanism for measuring the human rights performance of corporations. Human rights can be protected by using different angles of approach and various mechanisms. Innovative regulation should, in fact, be encouraged. The issue of mandatory certification for human rights performance is complex as it presents two major conceptual issues – namely, certification as a regulatory form and the dialectic between mandatory versus voluntary norms. The idea is not to select one instead of the other, but rather to critically evaluate the evolution and process of human rights certification in the fishery business under Ministerial Regulation No. 35/PERMEN-KP/2015 and Ministerial Regulation No. 2/PERMEN-KP/2017. This paper starts by discussing the development of certification prior to and after the adoption of Ministerial Regulation No. 35/PERMEN-KP/2015, and the impact of certification in the Indonesian context. In order to implement this system, this regulation requires much more detailed consideration.

2 DISCUSSION

2.1 Human rights certification in Indonesia

2.1.1 Prior to human rights certification in the fishery business

Identifying the influence of certification in Indonesia would necessitate a separate study. The purpose of this paper, instead, is simply to describe the difficulties faced by the certification process as a way to integrate human rights protections into day-to-day business policy.

Indonesia has mandatory and customary norms pertaining to work safety, labour rights, limited welfare, countering corruption, and environmental protection. In the context of corporate responsibility, there are some compulsory regulations to implement corporate social and environmental responsibility. Under Article 15 of the 2007 Investment Law No. 25, every corporation is obliged to implement corporate social and environmental responsibility. In contrast, Article 74 of the 2007 Limited Liability Corporation Law No. 40 only requires companies conducting their business activities in and or related to the field of natural resources to implement corporate social responsibility, which is elucidated as: 'the obligation of the company which is budgeted and calculated as the cost of the Company'. In other words, a company is obligated to allocate and spend 'obligatory funding' for implementing corporate social responsibility, which can be considered a corporate cost. Sanctions can be imposed for failure to comply with such an obligation. This differs from the 2007 Investment Law No. 25, which does not impose any sanction.

In the context of the fishery, different regulations have been passed to cover various aspects of the fishery business. The issuance of and requirements for fishing vessel licences, fishery management, fishing vessel safety regulations, and the prosecution of certain crimes such as illegal, unreported, and unregulated fishing are examples of how this business is heavily regulated in multiple legal formats from legislation to informal commitment.

In sum, these regulations provide boundaries regarding what corporations should or should not do, although their implementation is often problematic. In practice, weak law enforcement mechanisms – including the unavailability of judicial mechanisms to hold corporations responsible, prevalent corruption, and legal uncertainty due to overlapping norms on social and environmental issues – undermine this implementation. Moreover, conflict is common between economic interests and the protection of stakeholders in areas such as labour rights or environmental issues. Wishing to attract direct foreign investment, Indonesia, like any other developing country, often offers preferential treatment to corporations such as reduced taxes or looser environmental standards. Such behaviour undeniably impacts societal standards and the behaviour of the corporations.

The situation has been complicated by two controversial slavery cases in the town of Benjina, Maluku, as well as Ambon, which was considered the biggest case involving slavery in the twenty-first century. The cases were first highlighted by media revealing the allegation of human trafficking in the global fishery industry. The first case involved a Thai company based in Benjina on Aru Island in Maluku which had been treating its crew members as slaves. The crew members consisted of fishermen from Myanmar, Cambodia, Laos, and Thailand who claimed to have experienced brutal working conditions, including forced confinement and labour, non-payment of salaries, excessive working hours, and psychological and physical abuse amounting to torture. Many were abandoned and unable to return to their home countries. Some had been living in Benjina for more than ten years, and many thought they would never see their homes again. To respond to the case, the Indonesian government established a task force to carry out investigation and rescue efforts. Finally, the government, with the help of the International Organization for Migration (IOM), succeeded in evacuating more than 300 stranded fishermen from Benjina and repatriated them to their home countries. The police also investigated and brought more than four defendants before the court (*Jakarta Post*, 2018).

The second tragedy refers to the Ambon case. Similar to the case in Benjina, the slavery case in Ambon involved 365 crew members as victims of slavery, 105 of whom have been repatriated to their countries of origin. The Indonesian government investigated the case, and at least four captains (*tekong*) were named as suspects (Associated Press in Ambon, 2016). Trafficking in the fishing industry is not a new phenomenon, yet it has gone largely unpunished for too many years. The situations in Benjina and Ambon are just two examples of insidious trade in people, not only in Indonesia, but also globally. Many more cases like them represent the spread of transnational organized crime at sea. Hence, such cases brought several lessons.

First, the fishery industry is very fragile; various elements of the fishery industry lack regulation. In relation to human trafficking in the fishery business, it has yet to be comprehensively identified with a lack of adequate legal instruments. This has put workers on fishing boats under threats to acts of slavery. For that reason, Greenpeace has categorized the work on fishing boats as '3D' (dirty, dangerous, and demeaning) (International Organization for Migration, 2016). Moreover, there is an interdependency of crimes. The lack of monitoring mechanisms to address illegal fishing leads to other crimes such as people smuggling, drugs trafficking, and slavery.

Second, it is clear that a regulation should be equipped with a strong implementation mechanism. In this case, there are a lot of unsolved problems such as the inadequacy of advanced technology to monitor activities of vessels and the absence of comprehensive compliance monitoring mechanisms (compliance and deterrence approach). The fact that the enslavement was not detected and clarified for some time proves a lack of comprehensive compliance monitoring mechanisms.

Under this context, the Ministry of Maritime Affairs and Fishery issued two regulations. The first is Ministerial Regulation No. 35/PERMEN-KP/2005, passed in 2015 in response to two big slavery cases. This is one of the efforts of the special task force to combat human trafficking in the fishery business. The second regulation, Ministerial Regulation No. 2/PERMEN-KP/2017, was issued two years afterwards, focusing on the requirements and mechanisms of human rights certifications for fisheries. It was not really clear from the beginning why the form of assessment is certification. Unfortunately, there is no reference or argument behind the regulation, making it hard to analyse.

These two regulations were passed at different times. Ministerial Regulation No. 35/PERMEN-KP/2015 on Human Rights Systems and Certifications was adopted on 8 December 2015 and covered two main issues – namely, the human rights system and the mechanism to enforce the human rights system through certification. The human rights system refers to the corporate management system which is aimed to ensure the respect of human rights in the fishery industry; the certification is further defined as a process to assess and ensure the compliance of fishery businesses to implement the human rights system.

The second regulation – namely, Ministerial Regulation No. 2/PERMEN-KP/2017 – was passed two years after the first. This regulation specifically covers issues regarding the requirements and the mechanism of human rights certification for fisheries, including the issuance authority of certification, the requirements and issuance procedure of certification, assessment and training agencies, complaints, and funds. This regulation was intended to implement the previous regulation. As the implementing regulation was issued two years after the first, it is assumed that the certification process has not been initiated.

These two regulations cover several points:

1. Certification as a mandatory obligation
   These two regulations require every fishery business covering pre-production, production, processing, and marketing to have human rights certification from the Ministry of Maritime Affairs and Fishery. Such certification is valid for three years and is granted by the Ministry of Maritime Affairs and Fishery to every fishery corporation after the corporations pass the human rights certification assessment process. Failure to procure such certification will lead to administrative sanctions – namely, temporary and/or permanent revocation of the fishery's business licence, licence to fish sea fish, and/or fishing vessel licence, as well as a recommendation from the Ministry of Maritime Affairs and Fishery to the Ministry of Manpower to revoke the licence for employing crew/sailor/fisherman. Without these licences, fisheries cannot operate their business.

2. Certification process
   These two regulations specify the process all fishery businesses need to complete in order to obtain certification. First, all fishery businesses need to implement a human rights management system, which is defined as a company management system to ensure respect for human rights. Although neither regulation clarifies further what it means by a company management system to ensure human rights, they do refer to a due diligence test that a company should perform – namely, to identify and assess current and potential human rights violations caused by operations in the fishery business, to take necessary and effective actions, and to communicate the finding to stakeholders.
   The second step refers to the submission of the application for certification. Together with other documents, the findings of the due diligence test will be submitted to the head of the human rights team under the Ministry of Maritime Affairs and Fishery. Once all documents are received, the head of the team will assign an independent and accredited third party to assess the application – namely, to check and verify documentation, do field monitoring, and conduct an interview. The whole assessment process may take up to thirty days. In cases of potential human rights violations, a company is given two months to take necessary action. If the application is rejected, a company should reapply within several months, as mentioned in the regulation.

3. Human rights protection as a condition for granting certification
   The human rights management system as part of human rights certification only focuses on several human rights–related issues:
   a. the rights of workers, including the right to healthy and safe working conditions, the right to transparent and just crew recruitment processes, a minimum age requirement,

the implementation of sea crew working contracts, and the right to social welfare and insurance for all crew members;

b. corporate responsibility to improve social conditions, particularly in creating jobs and increasing the standard of living for local peoples;

c. the application of a human rights–based approach in the security policy, including training for security personnel;

d. an environmental management system; and

e. land transfer management.

All issues are further described in the annexes of these regulations.

4. Human rights team and third party/assessor

In order to grant human rights certification, the minister will establish a human rights team with a mandate to grant certification to all fishery businesses which pass the certification process. The team, led by its head, will also assign an accredited third party to provide independent assessment and oversee the whole certification process.

5. Public participants

Society's voice regarding a corporation can be heard through two mechanisms. The first mechanism is very limited – namely, during the verification period after the findings by an independent assessor have been completed, the findings will be reverified by the head of human rights and will involve various stakeholders, including the local government and a representative from civil society. The decision to involve civil society or not falls under the human rights team's discretion. Another platform of conveying the public voice is through a complaint mechanism. Here, the public may express complaints regarding the implementation of the certification in the fishery business.

This certification system has played a very significant role in institutionalizing human rights protection into corporate policy in Indonesia from at least four different perspectives. First, the mandatory nature of this certification strengthens the position of human rights protection as a legal obligation of corporations, although in this case it is limited to a certain business sector. Certification further provides a standard that should be applied throughout the fishery business in Indonesia. This, indeed, addresses the problem of diversity of standards and the questionable efficacy of soft voluntary mechanisms.

Second, the adoption of the certification process into law and/or policy enhances its reception and implementation in Indonesia by various stakeholders at a different level. At the corporate level, while the primary target may be corporations that directly or indirectly impact the fishery business, the certification's mandatory nature at the very least promotes a degree of awareness among other corporations to look beyond motives of profit maximization. It is conceivable that in the future, the same certification could be applied to all business sectors. At the grass-roots level, mandatory certification would encourage society to focus more attention on the monitoring of corporate behaviour regarding human rights.

Third, the mandatory nature of certification can serve as a preventive mechanism that keeps companies from unduly benefiting as a result of the system. Moreover, mandatory certification can be a complement to – not a replacement for – another remedial mechanism/punishment. For example, in the case of a fishing vessel owner who does not provide a written contract for his crew with the standards as regulated in Indonesian law, his fishing licence can be revoked temporarily until such a written contract is signed between the vessel owner and his crew. In this case, the certification may have the effect of actually strengthening and empowering these victims.

Fourth, while certification is probably not a favourable approach among states, it does lay down a precedent for other countries to take similar steps towards institutionalizing human rights in the form of legal obligations. The case of Indonesia also signifies a regulatory shift from the implicit to the explicit regulatory model of human rights protection and business.

While acknowledging all of these advantages, it is also important to observe that certification raises some pragmatic and conceptual questions. First, there is a lack of clarification on how to determine who should have certification. According to Article 5 of Regulation No. 35, all fisheries entrepreneurs are obligated to have a human rights certification. The regulation only defines the fisheries entrepreneur as everyone who conducts business in the fisheries field, covering all activities relating to the management and utilization of fish sources and their environment, commencing from pre-production, production, and processing to marketing.

This definition raises two points. First, no fisheries businesses, regardless of their legal formats, are except from this obligation. Some limitation is applied in Regulation No. 35 in that the obligation is imposed on four types of fishery business:

- An individual and/or corporation based on *grosse akta* or their representative who has a fishing vessel, including its licence;
- An individual, a group of individuals, and/or a corporation or representative which rents and/or manages a fishing vessel, including its licence;
- An individual, a group of individuals, or a corporation or representative which has a fishery management unit;
- An individual, a group of individuals, or a corporation or representative which produces fishery products in Indonesia.

In reality, the definition and its limitations are still not clear. The activities of fishery cover a broad range of endeavours. Should an individual fisherman, a person in the market selling fresh fish, or a corporation producing fish in a can be obligated to have a certification? The question remains unanswered. The representative from the Ministry of Maritime Affairs and Fishery in various interviews has stated that only a company which has or operates fishing vessels of thirty gross tons and upwards is required to have human rights assessment. Nevertheless, such a limitation is not reflected in the current regulations.

The second issue refers to the independent assessor or third party which will assess corporate behaviours. These two regulations clearly state that the head of the human rights team under the Ministry of Maritime Affairs and Fishery can assign an independent and accredited third party/ assessor. In a case where no assessor is available, the human rights team may carry out the assessment. The question is, who can be an assessor of human rights certification? Articles 19 and 20 stipulate that two mandatory capacities are required to be a human rights assessor: capacity as an auditor with management system certification, as well as an auditor with a technical training certificate in assessing human rights in the fishery business. At this point, very few auditors have both capacities, hence the selection of auditors is limited. Moreover, while there is already management system certification which has been commonly accepted by various institutions, as the issue is developing, there is no clear information regarding technical training on human rights in the fishery business which has been accepted as a common standard at the national, regional, or international level. In the absence of such training, the Ministry of Maritime Affairs and Fishery needs to develop it. While waiting for such training, the question is whether the certification process can start immediately. Pilot projects are needed in this case to showcase whether certification works.

Third, in order for such measure to be effective, there should be a monitoring and appeal mechanism which is transparent and accessible to oversee the process. The decision of an independent assessor is final and binding. No mechanism to challenge the decision is available under these regulations. Once an application for human rights certification is rejected, the proponent should reapply within one year. No clear information is available on whether a fishery corporation can still operate during that time.

Fourth, the issue of access to information is not addressed in the regulations. It is not clear who can access the findings of the certification process. Article 12 of Regulation No. 2 states that the results of the assessment held by a third party will be submitted to the head of the human rights team. For verification, the head of the human rights team can consult with stakeholders, including civil society or non-governmental organization representatives. Nevertheless, there is no reference to whether the public can access the findings of the assessment. Access to such documents will ensure transparency and accountability among different stakeholders.

These issues need to be tackled and elaborated prior to the implementation of certification. Having a pilot project may be a good way to assess whether this type of regulation works.

3 IMPLICATIONS

3.1 *Will a certification ensure the effectiveness of human rights protection?*

As mentioned earlier, certification is mainly conducted as a voluntary initiative. Making a certification mandatory is expected to ensure its effectivity. Nevertheless, the relationship between the legal character of a norm and compliance-effectiveness is complex; a mandatory norm does not necessarily coincide with corporate compliance. What constitutes a voluntary or obligatory norm is a matter of compromise and acceptance, and is dependent on the subject and the context. To be voluntary does not imply anything binding since many such norms function effectively to reshape, implement, interpret, or even substitute for mandatory norms (Chinkin, 2003, pp. 30–31). On the other hand, to be mandatory does not mean that the norm is automatically enforceable since it may be imprecise or lack a monitoring mechanism (Chinkin, 2003). Human rights certification serves as an example of the problems associated with such a norm, which may actually jeopardize its efficiency.

3.2 *Will the mandatory certification increase corporate costs?*

Before this question is answered, it is important to answer another question: does corporate social responsibility require expenses? In some cases, it may cost money and require the planning, recounting, and redistribution of sources, including profits. In other situations, it may cost very little, particularly in the context of energy saving. Having certification indeed involves an additional cost. This is apparently unavoidable. Article 27 of Ministerial Regulation No. 2 clearly reconfirms this. This regulation specifies three different costs which are borne by national as well as corporate budgets. First, the cost for technical assistance to the corporation to improve its conditions in order to pass the assessment should be covered by the national budget. Second, all costs needed to implement the tasks of the human rights team ought to be charged to the state budget and/or other legitimate sources. The last type of cost refers to the cost of the third party/auditing company/assessor. These two regulations require a corporation which has submitted a proposal to have a contract with the assigned assessor. While no financial implication is clearly mentioned in the regulations, it can be assumed that the any financial consequences caused by the contract between the third party and a corporation will be paid by the corporation. While large companies will be willing to participate in this mandatory initiative, high costs may be a big issue for small corporations to comply with such regulations.

On the other hand, will implementing this certification increase profits? Although many scholars, economists, and business actors point to the mutual relationship between human rights and profits, there is no definitive evidence regarding this relationship (Vogel, 2005, pp. 17–18). We are left with the question of whether such an obligation should still be required if this effort has the effect of obstructing business and creating additional burdens to corporations and shareholders. From a normative perspective, the answer is yes, based on two arguments. First, if these costs are a simple consequence of obeying the law, then the reduction of profits should be accepted. Second, if costly and, in some cases, destructive actions are necessary in order to achieve that profit, then social responsibility must also be taken into consideration by managers and directors.

4 CONCLUSION

Having a certification as a regulatory tool to integrate human rights protection into business, while having partly resolved the problems related to the diversity of standards and providing a greater degree of certainty around the practice of voluntary initiatives, also opens up new problems with respect to substance and procedure. The case of certification is no exception. The adoption of these two regulations, which impose sanctions for non-compliance, should be given credit where it is due. It serves to push forward the implementation of human rights protection and business in Indonesia and, moreover, may act as a preventive tool to keep corporations from further irresponsible behaviour. While a certification system offers hope to those trying to improve corporate practices in Indonesia, its successful implementation is still in doubt, for a number of reasons. First, the substance is too general, insubstantial, and ambiguous. The subject and object of human rights assessment are too broad. Second, it currently fails to provide any implementation mechanism and gives little clarity on how it would be implemented and monitored. Third, there is no clear direction from either the drafters or the government as to exactly what the certification is intended to achieve. Without a greater degree of clarification with respect to all of these details, a certification system will create more problems than it will offer solutions.

REFERENCES

Associated Press in Ambon. (2016). 'Five Jailed in Seafood Slavery Case'. *The Guardian*. 11 March. Accessed 23 April 2018. www.theguardian.com/world/2016/mar/11/seafood-slave-drivers-given.

Chinkin. (2003). 'Normative Development in the International Legal System'. In *Commitment and Compliance*. Edited by Diana Shelton. Oxford: Oxford University Press.

Deva, Surya. (2004). 'Relations of Corporate Observance of Human Rights to Competitive Advantage: An Economic Integration'. Sydney: University of Sydney.

Head, Katie Bird. (2010). 'It's Crystal Clear. No Certification to ISO 26000 Guidance Standard on Social Responsibility'. ISO. Accessed 23 April 2018. www.iso.org/cms/render/live/en/sites/isoorg/contents/news/2010/11/Ref1378.html.

International Organization for Migration. (2016). *Human Trafficking, Forced Labour and Fisheries Crime in the Indonesian Fishing Industry*. Jakarta: International Organization for Migration.

Jakarta Globe. 'At Least Six Suspects Arrested in Benjina Human Trafficking Case | Jakarta Globe'. Accessed 23 April 2018. http://jakartaglobe.id/law-and-order/at-least-six-suspects-arrested-in-benjina-human-trafficking-case/.

Jakarta Post. (2015). 'Seven Arrested in Benjina Case'. 5 December. Accessed 20 April 2018. www.thejakartapost.com/news/2015/05/12/seven-arrested-benjina-case.html.

'Minister Susi: Illegal Fishing Links to Other Crimes – Nasional Tempo.Co'. Accessed 23 April 2018. https://nasional.tempo.co/read/811082/minister-susi-illegal-fishing-links-to-other-crimes.

'Modern-Day Slavery: Indonesia Cracks Down on Brutal Conditions on Foreign Fishing Boats'. Accessed 23 April 2018. www.smh.com.au/world/modernday-slavery-indonesia-cracks-down-on-brutal-conditions-on-foreign-fishing-boats-20170124-gtxseo.html.

PECB. 'SRMS: Why Should Organizations Have Certified ISO 26000 Lead Implementers?' Accessed 23 April 2018. https://pecb.com/article/srms-why-should-organizations-have-certified-iso-26000-lead-implementers.

'Perusahaan Perikanan Wajib Kantongi Sertifikat HAM'. Accessed 23 April 2018. http://industri.kontan.co.id/news/perusahaan-perikanan-wajib-kantongi-sertifikat-ham.

Regulation of the Ministry of Marine Affairs and Fisheries of the Republic Indonesia No. 2/PERMEN-KP/2017 concerning Requirement and Mechanism of Human Rights Certification for Fisheries (2017).

Regulation of the Ministry of Marine Affairs and Fisheries of the Republic of Indonesia No. 35/PERMEN-KP/2017 concerning Human Rights System and Certification (2015).

'The Mandatory Corporate Social Responsibility in Indonesia: Problems and Implications'. *Journal of Business Ethics*, 2010. https://link.springer.com/article/10.1007/s10551-010-0587-x.

Vogel, David. (2005). *The Market for Virtue: The Potential and Limits of Corporate Social Responsibility*. 1st ed. Washington, DC: Brookings Institution Press.

Advancing Rule of Law in a Global Context – Susetyo, Rinwigati Waagstein & Budi Cahyono (eds)
© 2020 Taylor & Francis Group, London, ISBN 978-1-138-32782-5

Pros and cons of the government regulation in lieu of law no. 2 Year 2017 concerning mass organizations

Qurrata Ayuni
Faculty of Law, University of Indonesia, Indonesia

ABSTRACT: The birth of the Government Regulation in Lieu of Law No. 2 of 2017 Concerning Mass Organizations (Peraturan Pemerintah Pemerintah Pengganti Undang-Undang/Perppu Ormas) (Perppu) has pros and cons for the community. Not a few people think that the Perppu is a violation of freedom of association in Indonesia; others state that the Perppu is a tool to prevent the development of radical organizations in Indonesia. This paper conducts a critical study of this regulation, established by President Jokowi, from the perspective of constitutional law. The government must use its power to form laws accurately to guarantee the rule of law and human rights protections in Indonesia.

Keywords: Government Regulation in Lieu of Law No. 2 of 2017, Mass Organizations

1 INTRODUCTION

The issuance of the Government Regulation in Lieu of Law No. 2 of 2017 Concerning Community Organizations (hereinafter referred to as the Perppu Ormas) on July 10, 2017, has many pros and cons. This Perppu was intended to amend Law No. 17 of 2013 with the same title. Through this Perppu, the government can dissolve a mass organization using the principle of *contrarius actus*.

The principle of *contrarius actus* in state administrative law refers to the concept by which a state administrative body or official who issues state administrative decisions by itself is also authorized to cancel them (Hadjon & Djatmiati, 2009, p. 25). Through the use of this principle, the government as the party that gives administrative permission in the establishment of a mass organization can also unilaterally revoke that organization's license.

The Perppu Ormas has also abolished at least seventeen articles previously included in Law No. 17 of 2013 Concerning Social Affairs Organizations. The Perppu Ormas therefore eliminates mediation processes, administrative processes, and the process of involving the Supreme Court. The abolition of Articles 63 to 80 then triggered the assumption that the government had attempted to curb the freedom of association and the right to express opinions that were guaranteed in the 1945 Constitution. Many parties are of the belief that the dissolution of mass organizations by the government, based on the provisions of the Perppu Ormas, has meant the abandonment of the principles of due process of law, presumption of innocence, and equality before the law.

This is considered contrary to the constitutional rights granted under the 1945 Constitution based on the provisions of Article 28D paragraph (1), which the provisions of Article 80A of the Perppu Ormas threaten to ignore and suppress, so that there is no due process of law and there is no equal position between mass organizations and the government. The government can unilaterally dissolve mass organizations without going through the legal process of allowing them to defend themselves administratively.

These pros and cons then triggered a number of case requests at the Constitutional Court. Applications submitted to the Constitutional Court include cases numbers 41/PUU-XV/2017, 48/PUU-XV/2017, 50/PUU-XV/2017, 58/PUU-XV/2017, and 94/PUU-XV/2017. As stated in

Article 24C of the 1945 Constitution, the Constitutional Court has the authority to test the law against the 1945 Constitution. In several jurisprudences, the Constitutional Court has also tested government regulations in lieu of laws several times so that this application becomes one of the objects of judicial review in the Constitutional Court.

This paper is divided into five parts. The first part explains why community organizations present a lot of pros and cons for society. The second part discusses the position of Pancasila as a *staatsfundamentalnorm* for the country of Indonesia. The third part examines the concept of "forceful urgency" as stated in Article 22 of the 1945 Constitution. The fourth part studies the issue of criminal sanctions, which is a testing argument in the Constitutional Court, while the fifth part is the conclusion that summarizes this paper.

2 POSITION OF PANCASILA

The Perppu Ormas stemmed from the government's perceived urgent need to address the rise of radicalism in Indonesian society. The government believes that so far, many mass organizations have arisen that have not based their ideologies or activities on Pancasila and the 1945 Constitution. Law No. 17 of 2013 Concerning Mass Organizations (Law 17/2013) is still considered to save a lot of legal void in terms of effective application of sanctions. The government assesses the existence of community organizations that oppose Pancasila, which leads to radicalism.

For the Indonesian people, Pancasila is a *staatsfundamentalnorm*, as in Notonagoro's view (Asshiddiqie, 2006). Pancasila can also be seen as the ideal law (*rechtsidee*). This position requires the establishment of positive law to achieve ideas, and can be used to test positive law.

Pancasila can also be found in the opening of the 1945 Constitution, which contains five precepts – namely, the One God, just and civilized humanity, Indonesian unity and populism led by wisdom in representative deliberations, and social justice for all Indonesian people.

Pancasila, which holds an essential position as the fundamental norm of the state (*staatsfundamentalnorm*), is a term used by Hans Nawiasky in his theory of norms of law (*die theorie von stufenordnung der rechtsnormen*) as a development of Hans Kelsen's theory of norms (Hamidi, 2006, p. 59). Nawiasky uses a legal hierarchy with four levels:

1. *Staatsfundamentalnorm* in the form of basic norms of state or the source of all legal sources;
2. *Staatsgrundgezetze* in the form of basic law, which if stated in state documents becomes the constitution or *vervassung*;
3. *Formelegezetze* or formal law, which stipulates an imperative provision in terms of implementation and legal sanctions; and
4. *Verordnung en and satzungen autonome*, which are autonomous, implementing rules and regulations, born of both delegation and attribution (Darmodihardjo, 1999, p. 21).

Therefore all legal and social aspects of Indonesian law must come from Pancasila. So as a *staatsfundamentalnorm*, Pancasila must also be the main ideology in every mass organization in Indonesia so that no social organization should carry out activities that conflict with Pancasila.

Hamid S. Attamimi, in his essay entitled "The Pancasila of the Law in the Life of Indonesian Nation Law," discusses Pancasila from the point of view of legal philosophy. He deliberately does not use the term *ideology* in his essay because, according to him, the term *legal ideation* (*rechtsidee*) is more appropriate since *ideology* has the connotation of a sociopolitical program that tends to view other programs, including law, as a tool and therefore under its subordination. The legitimate aspiration is none other than Pancasila as the main embodiment of the ideals of Indonesian law (Attamimi, 1990).

Nine days after the publication of the Perppu Ormas, on July 19, 2017, the director general of public law administration, on behalf of the minister of law and human rights, issued an order to dissolve Hizb ut-Tahrir Indonesia (HTI) through Decree No. AHU-30.AH.01.08. In

2017, the government revoked the legal status of HTI, whose activities were considered contrary to Pancasila and the 1945 Constitution.

The revocation of HTI's legal status caused unrest in the community. Because the government tends only to supervise Islamic organizations, this cancellation of status raised concerns about the dissolution of HIT and called to mind similar incidents that occurred in the Soekarno era. At that time, an Islamic mass organization called Masyumi, which was often considered critical of the government, was dissolved.

The supporters of the dissolution of HTI argued that HTI adheres to the Islamic Caliphate, which wants to establish an Islamic state in Indonesia. This is considered contrary to the values of Pancasila and the 1945 Constitution. But for HTI, having confidence in the Islamic Caliphate is a belief value for Muslims so that it is in line with the first value of Pancasila, the One God.

Some parties contended that the dissolution of HTI without any legal defense before the state administrative court was a constitutional rights violation. Article 28E paragraph (3) of the 1945 Constitution states, "Everyone has the right to freedom of association, assembly and opinion." A more significant concern over the Perppu Ormas is the rebirth of an authoritarian regime in Indonesia.

According to Alexis de Tocqueville in his two-volume *Democracy in America*, a democratic state is a country in which people pursue shared goals that are applied for very many purposes (de Tocqueville, 2005, p. 116). The health of a mass organization also determines the quality of democracy in a country. The unilateral cancellation of permits without an administrative defense is contrary to human rights principles as they are understood in a proper democratic system.

The explanation of the Perppu Ormas states that restrictions on human rights are carried out in the framework of protecting human rights, and every citizen has an obligation to protect the human rights of others. The affirmation of the protection of human rights and human obligations is included in Article 28J, which reads:

1. Everyone must respect the human rights of others in orderly life in the community, nation, and state.
2. In exercising their rights and freedoms, each person must submit to the restrictions stipulated by law with the sole purpose of guaranteeing recognition and respect for the rights and freedoms of others and to fulfill just demands in accordance with moral considerations, values, religion, security, and public order in a democratic society.

From the provisions of Article 28J of the 1945 Constitution of the Republic of Indonesia, it can be concluded that the concept of human rights based on the 1945 Constitution is not absolute. This is in line with ASEAN's view in the first and second points of the Bangkok Declaration on Human Rights 1993.

First there is the matter of fair application: the approach to human rights has to be "balanced"; double standards in the implementation of human rights are to be avoided; "concern" is expressed about the priority accorded one category of rights; "economic, social, cultural, civil and political rights" are interdependent and indivisible and must therefore be "addressed in an integrated and balanced manner." The barely disguised subtext here is that civil and political rights (with their assertions of democratic and protest rights) have been wrongly prioritised by the supporters of human rights in the Global North with the result that the subject of human rights often appears exhausted once the issue of democratic freedom has been fully ventilated. In fact from the Bangkok perspective, social and economic rights are of at least equal importance.

Second the declaration introduces the notion of regional values as potentially in opposition to human rights. The diverse and rich cultures and traditions of Asia need to be better recognised. "Confrontation and the imposition of incompatible values" are to be avoided. Though "universal" in nature, human rights must, as the substance of the declaration went on to say, be considered in the context of a dynamic and evolving process of international norm-setting, bearing in mind the significance of national and regional particularities and various historical, cultural and religious backgrounds.

Based on the ASEAN Declaration of Human Rights in Bangkok, this quotation emphasizes that the Universal Declaration of Human Rights as applied in the ASEAN context must take into account regional and national specificities and various historical, cultural, and religious backgrounds.

The development of the protection of human rights as described from the national, regional, and international aspects has distinguished the protection of human rights under normal conditions and in emergency situations. In Indonesian national law, the government has enacted Law No. 39 of 1999 Concerning Human Rights, Law No. 26 of 2000 Concerning Human Rights Courts and several other laws related to the protection of human rights, and Law No. 23 of 1959 Concerning the State of Hazard, a condition that excluded the protection of human rights.

3   FORCEFUL URGENCY

Another criticism regarding the issuance of the public order Perppu is the form of legislation used. Many parties argue that the government is too hasty in setting up this Perppu because there are actually no real pressing forces. The rules that limit human rights should not be unilaterally determined by the government, but must be discussed in depth by the House of Representatives.

Article 22 paragraph (1) of the 1945 Constitution states, "In the case of compulsive matters, the President has the right to stipulate government regulations in lieu of the law." According to Decision of the Constitutional Court of the Republic of Indonesia Number 138/PUU-VII/2009, three conditions must be met in the event of a compulsive crisis:

1. There is an urgent need to resolve legal issues quickly based on the law.
2. The required laws do not yet exist, so there is a legal vacuum, or the current laws are inadequate.
3. The legal vacuum cannot be overcome by making law through the usual procedures because it will require a considerable amount of time while the urgent situation needs certainty to be resolved.

Those who oppose the Perppu Ormas argue that there is actually no urgent need to resolve legal issues. The state of Indonesia at the time of issuance of the Perppu Ormas was not in an emergency condition but under normal circumstances. The government can still use the mass organizations law and foster the organizations that are considered radical. Even in the dissolution of HTI, the government never carried out the previous coaching, which should have been carried out by good governance.

There is also no legal vacuum that does not regulate the dissolution of mass organizations. The Perppu Ormas contains a sanction mechanism in an effort to foster organizations in Indonesia. Sanctions can be given consisting of one to three written warnings. If the written notice is not heeded, the government can temporarily terminate the activities of the mass organization that is under investigation and potentially can even revoke its legal status after requesting consideration from the Supreme Court.

Unfortunately, the provisions for carrying out tiered sanctions were revoked by the Perppu Ormas, so the government can immediately revoke legal status when a mass organization is considered contrary to Pancasila. This raises concerns for the public that the government can unilaterally dissolve mass organizations that may conflict with the government's political interests.

Therefore, in the context of human rights restrictions, the appropriate legislation is the law, not the Perppu Ormas. The bill in the parliament using the law is in line with the human rights limitation clause in Article 28J paragraph (2) of the 1945 Constitution, which states, "In exercising their rights and freedoms, every person must submit to the restrictions stipulated by the law." Many critics have contended that the government should not issue the Perppu Ormas unilaterally without involving the DPR.

The government has special reasons why it finally chose to enact the Perppu Ormas. One of them is the theory put forward by Vernon Bogdanor, as quoted by Jimly Asshiddiqie, that there are at least three conditions of emergencies that can lead to a forceful compulsion – namely, war, civil emergency, and internal emergency. An internal emergency can be declared based on the subjective assessment of the president, which in turn can be a reason for the president to issue the Perppu Ormas.

The characteristics of "compulsive matters" are also in line with Article 4 of the International Covenant on Civil and Political Rights (ICCPR):

> In time of public emergency which threatens the life of the nation and the existence of which is officially proclaimed, the States Parties to the present Covenant may take measures derogating from their obligations under the present Covenant to the extent strictly required by the exigencies of the situation, provided that such measures are not inconsistent with their other obligations under international law and do not involve discrimination solely on the ground of race, colour, sex, language, religion or social origin.

Referring to Article 4 of the ICCPR, explain who questioned the "matters of urgency that support" including "threatening the life of the nation and the existence that was officially proclaimed (regarding the future of the life of the Indonesian people as well as the Republic of Indonesia 4) ICCPR and strengthened in Article 22 paragraph (1) of the 1945 Constitution of the Republic of Indonesia, allows the state to be able to improve it in relation to protection of rights, human problems with special reasons in these emergency situations.

The issuance of the Perppu Ormas is also necessary for the state to guarantee peace and security, and the state has five obligations when issuing it: (1) not to intervene in problems that occur in other countries; (2) not to cause upheaval in other countries; (3) not to forcefully relocate people who depend on its territory; (4) to guarantee its territory so as to guarantee safety and security; (5) to manage relations with allied countries in accordance with international law (Adolf, 1996, pp. 37–38).

The Unitary State of the Republic of Indonesia is based on Pancasila and the 1945 Constitution of the Republic of Indonesia, and it addresses certain civil society organizations that have taken hostile actions – including saying, advising, judging, or aspiring either verbally or in writing, through electronic media or not using media electronics, hatred toward certain groups and those who are included in the administration of the state. Such actions have the potential to cause social conflict between community members (Explanation of the Mass Organization in Lieu of Law, 2017).

The alleged violations of the mass organizations questioned by the government do not contradict Pancasila or the 1945 Constitution of the Republic of Indonesia; in essence, they are actions that are highly criticized by the organizers or by mass organizations that are part an agreement between the Unitary State of the Republic of Indonesia and the 1945 Constitution of the Republic of Indonesia. Such violations are manifestations of the intentions and the legal objectives of the registered mass organizations.

The purpose of the Government Regulation in Lieu of Law No. 2 of 2017 is to determine and simultaneously to protect approved civil society organizations and to hold them accountable based on Pancasila and the 1945 Constitution of the Republic of Indonesia. Government Regulation In Lieu of this Law has agreed to these two groups of Organizations and has been approved with extraordinary types of agreements and applications (Explanation of the Mass Organization In Lieu of Law, 2017).

Regarding the two views on the authority under the Perppu Ormas as used by the president, the authors consider this the authority of the president given by the constitution. In this case, the president can issue a presidential compilation module to examine the legal requirements that need to be addressed immediately. Jimly Asshiddiqie contends that Article 22 of the 1945 Constitution grants a presidential license for subjective judgments of the state (Asshiddiqie, 2007). This then leads to non-subjectivity. However, subjective judgments can be obtained through examination, and balance by means of legislative discussions of the Perppu Ormas.

The authority to issue the Perppu stipulated by this act was also granted by the DPR (Farida, 2007, p. 94). Article 52 of Law No. 12 of 2011 Concerning Formation of Laws and

Regulations states that the Perppu must be approved by the DPR at the time of the next trial. In this case, the DPR will discuss the Perppu and will either get approval or not receive support. If, in a plenary session, a Perppu does not get approval, the Perppu must be revoked. In the event that a Perppu is approved in a DPR plenary session, the Perppu becomes law.

The president can normatively issue a Perppu as a subjective view of a state condition that is considered urgent. In this case, the DPR must also immediately debate this Perppu as a form of checks and balances provided by the DPR members as legislators. The Perppu Ormas may not remain too long as a Perppu, but must be immediately decided upon by the DPR.

4   CRIMINAL SANCTION

Another issue petitioners have raised before the Constitutional Court is the imposition of criminal sanctions contained in Article 82A of the public order Perppu. This provision states that anyone who directly or indirectly engages in hostility on the grounds of ethnicity, religion, or race can be sentenced to a minimum of five years and a maximum of twenty years in prison.

The petitioner in case number 48/PUU-XV/2017 postulates that the existence of the Article 82A removes collateral for justice and legal certainty. This is due to the fact that such arrangements already exist in Law No. 40 of 2008 Concerning the Elimination of Racial and Ethnic Discrimination. Article 15 of this law states that the penalty for racial and ethnic differentiation is imprisonment of a maximum of one year and/or a fine of at most 100 million rupiah.

The law has also been regulated in Article 28 paragraph (2) of Law No. 11 of 2008 Concerning Electronic Information Technology, which states that anyone who intentionally disseminates information intended to cause hatred or individual hostility toward certain community groups based on ethnicity, religion, race, or intergroup characteristics can be punished with a sentence of imprisonment of a maximum of six years and/or a fine of at most 1 billion rupiah.

These provisions make a statement that the five-to-twenty-year sentence is very different from other penalties agreed upon by the DPR. The existence of this severe sentence will result in the loss of freedom for mass organizations due to a fear of being accused of hostility. The "directly or indirectly" phrase gives rise to multiple interpretations, and their consequences can be imposed on people who do not directly commit these criminal acts.

The concept of individual criminal acts as part of an organization has a different characteristic. In the life of an organization – both political and community organizations – there are always leaders and followers. Among the leaders are "ideologues" who formulate ideas and thoughts, doctrines and beliefs. Some even act as "organizers" who move the organization. There is also a "propagandist" whose duty is to disseminate the organization's ideology. Members or followers of an organization may lack in-depth understanding about its ideology or doctrine. In a corporate crime, the leader of the organization can be punished, not its followers.

Therefore, those who must get punishment are the main perpetrators who commit insults based on race, ethnicity, and religion. R. Soesilo, in his comments about criminal offense, argues that the object of insults must be individual human beings, not government agencies, administrators of an association, a group of residents, etc. (Soesilo, 1988, p. 225). The objective of criminal law is to determine whether a person's actions are criminal, and then to determine whether the person who committed them can be blamed (Moeljanto, 2008, pp. 10–11).

According to Bagir Manan, several principles must be considered under the legal concept of *lex specialis derogat legi generalis* – namely, (1) the provisions found in the general legal rules remain valid, except those specifically regulated in the special legal rules; (2) The provisions of *lex specialis* must be equal to the provisions of *lex generalis* (law with law); (3) The provisions of *lex specialis* must be in the same legal (regime) environment as the provisions of *lex generalis*. The Code of Commerce and the Civil Code are equally included in the civil law environment (Manan, 2001, p. 56).

In this regard, criminal provisions should indeed be discussed by the DPR first. The government should not impose criminal sanctions that exceed criminal provisions governing the provisions of the same in other laws. Therefore, the existence of the Perppu Ormas in terms of criminal sanctions must first be confirmed by the DPR. Even if the Perppu Ormas wants to regulate criminal provisions, it should not exceed the penalty on special laws that have regulated the same thing.

5 CONCLUSION

The issuance of the Government Regulation No. 2 of 2017 Concerning Community Organizations has many pros and cons. The issues include the absence of opportunities for organizations to defend themselves before being dissolved by the government. The issue that arises is the enactment of the principle of *contrarius actus*, which allows the government to unilaterally grant and revoke the status of legal entities. Another legal issue is the presence of criminal sanctions that exceed criminal provisions in other laws that regulate the same thing. The five-to-twenty-year sentence is more excessive than the penalty that should be applied to similar cases. Moreover, the phrase "directly or indirectly," which allows the prosecution of people who do not commit criminal acts directly, can also be ensnared in criminal sanctions.

However, for those who support the Perppu Ormas, it is considered a constitutional action that can be legitimately carried out by the president. It is based on Article 22 of the 1945 Constitution, which indeed gives the president the authority to subjectively weigh "compulsive matters." Safeguarding Pancasila as the ideology of all existing mass organizations in Indonesia is a legitimate legal, political policy.

REFERENCES

Adolf, Huala. (1996). *Country Aspects in International Law*. Jakarta: Raja Grafindo Persada.

Asshiddiqie, Jimly. (2006). *Introduction to Constitutional Law*. Jakarta: Secretariat General and Registrar's Office of the Constitutional Court.

Asshiddiqie, Jimly. (2007). *Emergency Constitutional Law*. Jakarta: Rajawali Press.

Attamimi, Hamid S. (1990). Role of the Decree of the President of the Republic of Indonesia in the Implementation of State Government (Dissertation). Jakarta: University of Indonesia.

Darmodihardjo, Dardji. (1999). *Principles of Legal Philosophy*. Indonesia: Main Gramedia.

de Tocqueville, Alexis. (2005). *About Revolution, Democracy and Society*. Jakarta: Yayasan Obor Indonesia.

Farida, Maria. (2006). *Legislation: Basics and Formation*. Indonesia: Kanisius.

Hadjon, Philipus M. and Tatiek Sri Djatmiati. (2009). *Legal Arguments*. Indonesia: Gadjah Mada University Press.

Manan, Bagir. (2001). *Welcoming Regional Autonomy*. Yogyakarta: Center for Legal Studies (PSH) Faculty of Law UII Yogyakarta.

Moeljanto. (2008). *Principles of Criminal Law*. Indonesia: Rineka Cipta.

Soesilo, R. (1988). *The Criminal Code*. Indonesia: Politea Bogor.

Advancing Rule of Law in a Global Context – Susetyo, Rinwigati Waagstein & Budi Cahyono (eds)
© 2020 Taylor & Francis Group, London, ISBN 978-1-138-32782-5

Awakening the geopolitical 'spirit': Transforming Indonesia's institutions to confront global challenges

Fakhridho S.B.P. Susilo & Kris Wijoyo Soepandji
Faculty of Law, University of Indonesia, Indonesia
Australia National University, Canberra, Australia

ABSTRACT: The study of global governance has never been so interesting than the current situation as we face multi-faceted global socio-economic and political challenges. The presupposed peace that the post-Cold War unipolar world order would entail, has never come into fruition. Rather, we see conflicts increasing with their multitudes of consequences which create further set of problems. Globalization that is supposed to bring welfare to the global populace, while to some extent benefitting the developing countries, exacerbate inequality and create socio-economic tensions not only in developing countries but also developed ones, which in turn fuels antipathy toward globalization and the open-access order of the 21st century. These, coupled with the rise of emerging powers such as Russia and China, point toward the uncertain and largely unknown future of the global order. The implications for these sets of challenges to the national front cannot be underestimated and must be met with institutional transformation that seeks to leverage on the national identity and character of that nation. That is a task that Indonesia still has to grapple with. Yet, the answer to that, as this paper proposes, may lie in the geopolitical and historical values that have rarely been examined and cultivated in the institutional transformation discourse in Indonesia.

Keywords: geopolitics, law, nationalism, governance, public policy, state, Indonesia

1   INTRODUCTION

The national dynamics cannot be completely disentangled from what occurs at the international or global level, and vice versa. This is a thesis that is best portrayed in the classical saying of the former Indonesian president, Soekarno: "*Internasionalisme tidak dapat hidup subur, kalau tidak berakar di dalam buminya nasionalisme. Nasionalisme tidak dapat hidup subur, kalau tidak hidup di taman-sarinya internasionalisme.*"[1] Yet this notion has never before resounded more profundly than in today's world in which we live, where boundaries between nation states are becoming more blurred due to socio-economic and technological progress. The interweaving of socio-cultural, political, and economic nodes between countries mean that events occuring in one country may almost asuredly create spillovers that are felt beyond the national border of such state.

'Resilience' is therefore the key in navigating the current global order, and it necessitates institutional arrangments that are conducive for governance to function effectively. How does Indonesia, both as a singular political entity and part of the global society cope in the new global context? More specifically, what implications do the current global challenges hold for Indonesia? And what kind of institutional arrangements are necessary for Indonesia to develop in confronting such challenges? What kind of reforms

---

1. Soekarno (1945 June 1) Speech at the *Dokuritu Zyunbi Tyoosakai known as Lahirnya Pancasila.*

or transformation are needed? These are some of the questions that this paper seek to address. It thus proceeds in the following fashion. In the first section, the author discusses some of the rising challenges facing the world today and how they impact Indonesia; while in the second section, the author argues for fundamental institutional transformation that is necessary for Indonesa to succesfully confront these challenges. The author then provides a concluding remarks.

2 RISING GLOBAL CHALLENGES AND THEIR IMPACTS ON INDONESIA

2.1 *Challenges in the socio-political and economic dimension*

In 1989, against the backdrop of a series of dramatic events that would eventually led to the collapse of the Soviet Empire and – with it – global communism as an ideology, Francis Fukyama self-assuredly predicted 'the end of history' with Western liberal democracy becoming the final form of government that nation states would embrace.[2] While his prediction resounded well with the contagious effect of the so-called 'third wave of democratization' propounded by Samuel Huntington,[3] the peace dividend that liberal democracy is supposed to yield has still yet to come into fruition in today's world, almost three decades after Fukyama's monumental analysis.

Indeed, rather than imbuing a complacent sense of security, the world nowadays has become an ever more uncertain place. While interstate war has declined markedly since the end of 20th century, the number of intrastate armed conflict has not declined significantly as depicted in Figure 1 below and becoming even deadlier as the ongoing conflicts

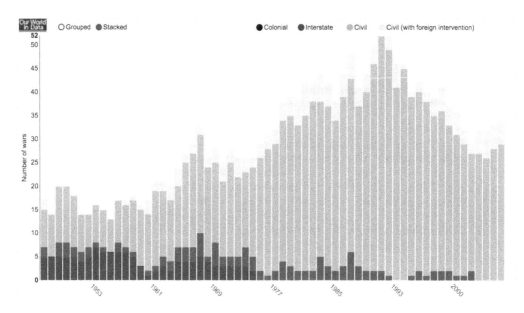

Figure 1.   Number of state-based armed conflicts by type, 1946-2007 (Source: Roser, 2017).

2. Francis Fukuyama. (1989). "The End of History?", The National Interest, Summer.; and Francis Fukuyama. (1992). *The End Of History And The Last Man*, New York: The Free Press.
3. Samuel Huntington. (1991). *The Third Wave: Democratization in the Late Twentieth Century*, Norman: University of Oklahoma Press.

in the Middle East have demonstrated.[4] The multitude spillovers of such conflict, often taking the form of forced migration, international terrorism, not to mention other humanitarian and social problems such as drugs and arms trafficking, and human rights abuses have created complex, interwoven challenges that the international society needs to deal with.

This is not to say that looming border conflicts or territorial tensions are implausible as the current situation in the South China Sea and the annexation of Crimea have recently suggested. Indeed, the two examples given relate to a phenomenon of the emergence of economically succesful illiberal democracy or soft-authoritarianism like China and Russia that have become increasingly assertive politically and militarily, which probably constitutes as the greatest challenge to the 'unipolar' liberal democratic world order that Francis Fukuyama envisioned.[5]

Such is also an argument echoed by sholar such as Danny Quah, who advocated for a global power shift to the 'East' to reflect global demographic reality and move in the world's economic center of gravity.[6] Much less certain though, is the implications of this power balance shift on global peace. While it is too early to predict a war on a global scale between China and the West happening, the breadth of cases from Punic Wars fought between Rome and Carthage from 264 BC to 146 BC, to the first and second World Wars in the 20th century, have taught mankind that the wrestling of power among hegemons rarely happened peacefully. Again, this reminds us on the fragile, and uncertain nature of the present geopolitical order.

'The new Asian hemisphere' as famously espoused by Kishore Mahbubani[7] is primarily propelled by the growing economic significance of many states in Asia, particularly

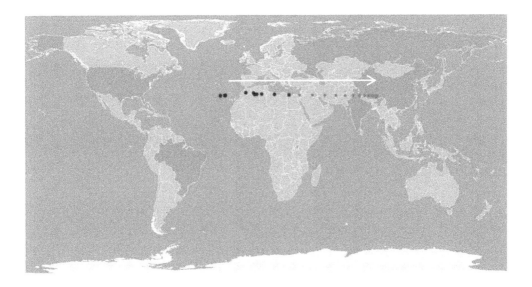

Figure 2.   Shifting economic gravity of the world (Source: Quah, 2017).

4. Richard Norton-Taylor (2015, May 20). "Global armed conflicts becoming more deadly, major study finds". Retrieved from https://www.theguardian.com/world/2015/may/20/armed-conflict-deaths-increase-syria-iraq-afghanistan-yemen
5. Robert Kagan. (2008). *The Return Of History And The End Of Dreams*, New York: Alfred A. Knopf.
6. Danny Quah, (2017, September 17) *Trade, Globalisation, and Sovereignty*, presentation at the Indonesia Update 2017: Indonesia in the New World, ANU Indonesia Project, Crawford School.
7. Kishore Mahbubani. (2009). The New Asian Hemisphere: The Irresistible Shift of Global Power to the East. New York: Public Affairs.

China, India, and Southeast Asia in the globalized economy. Sadly, yet, rapid growth in the region has not been accompanied by equity distribution of wealth, which resulted in widening gap between the rich and the poor. Based on a recent work by the ADB, out of the 37 economies in the Asia Pacific with available data in the 2000s, 14 had a Gini coefficient – which is a measure of income inequality – of or greater than 40, widely considered the threshold for 'high inequality.'[8]

In China, for example, Gini coefficient which is a measure of income inequality increased markedly from about 0.310 in 1981 to 0.491 in 2008, only declining marginally to 0.474 in 2012, making China one of the least equal 25 percent of countries worldwide.[9] China's disparity from a regional standpoint, between the rich coastal area and impoverished interior, has been attributed as a potential source of national disunity and intrastate conflict by George Friedman, leading ultimately to China's fragmentation and thwarted global aspiration.[10] The negative spillover of this event, if it occurs, can be pretty damaging to the rest of the world as many have come to rely on China's economic (and political) patronage.

But inequality, and many other socio-economic adversities exarcebated by globalization in recent times are not exclusive to the Asian domain. Even the West are experiencing economic downturn following the 2008 Global Financial Crisis, rising inequalities particularly in the United States, and loss of jobs and opportunities due to shifting capital and industrial base to the developing world. In the United States, for instance, the

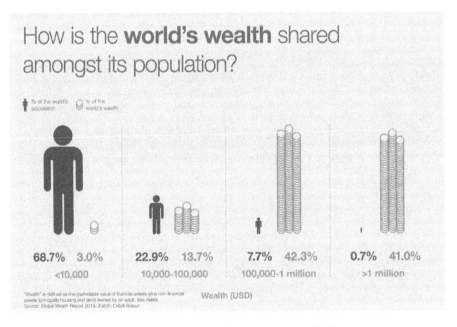

Figure 3. Share of the world's population and wealth (Source: WEF, 2014).

8. Juzhong Zhuang, Ravi Kanbur, and Dalisay Maligalig. (2014). "Asia's income inequalities: recent trends." in Ravi Kanbur, Changyong Rhee and Juzhong Zhuang (eds.). *Inequality in Asia and the Pacific: Trends, drivers, and policy implications.* New York: Routledge.
9. Yixiao Zhou and Ligang Song. (2016). *Income inequality in China: Causes and Policy Responsese.* Retrieved from: https://openresearch-repository.anu.edu.au/bitstream/1885/107259/2/01_Zhou_Income_Inequality_2016.pdf.
10. George Friedman. (2009). *The Next 100 Years: A Forecast For The 21st Century.* New York: Doubleday.

income disparity between the typical male worker (representing the majority of the population) and the top 1% earners has risen by more than 300% in the span of 32 years, from US$ 345.380 in 1978 to US$ 1.067.338 in 2010.[11] A recent figure produced by World Economic Forum below shows how, in a global context, 8,4% of the world's population unjustly control a combined 83,3% of the world's wealth.[12]

The rise in income inequality is truly a global phenomenon that may eventually trigger political polarization and horizontal conflicts which threaten to destroy the social fabric of humanity. Undeniably, an atmosphere of suspicion and skeptic attitudes toward globalisation and the open-access international order have accumulated in recent years. A recent survey by *The Economist* and YouGov, reveals how the West is increasingly turning its back on globalisation as less than half of respondents in America, Britain and France loses belief in globalisation as a 'force for good' and generally stands against immigration.[13] This phenomenon resonates with the chain of events such as Trump's victory in the United States and the new 'isolationist' spirit that it brings; the withdrawal of the United Kingdom from the European Union (i.e. 'Brexit'); and the rising popularity of anti-establishment and neo-nationalist parties across Europe.[14] All these trends point toward what may potentially seem like the unraveling of the established global order into a new 'unknowns'. The interesting question is: what implications do these hold for Indonesia?

2.2 *Implications on Indonesia's national dynamics*

The national dynamics of Indonesia as a geopolitical entity can never be detached from what happens in the world at large. Afterall, Indonesia geostrategically lies at the intersection of the world; a *totenkreuz* ('death crosss') where numerous global interests meet and clash, as A.R. Soehoed once alluded, citing a famous Austrian geopolitical professor in the 20th century, to remind Indonesia on the ever present danger and opportunity of being in such a strategically important location.[15] In the 21st century context, Kris Wijoyo Soepandji also brings into attention the need to pay attention to the 'global extraordinary networks' that shape the workings of the state and Indonesia's geostrategical position within such networks to ensure the nation's survival.[16] As such, it comes to no surprise that the global challenges that have been described above render several important implications for Indonesia.

In the socio-economic sphere, globalization and the rise of Asia brings as much opportunity as challenges for Indonesia. Numerous studies by global financial institutions and multilateral agencies have dubbed Indonesia as the 'next big thing' in terms of economic potential and relative size. One report by Goldman Sachs included Indonesia in the 'Next-11' countries along with the likes of Bangladesh, Iran, Mexico, Philippines, Vietnam, and five others with growth potential matching that of the 'BRICs' countries.[17] In a more optimistic tone, Standard Chartered projected Indonesia as one of

11. Robert Reich. (2013). *Inequality for All*. Retrieved from, http://inequalityforall.com/resources/#/1.

12. Amina Mohammad. (2014 November 7) "Why inequality is 2015's most worrying trend," Retrieved from, https://www.weforum.org/agenda/2014/11/inequality-2015s-worrying-trend/

13. "What the world thinks about globalisation", Retrieved from: https://www.economist.com/blogs/graphicde tail/2016/11/daily-chart-12.

14. Jon Henley, Helena Bengtsson and Caelainn Barr. (2016 May 25). "Across Europe, distrust of mainstream political parties is on the rise". Retrieved from https://www.theguardian.com/world/2016/may/25/across-europe-distrust-of-mainstream-political-parties-is-on-the-rise. and "Guide to nationalist parties challenging Europe", source: http://www.bbc.com/news/world-europe-36130006.

15. A.R. Soehoed. (2002). *Bunga Rampai Pembangunan: Antara Harapan dan Ancaman Masa Depan*, Jakarta: Puri Fadjar Mandiri.

16. Kris Wijoyo Soepandji. (2017). *Ilmu Negara: Perspektif Geopolitik Masa Kini*, Depok: Badan Penerbit FHUI.

17. Goldman Sachs Global Economics Group. (2012) *BRICs and Beyond*, The Goldman Sachs Group, Inc.

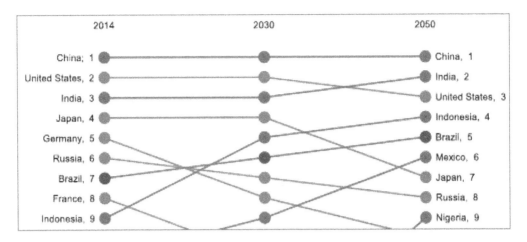

Figure 4.   The World in 2050 based on GDP at PPP rankings (Source: Hill & Pane, 2017).

the world's 10 largest economies in 2020 and among the top six in 2030.[18] In the recent 2017 Indonesia Update at the Australian National University, economists Hal Hill and Deasy Pane, based on data from IMF and PricewaterhouseCoopers, suggested even further that Indonesia will become the fourth biggest economies of the world in 2050.[19]

The problem with these – and indeed, many other – economic projections, however, is the 'hold all else constant' (*ceteris paribus*) assumption whereas we know that the global dynamics in the forthcoming decades is mired in uncertainty. This means that without deliberate efforts to eliminate impending challenges, all these predictions may as well be mere wishful thinking. More importantly, while all these projections center on economic growth, the 'quality' of such growth has often escaped attention, which brings into mind a set of socio-economic problems long embedded in Indonesia's economic growth.

One important source of socio-economic contention for Indonesia is the rate of inequality that has risen dramatically ever since the 1998 Reform as depicted in the adjacent graph. In fact, wealth concentration in Indonesia has become so high and increasing faster than in other countries that the richest 10 percent of Indonesians now own an estimated 77 percent of the country's wealth while half of the country's assets are owned by the richest 1 percent.[20] Another study by Oxfam finds that "the four richest men in Indonesia have more wealth than the combined total of the poorest 100 million people".[21] Meanwhile, according to Jeffrey Winters, as of 2010, 43.000 wealthiest Indonesians own a combined wealth equal to 25% of Indonesia's GDP (US$ 708 billion), while 50 wealthiest Indonesians have a net worth of 630.000 times more than ordinary people (as measured in GDP per capita).[22]

18. Dion Bisara. (2011 November 27) "Indonesia's Economy to Be in World's Top Six in 2030: Standard Chartered Transforming Institutions To Confront Global Challenges", http://jakartaglobe.id/archive/indonesias-economy-to-be-in-worlds-top-six-in-2030-standard-chartered/

19. Hal Hill & Deasy Pane. (2017 September 16). *Indonesia and the Global Economy: Missed Opportunities?*, presentation at The Indonesia Update 2017: Indonesia in the New World, ANU Indonesia Project, Crawford School.

20. Winarno Zain. (2016 June 8) "Behind the rise of income inequality in Indonesia". Retrieved from, http://www.thejakartapost.com/academia/2016/06/08/behind-the-rise-of-income-inequality-in-indonesia.html

21. Luke Gibson. (2017 February) *Towards a More Equal Indonesia: How the government can take action to close the gap between the richest and the rest*, Oxford: Oxfam International.

22. Jeffrey Winters. (2011). *Oligarchy*. Cambridge: Cambridge University Press; and Jeffrey Winters, "Who will tame the oligarchs?", *Inside Indonesia*, Edition 104: April-June 2011. Retrieved from http://www.insideindonesia.org/who-will-tame-the-oligarchs.

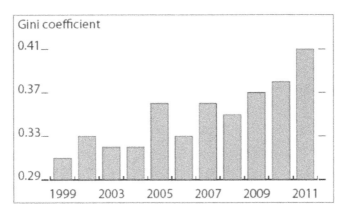

Figure 5. Gini coefficient in Indonesia, 1999-2011 (Source: Visually, proceesed from https://www. bps.go.id/).

All these problems of inequality point to another potentially explosive issue, namely social cohesion. The example of recent drama in the Jakarta's gubernatorial election where many elements of radical Islamic groups took to the streets along with hundreds of thousands of Indonesian moslems to protest the 'blasphemous action' of the then-incumbent, Basuki Tjahaja Purnama of Chinese descent, underscores a bigger problem rather than diminishing pluralism and the politicization of religious issues as the mainstream media and many commentators have suggested. Ian Wilson rightly points to the fundamental cause of gaping inequality that have spatially torn the city into deep social class division. In large parts of Jakarta, Wilson continues, where the poor has been victimized by state-endorsed development projects, elements of radical Islamic groups stepped in to leverage on the situation and gained ground; opposing evictions in some parts and providing services to the displaced in some others while whispering (their version of) religious messages.[23] Hence in the case of Jakarta we have seen how religious and ethnical sentiment have aggravated the pre-existing grave inequality, and hurts social cohesion.

The growing inequality in Indonesia is also tightly tied with the political dimension that is another focus of this paper. Sharp wealth concentration breeds political inequality, such is the thesis propounded by Jeffrey Winters, as 'oligarchic' actors occupying the highest stratum of material posession in a society strive to maintain their position by using all possible means of wealth defense.[24] Depending on the types of oligarchy in question, this can take various forms. In the case of post-*Reformasi* Indonesia where transition to an electoral democracy unfolded in a context of weak legal institutions that failed to replace the controls Suharto had imposed, untamed oligarchs trampled on the nascent democracy; stretching their arms ever wider in the bureaucracy, parliaments, law enforcement, the media and elements of the civils society.[25] Winters goes as far to suggest that political processes and choices too have been limited by the oligarchs in democratic Indonesia.[26]

23. Ian Wilson. (2017 April 19). 'Jakarta: inequality and the poverty of elite pluralism'. Retrieved from http://www.newmandala.org/jakarta-inequality-poverty-elite-pluralism/. Ian Wilson. (2016 November 3)" Making enemies out of friends". Retrieved from http://www.newmandala.org/making-enemies-friends/and Jewel Topsfield. (2017 February 3). "The real reason many poor Jakartans are opposing Ahok in the gubernatorial election", source: http://www.smh.com.au/world/the-real-reason-many-poor-jakartans-are-opposing-ahok-in-the-gubernatorial-election-20170203-gu52aj.html.
24. Jeffrey Winters, *Oligarchy*, op. cit.
25. Jeffrey Winters, ibid.
26. Jeffrey A. Winters. (October 2013). "Oligarchy and Democracy in Indonesia", *Indonesia,* No. 96, Special Issue: Wealth, Power, and Contemporary Indonesian Politics.

To some extent, the emergence and influence of the oligarchs cannot be completely disentangled from the 'global extraordinary networks', the intermingling of global financial networks and global multimedia networks with other networks in the political, cultural, and even criminal realms, that have been attributed by Soepandji as a powerful force in eroding the relative power of the state in many regions, particularly Indonesia.[27] Such argument resonates with Richard Robison and Vedi Hadiz, who place the emergence of oligarchic force in Indonesia in the context of global market capitalism.[28] Indeed, Joe Studwell in his book on 'Asian Godfather' describes to a great extent how many powerful conglomerates in Asia are intricately linked with one another, and, to the underworld.[29]

As Soepandji maintains, the 'disruptive' power of the global extraordinary networks must be approached with caution. In Soepandji's GETANKAS speech at the National Resilience Institute, Republic of Indonesia (*Lemhannas RI*), he asserts that, Soenarto Mertowardojo's concept of human's psyche in *Sasangka Jati* can be utilized to neutralize the negative effect of those global extraordinary networks, as those networks actually enforce their influnce through the human psyche.[30] Mertowardojo's concept of human's psyche states that human behaviour is conducted by the instruction of the mind (*Pangaribawa, Prabawa and Kamayan*) which is supported by the passions (*Muthmainah, Sufiah, Lauwamah and Amarah*). If the mind only delivers the passions' greed (*Lauwamah* as the leading force), hence human behaviour will tend to create conflict among society even though at certain stage the conflict can still be avoided. However, if the mind delivers not the passions' greed (*Muthmainah* as the leading force), but the belief in the Creator and act of humanity hence, the behaviour will find the path for lasting peace.[31] The act of terror which is based on the belief of Creator shall be seen as expression of covered passion's greed, because it is against humanity.

The fact that global extraordinary networks can limit policy choices and influence policy processes and state dynamics in general will have serious consequences on the outcomes of such process and dynamics. To prevent policies producing much negative externalities and institutions devolving into 'extractive' ones due to the influence of the global extraordinary networks, the state must become ever more vigilant and resilient, which means the state must prepare its citizens seriously to survive and gain supremacy in this current situation. More importantly, conscious efforts to transform the governance of political-economic institutions underlying the workings of society in order to produce greater welfare for the larger part of the public must be constantly pursued. This involves making hard collective choices, but choices that are otherwise detrimental to the success (or failure) of Indonesia as a nation-state.

3 INSTITUTIONAL TRANSFORMATION TO CONFRONT GLOBAL CHALLENGES: AWAKENING INDONESIA'S 'SPIRIT' OF THE LAWS

Any discussion of institutional transformation must begin with a discussion of the law, because institutions are essentially "the *rules* of the game of a society, or, more formally, the humanly

27. Kris Wijoyo Soepandji, op. cit.
28. Richard Robison and Vedi R. Hadiz. (2004). *Reorganising Power in Indonesia: The Politics of Oligarchy in an Age of Markets.* London: RoutledgeCurzon.
29. Joe Studwell. (2007) *Asian Godfathers: Money and Power in Hong Kong and Southeast Asia*, London: Grove/Atlantic Inc.
30. Kris Wijoyo Soepandji. (2018 May 16). *Orasi Ilmiah HUT-53 LEMHANNAS RI: Geopolitik, Ketahanan Nasional dan Kemerdekaan Sejati, GETANKAS.* Jakarta: LEMHANNAS RI.
31. *"angen-angenira luwih becik sira anggo niti tuna bathining uripira (nyawanira), saka anggonira candhak kulak lan dol tinuku kalawan hawa napsunira aneng pasaring kauripan donya, kanthi satiti anggonira ngulatake lebu wetuning napsu, kang sira bandhani kalawan nyawanira, supaya sira tansah andarbeni weweka, kang bisa nyumurupi mungguh pepilahane kang bener lan kang luput, kang iya lan kang dudu, lan sapanunggalane. Mangkono sarana kang bisa agawe bathining uripira, supaya sira bisa antuk nugrahaning Pangeran."* Soenarto Mertowardojo. (1954) *Sasangka Jati.* Surakarta: Paguyuban Ngesti Tunggal.

devised constraints that structure human interaction."[32] In a classical treatise called "De l'esprit des loix" (*The Spirit of the Laws*), Baron de Montesquieu talks to a great extent about how climate, geography, and culture interact with the people to produce norms and rules that prevail in and suit the 'spirit' of such societies.[33] According to him: *"Mankind are influenced by various causes; by the climate, by the religion, by the laws, by the maxims of government, by precedents, morals, and customs; from whence is formed a general spirit of nations."*[34] While most of what Montesquieu say, especially on climate and geography have nowdays been subjected to debates,[35] they cannot be completely irrelevant, at least in the idea of how laws come into being and evolve.[36]

In 'Ilmu Negara: Perspektif Geopolitik Masa Kini', Soepandji echoes the idea on the intrinsically connnectedness of geography with institutions and national character formation in Indonesia, based on which, there are at least three geographical realities that Indonesia needs to account for in its institutional transformation. Firstly, that Indonesia is the world's largest archipelagic state with rich endowment of natural resources; secondly, that Indonesia lies at the intersection of global trade routes and civilizations; and thirdly, that Indonesia is the only state to exist upon a territory with the highest volvanic (and tectonic) activities in the world.[37]

These geographical realities bring together a set of domestic institutional implications that Indonesia needs to grapple with. Firstly, Indonesia must build an 'open-access' or 'inclusive' legal order to tap on its geostrategical potential in the intersection of global maritime routes and its rich resources. At the same time, Indonesia must develop institutional arrangements that foster cautiousness or vigilance, both in the governance of the state and as a national trait in general. This is to anticipate the ever present danger of disasaters stemming from volcanic and tectonic activities due to Indonesia's location in the world's 'ring of fire', as well as the competing interests seeking to take advantage out of Indonesia's geostrategic position and natural resource endowment. In short, Indonesia's institutional development must proceed toward augmenting 'national resilience'.

Empirically speaking, these two institutional implications have not been very well reflected in Indonesia nowadays. While Indonesia has submitted itself to many international free trade and open market arrangements, its domestic institutions still fail to reflect an 'inclusive' or 'open-access' order, characterized by high degree of competition and innovation due to ease of entry of citizens into economic activities and the absence or minimum threat of political intrusion into economic activities.[38] Yet, as described in the previous section, the persistence of oligarchic actors, which, through their sheer material force hijacks political and legal institutions as part of their wealth defense

32. Douglas C. North. (1990). *Institutions, Institutional Change and Economic Performance*, Cambridge: Cambridge University Press.
33. These discussions are evident and most lively in several chapters of the treatise, such as Book XIV on Laws As Relative To The Nature of The Climate up to Book XIX on Laws, In Relation To The Principles Which Form The General Spirit, The Morals, And Customs, Of A Nation. See: Charles Louis de Secondat, Baron de Montesquieu, *The Spirit Of Laws*, 2 Vols. *In 1 (1777)*, Volumes I And II of The Complete Works Of M. De Montesquieu, Indiana: Liberty Fund, Inc., 2004.
34. Charles Louis de Secondat, Baron de Montesquieu, op. cit., p. 325.
35. Jared Diamond, (2012 June 7). "What Makes Countries Rich or Poor?", source: http://www.nybooks.com/articles/2012/06/07/what-makes-countries-rich-or-poor/?pagination=false&printpage=true.
36. The idea of 'Mare Clausum' versus 'Mare Liberum' in the international law of the sea evolved from the need to exert control over geographical terrain (i.e. the ocean) that is crucial to the political-economic activities of the states in Europe in the 17th century. For a complete discussion, see: Mónica Brito Vieira, "Mare Liberum vs. Mare Clausum: Grotius, Freitas, and Selden's Debate on Dominion over the Seas", *Journal of the History of Ideas*, Vol. 64, No. 3, Jul., 2003.
37. Kris Wijoyo Soepandji, op. cit.
38. Douglass C. North, et. al., *Limited Access Orders: Rethinking the Problems of Development and Violence*, January 25, 2011.

	China	Indonesia	Philippines	USA	Thailand	Malaysia	Hong Kong	Taiwan	South Korea	Singapore
Oligarchic Scale (bln)	$3.78	$2.13	$0.85	$16.66	$1.14	$1.61	$3.78	$2.32	$1.64	$1.36
Total of 40 (bln)	$151	$85	$34	$667	$45	$64	$151	$93	$66	$54
Rank 1 (bln)	$9.30	$14.00	$7.20	$59.00	$7.40	$12.40	$22.00	$8.80	$9.30	$8.90
Rank 40 (bln)	$1.86	$0.63	$0.09	$7.00	$0.20	$0.16	$0.95	$0.93	$0.53	$0.21
GDP/ capita	$5,203	$3,361	$2,082	$47,992	$5,151	$8,479	$33,667	$21,844	$23,804	$49,352
Oligarchic Intensity	2.2	10.2	15.7	4.4	13.1	26.0	62.3	18.4	5.7	20.4
Material Power Index	726,504	632,740	408,261	347,141	221,316	189,881	112,276	106,207	68,896	27,557

Figure 6. Oligarchic scale, intensity, and material power index in Indonesia and several other countries, 2011 (Source: Winters, 2013)[39].

strategy, nurture rent-seeking activities and establish barriers of entry for players outside their own circle, which then, arguably, disincentivize innovation. This can be seen in how countries with lower Material Power Index as described in the table below – with the exception of China – generally ranks higher in the Global Innovation Index.[40]

Moreover, Indonesian policy-makers often do not have a geopolitical acumen (and good sense of history), leading to the lack of cautiousness or vigilance in policy-making, which may in turn create policies that only provide short-term benefits rather than long-term strategic benefits for the society. A classic example of this from a local perspective is the proliferation of local regulations in decentralized Indonesia that conflict with many national regulations and create, as Simon Butt puts it, a 'legal disorder'.[41] The latest figures of such regulations that have been annulled by the central government amounted to 3.143,[42] which vary in content from local taxation and *levies* to public order regulations. According to Raden Pardede and Shirin Zahro, the existence of such regulations add to the complexities of economic governance and policy implementation, which may in turn hamper national development and reducing Indonesia's global competitiveness.[43]

39. Oligarchic scale (row 1) is the average wealth of the top 40 oligarchs (in US$ billions). Oligarchic intensity (row 6) is the total wealth of the top 40 oligarchs (row 2) as a percent of GDP. The Material Power Index (row 7) is the average wealth of the top 40 oligarchs (row 1) divided by GDP/capita (row 5). For further information, see: Jeffrey A. Winters, "Oligarchy and Democracy in Indonesia." Op. cit.
40. In 2017 Global Innovation Index, the respective rankings for the countries as stated in the table (Taiwan not included) is as follows: United States (#4); Singapore (#7); South Korea (#11); China (#22); Malaysia (#37); Thailand (#51); Philippines (#73); Indonesia (#87). For further information, see: Cornell University, INSEAD, and WIPO, *The Global Innovation Index 2017: Innovation Feeding the World*, Ithaca, Fontainebleau, and Geneva, 2017.
41. Simon Butt. (2010). "Regional Autonomy and Legal Disorder: The Proliferation of Local Laws in Indonesia," *Sydney Law Review*, Vol. 32, Issue 177.
42. Jurig Lembur. (2016 June 21). "Setelah 3.143 Perda, Peraturan Lain yang Bertentangan UU Akan Dihapus", source: https://news.detik.com/berita/d-3238902/setelah-3143-perda-peraturan-lain-yang-bertentangan-uu-akan-dihapus.
43. Raden Pardede & Shirin Zahro. (2017 September 15) "Survey of Recent Developments: The Effectiveness of Policy Reform in Indonesia", presentation at The Indonesia Update 2017: Indonesia in the New World. ANU Indonesia Project, Crawford School.

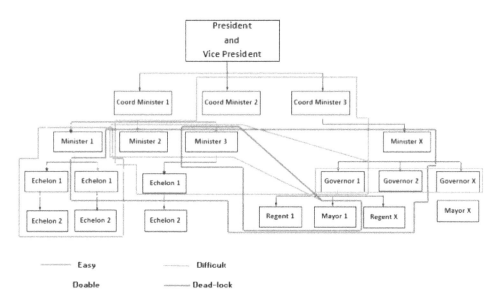

Figure 7. Illustration of the difficulty in inter-governmental coordination and implementation due to conflicting local regulations and policies (source: Pardede and Zahro, 2017).

From a more strategic point of view, another example is given by Soepandji pertaining to a statement given by a high ranking official at a foreign policy forum held in Jakarta in late 2000. In such forum, the official stated that Indonesia does not have 'history' before 1908 and has not yet given a masterpiece of architectural heritage to mankind in the likes of many of its Western counterparts. As Soepandji, maintains, this statement must be critically evaluated, because, even though the Indonesia's national consiousness as a political entity emerged in 1908 (when the national movement began with the establishment of Boedi Oetomo), the history of Indonesia as a geocultural entity can be traced as far back as the Tagaroa group of ocean people, as mentioned by John Rahasia in 1970 which includes kingdoms that have straddled the Indonesian archipelago. This would naturally include the Syailendra Dynasty that had built Borobudur Temple – one of the greatest heritage of mankind – in the 7th century.[44] From a realist viewpoint on the international interplay of power, this kind of lack of geopolitical and historical acumen will make Indonesia an easy prey to hostile narratives by actors that seek to weaken the nation's existence.

On the contrary, a contrasting example would be what the policymakers at the *Lemhannas RI* has accomplished in 2015 when, through scenario-planning, they developed a set of 'future' scenarios for Indonesia in 2045 to anticipate future global trends and threats facing the country.[45] This shows an awareness for geostrategic environment and its implications for Indonesia. As a matter of fact, mastery of historical and geopolitical knowledge is crucial to understand the broader political narratives within which conflicts are fought. Gao Hongwei and Tao Chun in a work on postmodern warfare tell of how the *Central Party School* and *National Defence College* of the People's Republic of China arrive upon the understanding that setting the narratives by winning information flow and leveraging on legal and ethical advantage are detrimental for

---

44. Kris W. Soepandji, op. cit.

45. (2016 March 14). "Lemhannas Luncurkan Buku 'Skenario Indonesia 2045'" Retrieved from http://sp.berita satu.com/home/lemhannas-luncurkan-buku-skenario-indonesia-2045/110973; and for a more detailed discussion of the range of scenarios, see: Lembaga Ketahanan Nasional Republik Indonesia. (2015). *Skenario Indonesia 2045*, Jakarta: Lembaga Ketahanan Nasional RI.

victory in the third and fourth generation (postmodern) warfare, respectively.[46] All these underscore the need for institutional transformation that is anchored on Indonesia's geopolitical and historical realities to increase resilience.

4 CONCLUSION AND RECOMMENDATIONS

It has been described to a great extent in the previous sections on how rising global challenges confront and have implications toward Indonesia's national dynamics and how institutional transformation is needed to cope with such challenges. Moreover, this paper has also submitted the idea of institutional transformation to be anchored on Indonesia's geopolitical and historical realties. We would like to now conclude by providing a three-point recommendations.

Firstly, in adressing institutional transformation we shall turn our attention to the often overlooked idea of human agency, of the collective choice that society's make that will ultimately chart its developmental course. In this point, the Mertowardojo's human psyche concept becomes very important, as it will help Indonesia state policy for effectively engaging its citizens to be active agents to transform the country positively, without sound belief in the Creator and act of humanity, the development of the country will easily lose its aim. The importance of human agency is also well demonstrated in a seminal work by Jared Diamond on the collapse of civilizations, as it is the collective failure of society to perceive and address impending fundamental – in his case, environmental – problems as the main cause of that society's collapse.[47] Indeed, looking back into the conception of institution given by North as stated in the preceding section, the agency of human actually occupies a central tenet by being both the subject and object of institutions; in the sense that 'they' devise the rules and norms that structure 'their' own interaction and exchanges.

This idea resounds with the argument made by economist Tim Harford where he prescribes three sets of recommendations in order for societies to succeed in overcoming fundamental problems, namely: first, to try new things, in the expectation that some will fail; second, to make failure survivable, because it will be common; and third, to make sure that we know when we have failed.[48] Hence, societies should always be audacious to challenge fundamental assumptions in the current order and to experiment with new sets of ideas, innovate, to arrive at institutional arrangments that work best and guarantee the long-term progress of that society.

Secondly, using the framework laid out in the first point, Indonesia should immediately address the fundamental problem of political inequality which impedes its 'open-access' and 'inclusive' order. This is an important question to address because inclusive socio-political and economic institutions, through their interlinkings, influence the long-term growth and progress of societies as the research of Daron Acemoglu and James Robinson have demonstrated.[49] And Indonesia should address this problem by experimenting with its democratic institutions. Tying the idea of geographical political representation with class' representation, or assigning a larger weight of electoral vote to the poor, for example, could be a starter.

The bottomline of this is to prevent Indonesia's democratic institutions from degenerating and succumbing to predatory elites. In this respect, Nial Fergusson's study provide important lessons to be learnt, in which he makes the case of how democratic

---

46. Gao Hongwei dan Tao Chun. (2008). *Post Modern Warfare and World New Order*, Hong Kong: Ming Pao Publishing Limited.
47. Jared Diamond. (2005). *Collapse: How Societies Choos to Fail or Succeed*, New Yotk: Penguin Group.
48. Tim Harford. (2011). *Adapt: Why Success Always Start With Failure*, London: Hachette Digital.
49. Daron Acemoglu and James Robinson. (2012). *Why Nations Fail? The Origins of Power, Prosperity, and Poverty*, New York: Crown Publishers.

institutions in the West have degenerated as democratic resource allocation has been overburdened by surging inter-generational debt; capitalism has subverted the vigor of civic activism; and the rule of law has devolved into rule of predatory lawyers, thereby causing many countries in the West to enter a 'stationary state' with very little economic growth.[50]

Thirdly, best practices work best. What have been done by *Lemhannas RI* to experiment with scenario-planning, in order to increase society's and the state's alertness on future geopolitical and global trends that shape the contour of Indonesia is a best practice that need to be institutionalized in the governing mechanism of Indonesia. In other words, more state agencies should adopt this kind of approach and the results of which endorsed by the President in policy-making processes. The objective of this, as proposed earlier in this paper, is to develop institutional arrangements that foster cautiousness or vigilance, both in the governance of the state and as a national trait in general.

---

50. Nial Fergusson. (2012). *The Great Degeneration: How Institutions Decay and Economies Die*, New York: Penguin Books.

Author index